# ANNUAL EDITIONS

# Environment

*Twenty-sixth Edition*

## 07/08

## EDITOR

**John L. Allen**
*University of Wyoming*

John L. Allen is professor and chair of geography at the University of Wyoming. He received his bachelor's degree in 1963 and his M.A. in 1964 from the University of Wyoming, and in 1969 he received his Ph.D. from Clark University. His special area of interest is the impact of contemporary human societies on environmental systems.

**Contemporary Learning Series**

2460 Kerper Blvd., Dubuque, IA 52001

Visit us on the Internet
*http://www.mhcls.com*

# Credits

1. **The Global Environment: An Emerging World View**
   Unit photo—Images produced by Hal Pierce, Laboratory for Atmospheres, NASA Goddard Space Flight Center/NOAA
2. **Population, Policy, and Economy**
   Unit photo—Royalty-Free/CORBIS
3. **Energy: Present and Future Problems**
   Unit photo—Kevin Phillips/Getty Images
4. **Biosphere: Endangered Species**
   Unit photo—U.S. Fish and Wildlife Service/Galen Rathburn
5. **Resources: Land and Water**
   Unit photo—Photo by Gene Alexander, USDA, Natural Resources Conservation Services
6. **The Hazards of Growth: Pollution and Climate Change**
   Unit photo—Flat Earth Images

# Copyright

Cataloging in Publication Data
Main entry under title: Annual Editions: Environment. 2007/2008.
1. Environment—Periodicals. I. Allen, John L., comp. II. Title: Environment.
ISBN-13: 978–0–07–351544–1   MHID-10: 0–07–351544–2   658'.05   ISSN 0272–9008

Twenty-sixth Edition

Cover image: Royalty-Free/CORBIS
Compositor: Laserwords Private Limited
Printed in the United States of America   1234567890QPDQPD987   Printed on Recycled Paper

# Editors/Advisory Board

Members of the Advisory Board are instrumental in the final selection of articles for each edition of ANNUAL EDITIONS. Their review of articles for content, level, currentness, and appropriateness provides critical direction to the editor and staff. We think that you will find their careful consideration well reflected in this volume.

# Preface

In publishing ANNUAL EDITIONS we recognize the enormous role played by the magazines, newspapers, and journals of the public press in providing current, first-rate educational information in a broad spectrum of interest areas. Many of these articles are appropriate for students, researchers, and professionals seeking accurate, current material to help bridge the gap between principles and theories and the real world. These articles, however, become more useful for study when those of lasting value are carefully collected, organized, indexed, and reproduced in a low-cost format, which provides easy and permanent access when the material is needed. That is the role played by ANNUAL EDITIONS.

At the beginning of our new millennium, environmental dilemmas long foreseen by natural and social scientists began to emerge in a number of guises: regional imbalances in numbers of people and the food required to feed them, international environmental crime, energy scarcity, acid rain, build-up of toxic and hazardous wastes, ozone depletion, water shortages, massive soil erosion, global atmospheric pollution, climate change, forest dieback and tropical deforestation, and the highest rates of plant and animal extinction the world has known in 65 million years. There are no longer valid scientific debates over these issues—they are real and measurable. Even worse, such environmental problems as climate change seem to be occurring more rapidly than expected and at least some scientists believe the Earth's environmental systems have reached a threshold or "tipping point" beyond which our ability to "fix things" diminishes rapidly.

These and other environmental problems continue to worsen in spite of an increasing amount of national and international attention to the issues surrounding them and increased environmental awareness and legislation at both global and national levels. The problems have resulted from centuries of exploitation and unwise use of resources, accelerated recently by the shortsighted public policies that have favored the short-term, expedient approach to problem-solving over longer-term economic and ecological good sense. In Africa, for example, the drive to produce enough food to support a growing population has caused the use of increasingly fragile and marginal resources, resulting in the dryland deterioration that brings famine to that troubled continent. Similar social and economic problems have contributed to massive deforestation in middle and South America and in Southeast Asia. And the refusal on the part of countries to address issues of greenhouse gas emissions has accelerated climate change that has manifested itself in increased melting of glacial ice and more extreme atmospheric events.

Part of the problem is that efforts to deal with environmental issues have been intermittent. During the decade of the 1980s, economic problems generated by resource scarcity caused the relaxation of environmental quality standards and contributed to the refusal of many of the world's governments and international organizations to develop environmentally sound protective measures, which were viewed as too costly. More recently, in the late 20th and early 21st century, as environmental protection policies were adopted, they were often cosmetic, designed for good press and TV sound bites, and—even worse—seemingly designed to benefit large corporations rather than to protect environmental systems. Even with these policies based more in public relations than in environmental ones, governments often lacked either the will or the means to implement them properly. The absence of effective environmental policy has been particularly apparent in those countries such as India or China that are striving to become economically developed. But even in the more highly developed nations, economic concerns tend to favor a loosening of environmental controls. In the United States, for example, the interests of maintaining jobs for the timber industry imperil many of the last areas of old-growth forests, and the desire to maintain agricultural productivity at all costs causes the continued use of destructive and toxic chemicals on the nation's farmlands. In addition, concerns over energy availability have created the need for foreign policy and military action to protect the developed nations' access to cheap oil and have prompted increasing reliance on technological quick fixes, as well as the development of environmentally sensitive areas to new energy resource exploration and exploitation. Yet the simpler measures of energy conservation do not seem to be important to policy makers.

Despite the recent tendency of the U.S. government to turn its back on environmental issues and refuse to participate in important international environmental accords, particularly those related to global warming, there is some reason to hope that a new environmental consciousness is awakening with the new global economic system. Unfortunately, increasing globalization of the economy has meant the increased spread of other things as well, such as conflict and infectious diseases, and economic development is increasing the release of greenhouse gases. The emergence of international terrorism as an instrument of national or quasi-national policy—particularly where terrorism may employ environmental contamination or disease as a weapon—has the potential to produce future environmental problems that are almost too frightening to think about. It has long been an accepted doctrine that we would all be better off when economic and political barriers were dropped. We are now learning that is not always the case.

In *Annual Editions: Environment 07/08* every effort has been made to choose articles that encourage an understanding of the nature of the environmental problems that beset us and how, with wisdom and knowledge and the proper perspective, they can be solved or at least mitigated. Accordingly, the selections in this book have been chosen more for their intellectual content than for their emotional tone. They have been arranged into an order of topics—the global environment; population, policy, and economy; energy; the biosphere; land and water resources; and pollution—that lends itself to a progressive understanding of the causes and effects of human modifications of Earth's environmental systems. We will not be protected against the ecological consequences of human actions by remaining ignorant of them.

Readers can have input into the next edition of *Annual Editions: Environment* by completing and returning the postpaid *article rating form* at the back of the book.

John L. Allen
*Editor*

# Contents

## UNIT 1
## The Global Environment: An Emerging World View

## UNIT 2
## Population, Policy, and Economy

The concepts in bold italics are developed in the article. For further expansion, please refer to the Topic Guide and the Index.

# UNIT 3
## Energy: Present and Future Problems

The concepts in bold italics are developed in the article. For further expansion, please refer to the Topic Guide and the Index.

# UNIT 4
## Biosphere: Endangered Species

# UNIT 5
## Resources: Land and Water

The concepts in bold italics are developed in the article. For further expansion, please refer to the Topic Guide and the Index.

# UNIT 6
# The Hazards of Growth: Pollution and Climate Change

The concepts in bold italics are developed in the article. For further expansion, please refer to the Topic Guide and the Index.

# Topic Guide

This topic guide suggests how the selections in this book relate to the subjects covered in your course. You may want to use the topics listed on these pages to search the Web more easily.

   On the following pages a number of Web sites have been gathered specifically for this book. They are arranged to reflect the units of this *Annual Edition*. You can link to these sites by going to the student online support site at *http://www.mhcls.com/online/*.

**ALL THE ARTICLES THAT RELATE TO EACH TOPIC ARE LISTED BELOW THE BOLD-FACED TERM.**

# Internet References

The following Internet sites have been carefully researched and selected to support the articles found in this reader. The easiest way to access these selected sites is to go to our student online support site at *http://www.mhcls.com/online/*.

# AE: Environment 07/08

The following sites were available at the time of publication. Visit our Web site—we update our student online support site regularly to reflect any changes.

## General Sources

### Britannica's Internet Guide
*http://www.britannica.com*

This site presents extensive links to material on world geography and culture, encompassing material on wildlife, human lifestyles, and the environment.

### CIA Factbook
*http://www.cia.gov/cia/publications/factbook*

This site is the United States government's official source for data on the population, production, resources, geography, political systems, and other important characteristics of each of the world's countries.

### EnviroLink
*http://www.envirolink.org*

One of the world's largest environmental information clearinghouses, EnviroLink is a grassroots nonprofit organization that unites organizations and volunteers around the world and provides up-to-date information and resources.

### Library of Congress
*http://www.loc.gov*

Examine this extensive Web site to learn about resource tools, library services/resources, exhibitions, and databases in many different subfields of environmental studies.

### The New York Times
*http://www.nytimes.com*

Browsing through the archives of the New York Times will provide a wide array of articles and information related to the different subfields of the environment.

### SocioSite: Sociological Subject Areas
*http://www.pscw.uva.nl/sociosite/TOPICS*

This huge sociological site from the University of Amsterdam provides many discussions and references of interest to students of the environment, such as the links to information on ecology and consumerism.

### U.S. Geological Survey
*http://www.usgs.gov*

This site and its many links are replete with information and resources in environmental studies, from explanations of El Niño to discussion of concerns about water resources.

## UNIT 1: The Global Envirwonment: An Emerging World View

### Alternative Energy Institute (AEI)
*http://www.altenergy.org*

The AEI will continue to monitor the transition from today's energy forms to the future in a "surprising journey of twists and turns." This site is the beginning of an incredible journey.

### Earth Science Enterprise
*http://www.earth.nasa.gov*

Information about NASA's Mission to Planet Earth program and its Science of the Earth System can be found here. Surf to learn about satellites, El Niño, and even "strategic visions" of interest to environmentalists.

### IISDnet
*http://www.iisd.org*

The International Institute for Sustainable Development, a Canadian organization, presents information through gateways entitled Business, Climate Change, Measurement and Assessment, and Natural Resources. IISD Linkages is its multimedia resource for environment and development policy makers.

### National Geographic Society
*http://www.nationalgeographic.com*

Links to National Geographic's huge archive are provided here. There is a great deal of material related to the atmosphere, the oceans, and other environmental topics.

### Research and Reference (Library of Congress)
*http://lcweb.loc.gov/rr*

This research and reference site of the Library of Congress will lead to invaluable information on different countries. It provides links to numerous publications, bibliographies, and guides in area studies that can be of great help to environmentalists.

### Solstice: Documents and Databases
*http://solstice.crest.org/index.html*

In this online source for sustainable energy information, the Center for Renewable Energy and Sustainable Technology (CREST) offers documents and databases on renewable energy, energy efficiency, and sustainable living. The site also offers related Web sites, case studies, and policy issues.

### United Nations
*http://www.unsystem.org*

Visit this official Web site Locator for the United Nations System of Organizations to get a sense of the scope of international environmental inquiry today. Various UN organizations concern themselves with everything from maritime law to habitat protection to agriculture.

### United Nations Environment Programme (UNEP)
*http://www.unep.ch*

Consult this home page of UNEP for links to critical topics of concern to environmentalists, including desertification, migratory species, and the impact of trade on the environment. The site will direct you to useful databases and global resource information.

### World Resources Institute (WRI)
*http://www.wri.org*

The World Resources Institute is committed to change for a sustainable world and believes that change in human behavior is urgently needed to halt the accelerating rate of environmental deterioration in some areas. It sponsors not only the general website above but also The Environmental Information Portal (www.earthtrends.wri.org) that provides a rich database on the interaction between human disease, pollution, and large-scale environmental, development, and demographic issues.

## UNIT 2: Population, Policy, and Economy

### The Hunger Project
*http://www.thp.org*

Browse through this nonprofit organization's site to explore the ways in which it attempts to achieve its goal: the sustainable end to global hunger through leadership at all levels of society. The Hunger Project contends that the persistence of hunger is at the heart of the major security issues that are threatening our planet.

### Poverty Mapping
*http://www.povertymap.net*

Poverty maps can quickly provide information on the spatial distribution of poverty. This site provides maps, graphics, data, publications, news, and links that provide the public with poverty mapping from the global to the subnational level.

### World Health Organization
*http://www.who.int*

The home page of the World Health Organization provides links to a wealth of statistical and analytical information about health and the environment in the developing world.

### World Population and Demographic Data
*http://geography.about.com/cs/worldpopulation*

On this site, information about world population and additional demographic data for all the countries of the world are provided.

### WWW Virtual Library: Demography & Population Studies
*http://demography.anu.edu.au/VirtualLibrary*

This is a definitive guide to demography and population studies. A multitude of important links to information about global poverty and hunger can be found here.

## UNIT 3: Energy: Present and Future Problems

### Alliance for Global Sustainability (AGS)
*http://globalsustainability.org*

The AGS is a cooperative venture seeking solutions to today's urgent and complex environmental problems. Research teams from four universities study large-scale, multidisciplinary environmental problems that are faced by the world's ecosystems, economies, and societies.

### Alternative Energy Institute, Inc.
*http://www.altenergy.org*

On this site created by a nonprofit organization, discover how the use of conventional fuels affects the environment. Also learn about research work on new forms of energy.

### Energy and the Environment: Resources for a Networked World
*http://zebu.uoregon.edu/energy.html*

An extensive array of materials having to do with energy sources—both renewable and nonrenewable—as well as other topics of interest to students of the environment is found on this site.

### Institute for Global Communication/EcoNet
*http://www.igc.org*

This environmentally friendly site provides links to dozens of governmental, organizational, and commercial sites having to do with energy sources. Resources address energy efficiency, renewable generating sources, global warming, and more.

### Nuclear Power Introduction
*http://library.thinkquest.org/17658/pdfs/nucintro.pdf*

Information regarding alternative energy forms can be accessed here. There is a brief introduction to nuclear power and a link to maps that show where nuclear power plants exist.

### U.S. Department of Energy
*http://www.energy.gov*

Scrolling through the links provided by this Department of Energy home page will lead to information about fossil fuels and a variety of sustainable/renewable energy sources.

## UNIT 4: Biosphere: Endangered Species

### Endangered Species
*http://www.endangeredspecie.com*

This site provides a wealth of information on endangered species anywhere in the world. Links providing data on the causes, interesting facts, law issues, case studies, and other issues on endangered species are available.

### Friends of the Earth
*http://www.foe.co.uk/index.html*

Friends of the Earth, a nonprofit organization based in the United Kingdom, pursues a number of campaigns to protect the Earth and its living creatures. This site has links to many important environmental sites, covering such broad topics as ozone depletion, soil erosion, and biodiversity.

### Natural Resources Defense Council
*http://nrdc.org*

The Natural Resources Defense Council (NRDC) uses law, science, and the support of more than 1 million members and activists to protect the planet's wildlife, plants, water, soils, and other resources. The site provides abundant information on global issues and political responses.

### Smithsonian Institution Web Site
*http://www.si.edu*

Looking through this site, which will provide access to many of the enormous resources of the Smithsonian, offers a sense of the biological diversity that is threatened by humans' unsound environmental policies and practices.

### World Wildlife Federation (WWF)
*http://www.wwf.org*

This home page of the WWF leads to an extensive array of information links about endangered species, wildlife management and preservation, and more. It provides many suggestions for how to take an active part in protecting the biosphere.

# www.mhcls.com/online/

## UNIT 5: Resources: Land and Water

### Global Climate Change
*http://www.puc.state.oh.us/consumer/gcc/index.html*

The goal of this PUCO (Public Utilities Commission of Ohio) site is to serve as a clearinghouse of information related to global climate change. Its extensive links provide an explanation of the science and chronology of global climate change, acronyms, definitions, and more.

### National Oceanic and Atmospheric Administration (NOAA)
*http://www.noaa.gov*

Through this home page of NOAA, you can find information about coastal issues, fisheries, climate, and more.

### National Operational Hydrologic Remote Sensing Center (NOHRSC)
*http://www.nohrsc.nws.gov*

Flood images are available at this site of the NOHRSC, which works with the U.S. National Weather Service to track weatherrelated information.

### Terrestrial Sciences
*http://www.cgd.ucar.edu/tss*

The Terrestrial Sciences Section (TSS) is part of the Climate and Global Dynamics (CGD) Division at the National Center for Atmospheric Research (NCAR) in Boulder, Colorado. Scientists in the section study land-atmosphere interactions, in particular surface forcing of the atmosphere, through model development, application, and observational analyses. Here, you'll find a link to VEMAP, The Vegetation/Ecosystem Modeling and Analysis Project.

## UNIT 6: The Hazards of Growth: Pollution and Climate Change

### Persistent Organic Pollutants (POP)
*http://www.chem.unep.ch/pops*

Visit this site to learn more about persistent organic pollutants (POPs) and the issues and concerns surrounding them.

### School of Labor and Industrial Relations (SLIR): Hot Links
*http://www.lir.msu.edu/hotlinks*

Michigan State University's SLIR page connects to industrial relations sites throughout the world. It has links to U.S. government statistics, newspapers and libraries, international intergovernmental organizations, and more.

### Space Research Institute
*http://arc.iki.rssi.ru/eng/index.htm*

For a change of pace, browse through this home page of Russia's Space Research Institute for information on its Environment Monitoring Information Systems, the IKI Satellite Situation Center, and its Data Archive.

### Worldwatch Institute
*http://www.worldwatch.org*

The Worldwatch Institute, dedicated to fostering the evolution of an environmentally sustainable society, presents this site with access to World Watch Magazine and State of the World 2000. Click on In the News and Press Releases for discussions of current problems.

**We highly recommend that you review our Web site for expanded information and our other product lines. We are continually updating and adding links to our Web site in order to offer you the most usable and useful information that will support and expand the value of your Annual Editions. You can reach us at:** http://www.mhcls.com/annualeditions.

# UNIT 1

# The Global Environment: An Emerging World View

## Unit Selections

## Key Points to Consider

- What are the connections between the attempts to develop sustainable systems and the quantity and quality of environmental data? Are there also relationships between data and the role of technology and economic systems in shaping the environmental future?

- What are some of the key "meta-trends" produced by increasing globalization of economic and other human systems? How can human societies and cultures adapt to such trends in order to prevent significant environmental disruption?

- How has the process of "globalization" altered the cultural and environmental patterns of the world? What kinds of changes brought about by an increasingly global economy have been unforeseen?

- What is the relationship between human attitudes and behavior and the attempts to develop systems of economic development that can be environmentally sustainable?

- What are some of the demographic trends now observable that were unpredicted a quarter of a century ago? Are there links between population changes and changes in economic and political systems?

## Student Web Site

www.mhcls.com/online

## Internet References

Further information regarding these Web sites may be found in this book's preface or online.

**Alternative Energy Institute (AEI)**
*http://www.altenergy.org*

**Earth Science Enterprise**
*http://www.earth.nasa.gov*

**IISDnet**
*http://www.iisd.org*

**National Geographic Society**
*http://www.nationalgeographic.com*

**Research and Reference (Library of Congress)**
*http://lcweb.loc.gov/rr*

**Solstice: Documents and Databases**
*http://solstice.crest.org/index.html*

**United Nations**
*http://www.unsystem.org*

**United Nations Environment Programme (UNEP)**
*http://www.unep.ch*

**World Resources Institute (WRI)**
*http://www.wri.org*

Nearly four decades after the celebration of the first Earth Day in 1970, public apprehension over the environmental future of the planet has reached levels unprecedented—even during the late 1960s and early 1970s in the "Age of Aquarius." No longer are those concerned about the environment dismissed as "ecofreaks" and "tree-huggers." Most serious scientists have joined the rising clamor for environmental protection, as have the more traditional environmentally conscious public-interest groups. There are a number of reasons for this increased environmental awareness. Some of these reasons arise from environmental events. The dramatic hurricane season of 2005 and its links to warmer ocean waters have, for example, made it increasingly difficult to deny the effects of global warming. Atmospheric scientists are nearly unanimous in their attribution of human agencies as at least a partial cause for increasing global temperatures.

But more reasons for environmental awareness arise simply from the process of globalization: the increasing unity of the world's economic, social, and information systems. Hailed by many as the salvation of the future, globalization has done little to make the world a better or safer place. Diseases once defined as "endemic" or confined to specific regions now have increased capacity for widespread dissemination and alarms have been sounded about new pandemic diseases—such as strains of flu with worldwide distribution. In addition, increased human mobility and the ease of travel has allowed human-caused disruptions to political, cultural, and economic systems to spread, and acts of terrorism now take place in locations once thought safe from such manifestations of hatred and despair. On the more positive side, the expansion of global information systems has fostered a maturation of concepts about the global nature of environmental processes.

Much of what has been learned through this increased information flow over the last two decades, particularly by American observers, has been of the environmentally ravaged world behind the old Iron Curtain—a chilling forecast of what other existing industrialized regions and rapidly industrializing countries such as India and China can become in the near future unless strict international environmental measures are put in place. For perhaps the first time ever, countries are beginning to recognize that environmental problems have no boundaries and that international cooperation is the only way to solve them.

The subtitle of this first unit, "An Emerging World View," is an optimistic assessment of the future: a future in which less money is spent on defense and more on environmental protection and cleanup—a new world order in which political influence might be based more on leadership in environmental and economic issues than on military might. It is probably far too early to make such optimistic predictions, to conclude that the world's nations—developed, developing, and underdeveloped—will begin to recognize that Earth's environment is a single unit. Thus far those nations have shown no tendency to recognize that humankind is a single unit and that what harms one harms all. The recent emergence of wide-scale terrorism and military action as a means of political and social policy is evidence of such a failure of recognition. Nevertheless, there is a growing international realization—aided by the information superhighway—that we are all, as environmental activists have been saying for decades, inhabitants of Spaceship Earth and will survive or succumb together.

The articles selected for this unit have been chosen to illustrate the increasingly global perspective on environmental problems and the degree to which their solutions must be linked to political, economic, and social problems and solutions. In the lead piece of the unit, "How Many Planets?" the editors of The Economist attempt an analysis of what they admit is a very slippery subject by beginning with the observation that "it comes as a shock to discover how little information there is on the environment." They note the lip service paid everywhere to the concept of sustainability and acknowledge that economic growth and environmental health are not mutually inconsistent but that a great deal more work is necessary to make them compatible. They also conclude that governments, corporations, and individuals are more prepared now to think about how to use the planet than they were even 10 years ago.

Issues surrounding globalization form the subject of the next selection in the unit. In "Five Meta-Trends Changing the World," David Snyder, lifestyles editor of The Futurist, recognizes the "meta-trends"—multidimensional and evolutionary trends—in the human-environmental systems are occurring as a result of a series of simultaneous demographic, economic, and technological trends. Snyder identifies these "meta-trends" as: (1) Cultural modernization, referring to the increasing "Westernization" of cultures in technological, economic, and demographic terms; (2) Economic globalization, meaning increasingly global competition for workers and resources and markets; (3) Universal connectivity or what the editor of The Economist has called "the death of distance," a recognition that the cell phone, the Internet, and other connective technologies have truly made the world smaller by linking people together on a virtually-instantaneous basis; (4) Transactional transparency, in which corporate integrity and openness will be forced to grow as watchdog groups and citizens demand a new transparency of business that will allow closer public scrutiny of business practices that impact the environment; and (5) Social adaptation, or responses to new medical and other technologies that will allow people to remain productive in the workplace for longer periods of time and may even produce a return to the multigenerational family of children, parents, and grandchildren that formed society's safety net prior to the onset of industrialization. Each of these trends has profound implications for the ways in which human societies around the world relate to the environmental systems they inhabit.

Some of the environmental consequences of the trend to globalization are discussed in the third article in this unit. Jo Kwong, Director of the Atlas Economic Research Foundation, notes that globalization is simply about the removal of barriers, allowing the free movement of goods, service, people, and ideas throughout the world. While this used to be viewed as a good thing, globalization has, in fact, raised many problems of both an economic and environmental nature. The author, however, tends to see the environmental issues—such as accelerated global warming through increasing economic development of the Third World and the concomitant increase in fossil fuel consumption—as producing a heightened global awareness and dialogue on global environmental issues. Globalization, Kwong notes, "can be a means to accelerate learning about the importance of market institutions to economic growth." It follows, she concludes, that environmental protection is one of the potential benefits of increasing markets and globalization.

The final two selections in the opening section deal with those issues of human intellectual and political response to the global changes in society, economy, and environmental relations that are part of today's world. In "Do Global Attitudes and Behaviors Support Sustainable Development?" a trio of leading experts on environmental issues and human behavior note that there are relatively few signs that societies around the world are prepared to embrace economic development that is environmentally sustainable. Anthony Leiserowitz, Robert Kates, and Thomas Parris also suggest that the types of attitudes that prevent ready acceptance of sustainable development are often highly resistant to change. They suggest, however, that by leveraging existing values such as self-interest and the desire for health and well-being, advocates of sustainable development can begin to turn things around. Long term changes in human attitudes and behavior, however, will require major shifts in human thinking about such things as race, environmental degradation, and human rights. A similar chord is struck by George Musser, staff writer and editor of *Scientific American.* In "The Climax of Humanity," Musser contends that the present is not what it was predicted to be a quarter-century ago, largely because of human attitudes toward resource management and environmental issues. Instead of flying cars we have diseases transmitted by air travel; instead of shorter work-weeks, most people are working more; instead of portable fusion energy, we have stratospheric gas prices brought on by increasingly short supplies of fossil fuel and economic and political manipulation. While prosperity is spreading, so are carbon dioxide emissions. Musser contends that many of the solutions to the environmental problems facing the world are fairly prosaic ones: management of fisheries for sustained yield; deciding to have fewer children; insulating homes more efficiently. The fact that we do not have—as was predicted—space colonies is less important than learning to manage the space that we occupy.

All the articles in this opening section deal with environmental problems that were once confined to specific locales but—as the world has grown increasingly smaller through advanced transportation, communication, and other technologies, and as a truly global economy has developed—have become problems on a global scale. While the potential for world-wide collapses of environmental systems is still a threat, the development of a global community with increasing awareness of the fragility of both economic and environmental systems provides a promise for the future.

# How Many Planets?

## *A Survey of the Global Environment*

## *The Great Race*

**Growth need not be the enemy of greenery. But much more effort is required to make the two compatible, says Vijay Vaitheeswaran**

Sustainable development is a dangerously slippery concept. Who could possibly be against something that invokes such alluring images of untouched wildernesses and happy creatures? The difficulty comes in trying to reconcile the "development" with the "sustainable" bit: look more closely, and you will notice that there are no people in the picture.

That seems unlikely to stop a contingent of some of 60,000 world leaders, businessmen, activists, bureaucrats and journalists from travelling to South Africa next month for the UN-sponsored World Summit on Sustainable Development in Johannesburg. Whether the summit achieves anything remains to be seen, but at least it is asking the right questions. This survey will argue that sustainable development cuts to the heart of mankind's relationship with nature—or, as Paul Portney of Resources for the Future, an American think-tank, puts it, "the great race between development and degradation". It will also explain why there is reason for hope about the planet's future.

The best way known to help the poor today—economic growth—has to be handled with care, or it can leave a degraded or even devastated natural environment for the future. That explains why ecologists and economists have long held diametrically opposed views on development. The difficult part is to work out what we owe future generations, and how to reconcile that moral obligation with what we owe the poorest among us today.

It is worth recalling some of the arguments fielded in the run-up to the big Earth Summit in Rio de Janeiro a decade ago. A publication from UNESCO, a United Nations agency, offered the following vision of the future: "Every generation should leave water, air and soil resources as pure and unpolluted as when it came on earth. Each generation should leave undiminished all the species of animals it found existing on earth." Man, that suggests, is but a strand in the web of life, and the natural order is fixed and supreme. Put earth first, it seems to say.

Robert Solow, an economist at the Massachusetts Institute of Technology, replied at the time that this was "fundamentally the wrong way to go", arguing that the obligation to the future is "not to leave the world as we found it in detail, but rather to leave the option or the capacity to be as well off as we are." Implicit in that argument is the seemingly hard-hearted notion of "fungibility": that natural resources, whether petroleum or giant pandas, are substitutable.

## Rio's Fatal Flaw

Champions of development and defenders of the environment have been locked in battle ever since a UN summit in Stockholm launched the sustainable-development debate three decades ago. Over the years, this debate often pitted indignant politicians and social activists from the poor world against equally indignant politicians and greens from the rich world. But by the time the Rio summit came along, it seemed they had reached a truce. With the help of a committee of grandees led by Gro Harlem Brundtland, a former Norwegian prime minister, the interested parties struck a deal in 1987: development and the environment, they declared, were inextricably linked. That compromise generated a good deal of euphoria. Green groups grew concerned over poverty, and development charities waxed lyrical about greenery. Even the World Bank joined in. Its World Development Report in 1992 gushed about "win-win" strategies, such as ending environmentally harmful subsidies, that would help both the economy and the environment.

By nearly universal agreement, those grand aspirations have fallen flat in the decade since that summit. Little headway has been made with environmental problems such as climate change and loss of biodiversity. Such progress as has been achieved has been largely due to three factors that this survey will explore in later sections: more decision-making at local level, technological innovation, and the rise of market forces in environmental matters.

The main explanation for the disappointment—and the chief lesson for those about to gather in South Africa—is that Rio overreached itself. Its participants were so anxious to reach a political consensus that they agreed to the Brundtland definition of sustainable development, which Daniel Esty of Yale University thinks has turned into "a buzz-word largely devoid of content". The biggest mistake, he reckons, is that it slides over the difficult trade-offs between environment and development in the real world. He is careful to note that there are plenty of cases where those goals are linked—but also many where they are not: "Environmental and economic policy goals are distinct, and the actions needed to achieve them are not the same."

## No Such Thing as Win-Win

To insist that the two are "impossible to separate", as the Brundtland commission claimed, is nonsense. Even the World Bank now accepts that its much-trumpeted 1992 report was much too optimistic. Kristalina Georgieva, the Bank's director for the environment, echoes comments from various colleagues when she says: "I've never seen a real win-win in my life. There's always somebody, usually an elite group grabbing rents, that loses. And we've learned in the past decade that those losers fight hard to make sure that technically elegant win-win policies do not get very fat."

So would it be better to ditch the concept of sustainable development altogether? Probably not. Even people with their feet firmly planted on the ground think one aspect of it is worth salvaging: the emphasis on the future.

Nobody would accuse John Graham of jumping on green bandwagons. As an official in President George Bush's Office of Management and Budget, and previously as head of Harvard University's Centre for Risk Analysis, he has built a reputation for evidence-based policymaking. Yet he insists sustainable development is a worthwhile concept: "It's good therapy for the tunnel vision common in government ministries, as it forces integrated policymaking. In practical terms, it means that you have to take economic cost-benefit trade-offs into account in environmental laws, and keep environmental trade-offs in mind with economic development."

Jose Maria Figueres, a former president of Costa Rica, takes a similar view. "As a politician, I saw at first hand how often policies were dictated by short-term considerations such as elections or partisan pressure. Sustainability is a useful template to align short-term policies with medium- to long-term goals."

It is not only politicians who see value in saving the sensible aspects of sustainable development. Achim Steiner, head of the International Union for the Conservation of Nature, the world's biggest conservation group, puts it this way: "Let's be honest: greens and businesses do not have the same objective, but they can find common ground. We look for pragmatic ways to save species. From our own work on the ground on poverty, our members—be they bird watchers or passionate ecologists—have learned that 'sustainable use' is a better way to conserve."

Sir Robert Wilson, boss of Rio Tinto, a mining giant, agrees. He and other business leaders say it forces hard choices about the future out into the open: "I like this concept because it frames the trade-offs inherent in a business like ours. It means that single-issue activism is simply not as viable."

Kenneth Arrow and Larry Goulder, two economists at Stanford University, suggest that the old ideological enemies are converging: "Many economists now accept the idea that natural capital has to be valued, and that we need to account for ecosystem services. Many ecologists now accept that prohibiting everything in the name of protecting nature is not useful, and so are being selective." They think the debate is narrowing to the more empirical question of how far it is possible to substitute natural capital with the man-made sort, and specific forms of natural capital for one another.

## The Job for Johannesburg

So what can the Johannesburg summit contribute? The prospects are limited. There are no big, set-piece political treaties to be signed as there were at Rio. America's acrimonious departure from the Kyoto Protocol, a UN treaty on climate change, has left a bitter taste in many mouths. And the final pre-summit gathering, held in early June in Indonesia, broke up in disarray. Still, the gathered worthies could usefully concentrate on a handful of areas where international co-operation can help deal with environmental problems. Those include improving access for the poor to cleaner energy and to safe drinking water, two areas where concerns about human health and the environmental overlap. If rich countries want to make progress, they must agree on firm targets and offer the money needed to meet them. Only if they do so will poor countries be willing to cooperate on problems such as global warming that rich countries care about.

That seems like a modest goal, but it just might get the world thinking seriously about sustainability once again. If the Johannesburg summit helps rebuild a bit of faith in international environmental cooperation, then it will have been worthwhile. Minimising the harm that future economic growth does to the environment will require the rich world to work hand in glove with the poor world—which seems nearly unimaginable in today's atmosphere poisoned by the shortcomings of Rio and Kyoto.

To understand why this matters, recall that great race between development and degradation. Mankind has stayed comfortably ahead in that race so far, but can it go on doing so? The sheer magnitude of the economic growth that is hoped for in the coming decades makes it seem inevitable that the clashes between mankind and nature will grow worse. Some are now asking whether all this economic growth is really necessary or useful in the first place, citing past advocates of the simple life.

"God forbid that India should ever take to industrialism after the manner of the West. . . It took Britain half the resources of the planet to achieve this prosperity. How many planets will a country like India require?", Mahatma Gandhi asked half a century ago. That question encapsulated the bundle of worries that haunts the sustainable-development debate to this day. Today, the vast majority of Gandhi's countrymen are still living the simple life—full of simple misery, malnourishment and material want. Grinding poverty, it turns out, is pretty sustainable.

If Gandhi were alive today, he might look at China next door and find that the country, once as poor as India, has been transformed beyond recognition by two decades of roaring economic growth. Vast numbers of people have been lifted out of poverty and into middle-class comfort. That could prompt him to reframe his question: how many planets will it take to satisfy China's needs if it ever achieves profligate America's affluence? One green group reckons the answer is three. The next section looks at the environmental data that might underpin such claims. It makes for alarming reading—though not for the reason that first springs to mind.

# Flying Blind

## It comes as a shock to discover how little information there is on the environment

What is the true state of the planet? It depends from which side you are peering at it. "Things are really looking up," comes the cry from one corner (usually overflowing with economists and technologists), pointing to a set of rosy statistics. "Disaster is nigh," shouts the other corner (usually full of ecologists and environmental lobbyists), holding up a rival set of troubling indicators.

According to the optimists, the 20th century marked a period of unprecedented economic growth that lifted masses of people out of abject poverty. It also brought technological innovations such as vaccines and other advances in public health that tackled many preventable diseases. The result has been a breathtaking enhancement of human welfare and longer, better lives for people everywhere on earth.

At this point, the pessimists interject: "Ah, but at what ecological cost?" They note that the economic growth which made all these gains possible sprang from the rapid spread of industrialisation and its resource-guzzling cousins, urbanisation, motorisation and electrification. The earth provided the necessary raw materials, ranging from coal to pulp to iron. Its ecosystems—rivers, seas, the atmosphere—also absorbed much of the noxious fallout from that process. The sheer magnitude of ecological change resulting directly from the past century's economic activity is remarkable.

To answer that Gandhian question about how many planets it would take if everybody lived like the West, we need to know how much—or how little—damage the West's transformation from poverty to plenty has done to the planet to date. Economists point to the remarkable improvement in local air and water pollution in the rich world in recent decades. "It's Getting Better All the Time", a cheerful tract co-written by the late Julian Simon, insists that: "One of the greatest trends of the past 100 years has been the astonishing rate of progress in reducing almost every form of pollution." The conclusion seems unavoidable: "Relax! If we keep growing as usual, we'll inevitably grow greener."

The ecologically minded crowd takes a different view. "GEO3", a new report from the United Nations Environment Programme, looks back at the past few decades and sees much reason for concern. Its thoughtful boss, Klaus Töpfer (a former German environment minister), insists that his report is not "a document of doom and gloom". Yet, in summing it up, UNEP decries "the declining environmental quality of planet earth", and wags a finger at economic prosperity: "Currently, one-fifth of the world's population enjoys high, some would say excessive, levels of affluence." The conclusion seems unavoidable: "Panic! If we keep growing as usual, we'll inevitably choke the planet to death."

"People and Ecosystems", a collaboration between the World Resources Institute, the World Bank and the United Nations, tried to gauge the condition of ecosystems by examining the goods and services they produce—food, fibre, clean water, carbon storage and so on—and their capacity to continue producing them. The authors explain why ecosystems matter: half of all jobs worldwide are in agriculture, forestry and fishing, and the output from those three commodity businesses still dominates the economies of a quarter of the world's countries.

The report reached two chief conclusions after surveying the best available environmental data. First, a number of ecosystems are "fraying" under the impact of human activity. Second, ecosystems in future will be less able than in the past to deliver the goods and services human life depends upon, which points to unsustainability. But it took care to say: "It's hard, of course, to know what will be truly sustainable." The reason this collection of leading experts could not reach a firm conclusion was that, remarkably, much of the information they needed was incomplete or missing altogether: "Our knowledge of ecosystems has increased dramatically, but it simply has not kept pace with our ability to alter them."

Another group of experts, this time organised by the World Economic Forum, found itself similarly frustrated. The leader of that project, Daniel Esty of Yale, exclaims, throwing his arms in the air: "Why hasn't anyone done careful environmental measurement before? Businessmen always say, 'what matters gets measured.' Social scientists started quantitative measurement 30 years ago, and even political science turned to hard numbers 15 years ago. Yet look at environmental policy, and the data are lousy."

# Gaping Holes

At long last, efforts are under way to improve environmental data collection. The most ambitious of these is the Millennium Ecosystem Assessment, a joint effort among leading development agencies and environmental groups. This four-year effort is billed as an attempt to establish systematic data sets on all environmental matters across the world. But one of the researchers involved grouses that it "has very, very little new money to collect or analyse new data". It seems astonishing that governments have been making sweeping decisions on environmental policy for decades without such a baseline in the first place.

One positive sign is the growing interest of the private sector in collecting environmental data. It seems plain that leaving the task to the public sector has not worked. Information on the environment comes far lower on the bureaucratic pecking order than data on education or social affairs, which tend to be overseen by ministries with bigger budgets and more political clout. A number of countries, ranging from New Zealand to Austria, are now looking to the private sector to help collect and manage data in areas such as climate. Development banks are also considering using private contractors to monitor urban air quality, in part to get around the corruption and apathy in some city governments.

"I see a revolution in environmental data collection coming because of computing power, satellite mapping, remote sensing and other such information technologies," says Mr Esty. The arrival of hard data in this notoriously fuzzy area could cut down on environmental disputes by reducing uncertainty. One example is the long-running squabble between America's midwestern states, which rely heavily on coal, and the north-eastern states, which suffer from acid rain. Technology helped disprove claims by the mid-western states that New York's problems all resulted from home-grown pollution.

The arrival of good data would have other benefits as well, such as helping markets to work more robustly: witness America's pioneering scheme to trade emissions of sulphur dioxide, made possible by fancy equipment capable of monitoring emissions in real time. Mr Esty raises an even more intriguing possibility: "Like in the American West a hundred years ago, when barbed wire helped establish rights and prevent overgrazing, information technology can help establish 'virtual barbed wire' that secures property rights and so prevents overexploitation of the commons." He points to fishing in the waters between Australia and New Zealand, where tracking and monitoring devices have reduced over-exploitation.

Best of all, there are signs that the use of such fancy technology will not be confined to rich countries. Calestous Juma of Harvard University shares Mr Esty's excitement about the possibility of such a technology-driven revolution even in Africa: "In the past, the only environmental 'database' we had in Africa was our grandmothers. Now, with global information systems and such, the potential is enormous." Conservationists in Namibia, for example, already use satellite tracking to keep count of their elephants. Farmers in Mali receive satellite updates about impending storms on hand-wound radios. Mr Juma thinks the day is not far off when such technology, combined with ground-based monitoring, will help Africans measure trends in deforestation, soil erosion and climate change, and assess the effects on their local environment.

# Make a Start

That is at once a sweeping vision and a modest one. Sweeping, because it will require heavy investment in both sophisticated hardware and nuts-and-bolts information infrastructure on the ground to make sense of all these new data. As the poor world clearly cannot afford to pay for all this, the rich world must help—partly for altruistic reasons, partly with the selfish aim of discovering in good time whether any global environmental calamities are in the making. A number of multilateral agencies now say they are willing to invest in this area as a "neglected global public good"—neglected especially by those agencies themselves. Even President Bush's administration has recently indicated that it will give environmental satellite data free to poor countries.

But that vision is also quite a modest one. Assuming that this data "revolution" does take place, all it will deliver is a reliable assessment of the health of the planet today. We will still not be able to answer the broader question of whether current trends are sustainable or not.

To do that, we need to look more closely at two very different sorts of environmental problems: global crises and local troubles. The global sort is hard to pin down, but can involve irreversible changes. The local kind is common and can have a big effect on the qualify of life, but is usually reversible. Data on both are predictably inadequate. We turn first to the most elusive environmental problem of all, global warming.

# *Blowing Hot and Cold*

## Climate change may be slow and uncertain, but that is no excuse for inaction

What would Winston Churchill have done about climate change? Imagine that Britain's visionary wartime leader had been presented with a potential time bomb capable of wreaking global havoc, although not certain to do so. Warding it off would require concerted global action and economic sacrifice on the home front. Would he have done nothing?

Not if you put it that way. After all, Churchill did not dismiss the Nazi threat for lack of conclusive evidence of Hitler's evil intentions. But the answer might be less straightforward if the following provisos had been added: evidence of this problem would remain cloudy for decades; the worst effects might not be felt for a century; but the costs of tackling the problem would start biting immediately. That, in a nutshell, is the dilemma of climate change. It is asking a great deal of politicians to take action on behalf of voters who have not even been born yet.

One reason why uncertainty over climate looks to be with us for a long time is that the oceans, which absorb carbon from the atmosphere, act as a time-delay mechanism. Their massive thermal inertia means that the climate system responds only very slowly to changes in the composition of the atmosphere. Another complication arises from the relationship between carbon dioxide ($CO_2$), the principal greenhouse gas (GHG), and sulphur dioxide ($SO_2$), a common pollutant. Efforts to reduce man-made emissions of GHGs by cutting down on fossil-fuel use will reduce emissions of both gases. The reduction in $CO_2$ will cut warming, but the concurrent $SO_2$ cut may mask that effect by contributing to the warming.

There are so many such fuzzy factors—ranging from aerosol particles to clouds to cosmic radiation—that we are likely to see disruptions to familiar climate patterns for many years without knowing why they are happening or what to do about them. Tom Wigley, a leading climate scientist and member of the UN's Intergovernmental Panel on Climate Change (IPCC), goes further. He argues in an excellent book published by the Aspen Institute, "US Policies on Climate Change: What Next?", that whatever policy changes governments pursue, scientific uncertainties will "make it difficult to detect the effects of such changes, probably for many decades."

As evidence, he points to the negligible short- to medium-term difference in temperature resulting from an array of emissions "pathways" on which the world could choose to embark if it decided to tackle climate change. He plots various strategies for reducing GHGs (including the Kyoto one) that will lead in the next century to the stabilisation of atmospheric concentrations of $CO_2$ at 550 parts per million (ppm). That is roughly double the level which prevailed in pre-industrial times, and is often mooted by climate scientists as a reasonable target. But even by 2040, the temperature differences between the various options will still be tiny—and certainly within the magnitude of natural climatic variance. In short, in another four decades we will probably still not know if we have over- or undershot.

## Ignorance Is Not Bliss

However, that does not mean we know nothing. We do know, for a start, that the "greenhouse effect" is real: without the heat-trapping effect of water vapour, $CO_2$, methane and other naturally occurring GHGs, our planet would be a lifeless 30° C or so colder. Some of these GHG emissions are captured and stored by "sinks", such as the oceans, forests and agricultural land, as part of nature's carbon cycle.

We also know that since the industrial revolution began, mankind's actions have contributed significantly to that greenhouse effect. Atmospheric concentrations of GHGs have risen from around 280ppm two centuries ago to around 370ppm today, thanks chiefly to mankind's use of fossil fuels and, to a

lesser degree, to deforestation and other land-use changes. Both surface temperatures and sea levels have been rising for some time.

There are good reasons to think temperatures will continue rising. The IPCC has estimated a likely range for that increase of 1.4° C–5.8° C over the next century, although the lower end of that range is more likely. Since what matters is not just the absolute temperature level but the rate of change as well, it makes sense to try to slow down the increase.

The worry is that a rapid rise in temperatures would lead to climate changes that could be devastating for many (though not all) parts of the world. Central America, most of Africa, much of south Asia and northern China could all be hit by droughts, storms and floods and otherwise made miserable. Because they are poor and have the misfortune to live near the tropics, those most likely to be affected will be least able to adapt.

The colder parts of the world may benefit from warming, but they too face perils. One is the conceivable collapse of the Atlantic "conveyor belt", a system of currents that gives much of Europe its relatively mild climate; if temperatures climb too high, say scientists, the system may undergo radical changes that damage both Europe and America. That points to the biggest fear: warming may trigger irreversible changes that transform the earth into a largely uninhabitable environment.

Given that possibility, extremely remote though it is, it is no comfort to know that any attempts to stabilise atmospheric concentrations of GHGs at a particular level will take a very long time. Because of the oceans' thermal inertia, explains Mr Wigley, even once atmospheric concentrations of GHGs are stabilised, it will take decades or centuries for the climate to follow suit. And even then the sea level will continue to rise, perhaps for millennia.

This is a vast challenge, and it is worth bearing in mind that mankind's contribution to warming is the only factor that can be controlled. So the sooner we start drawing up a long-term strategy for climate change, the better.

What should such a grand plan look like? First and foremost, it must be global. Since $CO_2$ lingers in the atmosphere for a century or more, any plan must also extend across several generations.

The plan must recognise, too, that climate change is nothing new: the climate has fluctuated through history, and mankind has adapted to those changes—and must continue doing so. In the rich world, some of the more obvious measures will include building bigger dykes and flood defences. But since the most vulnerable people are those in poor countries, they too have to be helped to adapt to rising seas and unpredictable storms. Infrastructure improvements will be useful, but the best investment will probably be to help the developing world get wealthier.

It is essential to be clear about the plan's long-term objective. A growing chorus of scientists now argues that we need to keep temperatures from rising by much more than 2–3° C in all. That will require the stabilisation of atmospheric concentrations of GHGs. James Edmonds of the University of Maryland points out that because of the long life of $CO_2$, stabilisation of $CO_2$ concentrations is not at all the same thing as stabilisation of $CO_2$ emissions. That, says Mr Edmonds, points to an unavoidable conclusion: "In the very long term, global net $CO_2$ emissions must eventually peak and gradually decline toward zero, regardless of whether we go for a target of 350ppm or 1,000ppm."

# A Low-Carbon World

That is why the long-term objective for climate policy must be a transition to a low-carbon energy system. Such a transition can be very gradual and need not necessarily lead to a world powered only by bicycles and windmills, for two reasons that are often overlooked.

One involves the precise form in which the carbon in the ground is distributed. According to Michael Grubb of the Carbon Trust, a British quasi-governmental body, the long-term problem is coal. In theory, we can burn all of the conventional oil and natural gas in the ground and still meet the most ambitious goals for tackling climate change. If we do that, we must ensure that the far greater amounts of carbon trapped as coal (and unconventional resources like tar sands) never enter the atmosphere.

The snag is that poor countries are likely to continue burning cheap domestic reserves of coal for decades. That suggests the rich world should speed the development and diffusion of "low carbon" technologies using the energy content of coal without releasing its carbon into the atmosphere. This could be far off, so it still makes sense to keep a watchful eye on the soaring carbon emissions from oil and gas.

The other reason, as Mr Edmonds took care to point out, is that it is net emissions of $CO_2$ that need to peak and decline. That leaves scope for the continued use of fossil fuels as the main source of modern energy if only some magical way can be found to capture and dispose of the associated $CO_2$. Happily, scientists already have some magic in the works.

One option is the biological "sequestration" of carbon in forests and agricultural land. Another promising idea is capturing and storing $CO_2$—underground, as a solid or even at the bottom of the ocean. Planting "energy crops" such as switch-grass and using them in conjunction with sequestration techniques could even result in negative net $CO_2$ emissions, because such plants use carbon from the atmosphere. If sequestration is combined with techniques for stripping the hydrogen out of this hydrocarbon, then coal could even offer a way to sustainable hydrogen energy.

But is anyone going to pay attention to these long-term principles? After all, over the past couple of years all participants in the Kyoto debate have excelled at producing short-sighted, selfish and disingenuous arguments. And the political rift continues: the EU and Japan pushed ahead with ratification of the Kyoto treaty a month ago, whereas President Bush reaffirmed his opposition.

However, go back a decade and you will find precisely those principles enshrined in a treaty approved by the elder George Bush and since reaffirmed by his son: the UN Framework Convention on Climate Change (FCCC). This treaty was perhaps the most important outcome of the Rio summit, and it remains the basis for the international climate-policy regime, including Kyoto.

The treaty is global in nature and long-term in perspective. It commits signatories to pursuing "the stabilisation of GHG concentrations in the atmosphere at a level that would prevent dangerous interference with the climate system." Note that the agreement covers GHG concentrations, not merely emissions. In effect, this commits even gas-guzzling America to the goal of declining emissions.

# Better than Kyoto

Crucially, the FCCC treaty not only lays down the ends but also specifies the means: any strategy to achieve stabilisation of GHG concentrations, it insists, "must not be disruptive of the global economy". That was the stumbling block for the Kyoto treaty, which is built upon the FCCC agreement: its targets and timetables proved unrealistic.

Any revised Kyoto treaty or follow-up accord (which must include the United States and the big developing countries) should rest on the three basic pillars. First, governments everywhere (but especially in Europe) must understand that a reduction in emissions has to start modestly. That is because the capital stock involved in the global energy system is vast and long-lived, so a dash to scrap fossil-fuel production would be hugely expensive. However, as Mr Grubb points out, that pragmatism must be flanked by policies that encourage a switch to low-carbon technologies when replacing existing plants.

Second, governments everywhere (but especially in America) must send a powerful signal that carbon is going out of fashion. The best way to do this is to levy a carbon tax. However, whether it is done through taxes, mandated restrictions on GHG emissions or market mechanisms is less important than that the signal is sent clearly, forcefully and unambiguously. This is where President Bush's mixed signals have done a lot of harm: America's industry, unlike Europe's, has little incentive to invest in low-carbon technology. The irony is that even some coal-fired utilities in America are now clamouring for $CO_2$ regulation so that they can invest in new plants with confidence.

The third pillar is to promote science and technology. That means encouraging basic climate and energy research, and giving incentives for spreading the results. Rich countries and aid agencies must also find ways to help the poor world adapt to climate change. This is especially important if the world starts off with small cuts in emissions, leaving deeper cuts for later. That, observes Mr Wigley, means that by mid-century "very large investments would have to have been made—and yet the 'return' on these investments would not be visible. Continued investment is going to require more faith in climate science than currently appears to be the case."

Even a visionary like Churchill might have lost heart in the face of all this uncertainty. Nevertheless, there is a glimmer of hope that today's peacetime politicians may rise to the occasion.

# Miracles Sometimes Happen

Two decades ago, the world faced a similar dilemma: evidence of a hole in the ozone layer. Some inconclusive signs suggested that it was man-made, caused by the use of chlorofluorocarbons (CFCS).

There was the distant threat of disaster, and the knowledge about a concerted global response was required. Industry was reluctant at first, yet with leadership from Britain and America the Montreal Protocol was signed in 1987. That deal has proved surprisingly successful. The manufacture of CFCs is nearly phased out, and there are already signs that the ozone layer is on the way to recovery.

This story holds several lessons for the admittedly far more complex climate problem. First, it is the rich world which has caused the problem and which must lead the way in solving it. Second, the poor world must agree to help, but is right to insist on being given time—as well as money and technology—to help it adjust. Third, industry holds the key: in the ozone-depletion story, it was only after DuPont and ICI broke ranks with the rest of the CFC manufacturers that a deal became possible. On the climate issue, BP and Shell have similarly broken ranks with Big Oil, but the American energy industry—especially the coal sector—remains hostile.

The final lesson is the most important: that the uncertainty surrounding a threat such as climate change is no excuse for inaction. New scientific evidence shows that the threat from ozone depletion had been much deadlier than was thought at the time when the world decided to act. Churchill would surely have approved.

# Local Difficulties

## Greenery is for the poor too, particularly on their own doorstep

Why should we care about the environment? Ask a European, and he will probably point to global warming. Ask the two little boys playing outside a newsstand in Da Shilan, a shabby neighbourhood in the heart of Beijing, and they will tell you about the city's notoriously foul air: "It's bad—like a virus!"

Given all the media coverage in the rich world, people there might believe that global scares are the chief environmental problems facing humanity today. They would be wrong. Partha Dasgupta, an economics professor at Cambridge University, thinks the current interest in global, future-oriented problems has "drawn attention away from the economic misery and ecological degradation endemic in large parts of the world today. Disaster is not something for which the poorest have to wait; it is a frequent occurrence."

Every year in developing countries, a million people die from urban air pollution and twice that number from exposure to stove smoke inside their homes. Another 3m unfortunates die prematurely every year from water-related diseases. All told, premature deaths and illnesses arising from environmental factors account for about a fifth of all diseases in poor countries, bigger than any other preventable factor, including malnutrition. The problem is so serious that Ian Johnson, the World Bank's vice-president for the environment, tells his colleagues, with a touch of irony, that he is really the bank's vice-president for health: "I say tackling the underlying environmental causes of health problems will do a lot more good than just more hospitals and drugs."

The link between environment and poverty is central to that great race for sustainability. It is a pity, then, that several powerful fallacies keep getting in the way of sensible debate. One popular myth is that trade and economic growth make poor countries' environmental problems worse. Growth, it is said, brings with it urbanisation, higher energy consumption and industrialisation—all factors that contribute to pollution and pose health risks.

In a static world, that would be true, because every new factory causes extra pollution. But in the real world, economic growth unleashes many dynamic forces that, in the longer run, more than offset that extra pollution. Traditional environmental risks (such as water-borne diseases) cause far more health problems in poor countries than modern environmental risks (such as industrial pollution).

## Rigged Rules

However, this is not to say that trade and economic growth will solve all environmental problems. Among the reasons for doubt are the "perverse" conditions under which world trade is carried on, argues Oxfam. The British charity thinks the rules of trade are "unfairly rigged against the poor", and cites in evidence the enormous subsidies lavished by rich countries on industries such as agriculture, as well as trade protection offered to manufacturing industries such as textiles. These measurements hurt the environment because they force the world's poorest countries to rely heavily on commodities—a particularly energy-intensive and ungreen sector.

Mr Dasgupta argues that this distortion of trade amounts to a massive subsidy of rich-world consumption paid by the world's poorest people. The most persuasive critique of all goes as follows: "Economic growth is not sufficient for turning environmental degradation around. If economic incentives facing producers and consumers do not change with higher incomes, pollution will continue to grow unabated with the growing scale of economic activity." Those words come not from some anti-globalist green group, but from the World Trade Organisation.

Another common view is that poor countries, being unable to afford greenery, should pollute now and clean up later. Certainly poor countries should not be made to adopt American or European environmental standards. But there is evidence to suggest that poor countries can and should try to tackle some environmental problems now, rather than wait till they have become richer.

This so-called "smart growth" strategy contradicts conventional wisdom. For many years, economists have observed that as agrarian societies industrialised, pollution increased at first, but as the societies grew wealthier it declined again. The trouble is that this applies only to some pollutants, such as sulphur dioxide, but not to others, such as carbon dioxide. Even more troublesome, those smooth curves going up, then down, turn out to be misleading. They are what you get when you plot data for poor and rich countries together at a given moment in time, but actual levels of various pollutants in any individual country plotted over time wiggle around a lot more. This suggests that the familiar bell-shaped curve reflects no immutable law, and that intelligent government policies might well help to reduce pollution levels even while countries are still relatively poor.

Developing countries are getting the message. From Mexico to the Philippines, they are now trying to curb the worst of the air and water pollution that typically accompanies industrialisation. China, for example, was persuaded by outside experts that it was losing so much potential economic output through health troubles caused by pollution (according to one World Bank study, somewhere between 3.5% and 7.7% of GDP) that tackling it was cheaper than ignoring it.

One powerful—and until recently ignored—weapon in the fight for a better environment is local people. Old-fashioned paternalists in the capitals of developing countries used to argue that poor villagers could not be relied on to look after natural resources. In fact, much academic research has shown that the poor are more often victims than perpetrators of resource depletion: it tends to be rich locals or outsiders who are responsible for the worst exploitation.

Local people usually have a better knowledge of local ecological conditions than experts in faraway capitals, as well as a direct interest in improving the quality of life in their village. A good example of this comes from the bone-dry state of Rajasthan in India, where local activism and indigenous know-how about rainwater "harvesting" provided the people with reliable water supplies—something the government had failed to do. In Bangladesh, villages with active community groups or concerned mullahs proved greener than less active neighbouring villages.

Community-based forestry initiatives from Bolivia to Nepal have shown that local people can be good custodians of nature. Several hundred million of the world's poorest people live in and around forests. Giving those villagers an incentive to preserve forests by allowing sustainable levels of harvesting, it turns out, is a far better way to save those forests than erecting tall fences around them.

To harness local energies effectively, it is particularly important to give local people secure property rights, argues Mr Dasgupta. In most parts of the developing world, control over resources at the village level is ill-defined. This often means that local elites usurp a disproportionate share of those resources, and that individuals have little incentive to maintain and upgrade forests or agricultural land. Authorities in Thailand tried to remedy this problem by distributing 5.5m land titles over a 20-year period. Agricultural output increased, access to credit improved and the value of the land shot up.

# Name and Shame

Another powerful tool for improving the local environment is the free flow of information. As local democracy flourishes, ordinary people are pressing for greater environmental disclosure by companies. In some countries, such as Indonesia, governments have adopted a "sunshine" policy that involves naming and shaming companies that do not meet environmental regulations. It seems to achieve results.

Bringing greenery to the grass roots is good, but on its own it will not avert perceived threats to global "public goods" such as the climate or biodiversity. Paul Portney of Resources for the Future explains: "Brazilian villagers may think very carefully and unselfishly about their future descendants, but there's no reason for them to care about and protect species or habitats that no future generation of Brazilians will care about."

That is why rich countries must do more than make pious noises about global threats to the environment. If they believe that scientific evidence suggests a credible threat, they must be willing to pay poor countries to protect such things as their tropical forests. Rather than thinking of this as charity, they should see it as payment for environmental services (say, for carbon storage) or as a form of insurance.

In the case of biodiversity, such payments could even be seen as a trade in luxury goods: rich countries would pay poor countries to look after creatures that only the rich care about. Indeed, private green groups are already buying up biodiversity "hot spots" to protect them. One such initiative, led by Conservation International and the International Union for the Conservation of Nature (IUCN), put the cost of buying and preserving 25 hot spots exceptionally rich in species diversity at less than $30 billion. Sceptics say it will cost more, as hot spots will need buffer zones of "sustainable harvesting" around them. Whatever the right figure, such creative approaches are more likely to achieve results than bullying the poor into conservation.

It is not that the poor do not have green concerns, but that those concerns are very different from those of the rich. In Beijing's Da Shilan, for instance, the air is full of soot from the many tiny coal boilers. Unlike most of the neighbouring districts, which have recently converted from coal to natural gas, this area has been considered too poor to make the transition. Yet ask Liu Shihua, a shopkeeper who has lived in the same spot for over 20 years, and he insists he would readily pay a bit more for the cleaner air that would come from using natural gas. So would his neighbours.

To discover the best reason why poor countries should not ignore pollution, ask those two little boys outside Mr Liu's shop what colour the sky is. "Grey!" says one tyke, as if it were the most obvious thing in the world. "No, stupid, it's blue!" retorts the other. The children deserve blue skies and clean air. And now there is reason to think they will see them in their lifetime.

# Working Miracles

## Can technology save the planet?

"Nothing endures but change." That observation by Heraclitus often seems lost on modern environmental thinkers. Many invoke scary scenarios assuming that resources—both natural ones, like oil, and man-made ones, like knowledge—are fixed. Yet in real life man and nature are entwined in a dynamic dance of development, scarcity, degradation, innovation and substitution.

The nightmare about China turning into a resource-guzzling America raises two questions: will the world run out of resources? And even if it does not, could the growing affluence of developing nations lead to global environmental disaster?

The first fear is the easier to refute; indeed, history has done so time and again. Malthus, Ricardo and Mill all worried that scarcity of resources would snuff out growth. It did not. A few decades ago, the limits-to-growth camp raised worries that the world might soon run out of oil, and that it might not be able to feed the world's exploding population. Yet there are now more proven reserves of petroleum than three decades ago; there is more food produced than ever; and the past decade has seen history's greatest economic boom.

What made these miracles possible? Fears of oil scarcity prompted investment that led to better ways of producing oil, and to more efficient engines. In food production, technological advances have sharply reduced the amount of land required to feed a person in the past 50 years. Jesse Ausubel of Rockefeller University calculates that if in the next 60 to 70 years the world's average farmer reaches the yield of today's average (not best) American maize grower, then feeding 10 billion people will require just half of today's cropland. All farmers need to do is maintain the 2%-a-year productivity gain that has been the global norm since 1960.

"Scarcity and Growth", a book published by Resources for the Future, sums it up brilliantly: "Decades ago Vermont granite was only building and tombstone material; now it is a potential fuel, each ton of which has a usable energy content (uranium) equal to 150 tons of coal. The notion of an absolute limit to natural resource availability is untenable when the definition of resources changes drastically and unpredictably over time." Those words were written by Harold Barnett and Chandler Morse in 1963, long before the limits-to-growth bandwagon got rolling.

## Giant Footprint

Not so fast, argue greens. Even if we are not going to run out of resources, guzzling ever more resources could still do irreversible damage to fragile ecosystems.

WWF, an environmental group, regularly calculates mankind's "ecological footprint", which it defines as the "biologically productive land and water areas required to produce the resources consumed and assimilate the wastes generated by a given population using prevailing technology." The group reckons the planet has around 11.4 billion "biologically productive" hectares of land available to meet continuing human needs. WWF thinks mankind has recently been using more than that. This is possible because a forest harvested at twice its regeneration rate, for example, appears in the footprint accounts at twice its area—an unsustainable practice which the group calls "ecological overshoot."

Any analysis of this sort must be viewed with scepticism. Everyone knows that environmental data are incomplete. What is more, the biggest factor by far is the land required to absorb $CO_2$ emissions of fossil fuels. If that problem could be managed some other way, then mankind's ecological footprint would look much more sustainable.

Even so, the WWF analysis makes an important point: if China's economy were transformed overnight into a clone of America's, an ecological nightmare could ensue. If a billion eager new consumers were suddenly to produce $CO_2$ emissions at American rates, they would be bound to accelerate global warming. And if the whole of the developing world were to adopt an American lifestyle tomorrow, local environmental crises such as desertification, aquifer depletion and topsoil loss could make humans miserable.

So is this cause for concern? Yes, but not for panic. The global ecological footprint is determined by three factors: population size, average consumption per person and technology. Fortunately, global population growth now appears to be moderating. Consumption per person in poor countries is rising as they become better off, but there are signs that the rich world is reducing the footprint of its consumption (as this survey's final section explains). The most powerful reason for hope—innovation—was foreshadowed by WWF's own definition. Today's "prevailing technologies" will, in time, be displaced by tomorrow's greener ones.

"The rest of the world will not live like America," insists Mr Ausubel. Of course poor people around the world covet the creature comforts that Americans enjoy, but they know full well that the economic growth needed to improve their lot will take time. Ask Wu Chengjian, an environmental official in booming Shanghai, what he thinks of the popular notion that his city might become as rich as today's Hong Kong by 2020: "Impossible—that's just not enough time." And that is Shanghai, not the impoverished countryside.

## Leaps of Faith

This extra time will allow poor countries to embrace new technologies that are more efficient and less environmentally damaging. That still does not guarantee a smaller ecological footprint for China in a few decades' time than for America now, but it greatly improves the chances. To see why, consider the history of "dematerialisation" and "decarbonisation". Viewed across very long spans of time, productivity improvements allow economies to use ever fewer material inputs—and to emit ever fewer pollutants—per unit of economic output. Mr Ausubel concludes: "When China has today's American mobility, it will not have today's American cars," but the cleaner and more efficient cars of tomorrow.

The snag is that consumers in developing countries want to drive cars not tomorrow but today. The resulting emissions have led many to despair that technology (in the form of vehicles) is making matters worse, not better.

Can they really hope to "leapfrog" ahead to cleaner air? The evidence from Los Angeles—a pioneer in the fight against air pollution—suggests the answer is yes. "When I moved to Los Angeles in the 1960s, there was so much soot in the air that it felt like there was a man standing on your chest most of the time," says Ron Loveridge, the mayor of Riverside, a city to the east of LA that suffers the worst of the region's pollution. But, he says, "We have come an extraordinary distance in LA."

Four decades ago, the city had the worst air quality in America. The main problem was the city's infamous "smog" (an amalgam of "smoke" and "fog"). It took a while to figure out that this unhealthy ozone soup developed as a result of complex chemical reactions between nitrogen oxides and volatile organic compounds that need sunlight to trigger them off.

Arthur Winer, an atmospheric chemist at the University of California at Los Angeles, explains that tackling smog required tremendous perseverance and political will. Early regulatory efforts met stiff resistance from business interests, and began to falter when they failed to show dramatic results.

Clean-air advocates like Mr Loveridge began to despair: "We used to say that we needed a 'London fog' [a reference to an air-pollution episode in 1952 that may have killed 12,000 people in that city] here to force change." Even so, Californian officials forged ahead with an ambitious plan that combined regional regulation with stiff mandates for cleaner air. Despite uncertainties about the cause of the problem, the authorities introduced a sequence of controversial measures: unleaded and low-sulphur petrol, on-board diagnostics for cars to minimise emissions, three-way catalytic converters, vapour-recovery attachments for petrol nozzles and so on.

As a result, the city that two decades ago hardly ever met federal ozone standards has not had to issue a single alert in the past three years. Peak ozone levels are down by 50% since the 1960s. Though the population has shot up in recent years, and the vehicle-miles driven by car-crazy Angelenos have tripled, ozone levels have fallen by two-thirds. The city's air is much cleaner than it was two decades ago.

"California, in solving its air-quality problem, has solved it for the rest of the United States and the world—but it doesn't get credit for it," says Joe Norbeck of the University of California at Riverside. He is adamant that the poor world's cities can indeed leapfrog ahead by embracing some of the cleaner technologies developed specifically for the Californian market. He points to

China's vehicle fleet as an example: "China's typical car has the emissions of a 1974 Ford Pinto, but the new Buicks sold there use 1990s emissions technology." The typical car sold today produces less than a tenth of the local pollution of a comparable model from the 1970s.

That suggests one lesson for poor cities such as Beijing that are keen to clean up: they can order polluters to meet high emissions standards. Indeed, from Beijing to Mexico city, regulators are now imposing rich-world rules, mandating new, cleaner technologies. In China's cities, where pollution from sooty coal fires in homes and industrial boilers had been a particular hazard, officials are keen to switch to natural-gas furnaces.

However, there are several reasons why such mandates—which worked wonders in LA—may be trickier to achieve in impoverished or politically weak cities. For a start, city officials must be willing to pay the political price of reforms that raise prices for voters. Besides, higher standards for new cars, useful though they are, cannot do the trick on their own. Often, clean technologies such as catalytic converters will require cleaner grades of petrol too. Introducing cleaner fuels, say experts, is an essential lesson from LA for poor countries. This will not come free either.

There is another reason why merely ordering cleaner new cars is inadequate: it does nothing about the vast stock of dirty old ones already on the streets. In most cities of the developing world, the oldest fifth of the vehicles on the road is likely to produce over half of the total pollution caused by all vehicles taken together. Policies that encourage a speedier turnover of the fleet therefore make more sense than "zero emissions" mandates.

## Policy Matters

In sum, there is hope that the poor can leapfrog at least some environmental problems, but they need more than just technology. Luisa and Mario Molina of the Massachusetts Institute of Technology, who have studied such questions closely, reckon that technology is less important than the institutional capacity, legal safeguards and financial resources to back it up: "The most important underlying factor is political will." And even a techno-optimist such as Mr Ausubel accepts that: "There is nothing automatic about technological innovation and adoption; in fact, at the micro level, it's bloody."

Clearly innovation is a powerful force, but government policy still matters. That suggests two rules for policymakers. First, don't do stupid things that inhibit innovation. Second, do sensible things that reward the development and adoption of technologies that enhance, rather than degrade, the environment.

The greatest threat to sustainability may well be the rejection of science. Consider Britain's hysterical reaction to genetically modified crops, and the European Commission's recent embrace of a woolly "precautionary principle". Precaution applied case-by-case is undoubtedly a good thing, but applying any such principle across the board could prove disastrous.

Explaining how not to stifle innovation that could help the environment is a lot easier than finding ways to encourage it. Technological change often goes hand-in-hand with greenery by saving resources, as the long history of dematerialisation shows—but not always. Sports utility vehicles, for instance, are technologically innovative, but hardly green. Yet if those SUVs were to come with hydrogen-powered fuel cells that emit little pollution, the picture would be transformed.

The best way to encourage such green innovations is to send powerful signals to the market that the environment matters. And there is no more powerful signal than price, as the next section explains.

# The Invisible Green Hand

## Markets could be a potent force for greenery— if only greens could learn to love them

"**M**andate, regulate and litigate." That has been the environmentalists' rallying cry for ages. Nowhere in the green manifesto has there been much mention of the market. And, oddly, it was market-minded America that led the dirigiste trend. Three decades ago, Congress passed a sequence of laws, including the Clean Air Act, which set lofty goals and generally set rigid technological standards. Much of the world followed America's lead.

This top-down approach to greenery has long been a point of pride for groups such as the Natural Resources Defence Council (NRDC), one of America's most influential environmental outfits. And with some reason, for it has had its successes: the air and water in the developed world is undoubtedly cleaner than it was three decades ago, even though the rich world's economies have grown by leaps and bounds. This has convinced such groups stoutly to defend the green status quo.

But times may be changing. Gus Speth, now head of Yale University's environment school and formerly head of the World Resources Institute and the UNDP, as well as one of the founders of the NRDC, recently explained how he was converted to market economics: "Thirty years ago, the economists at Resources for the Future were pushing the idea of pollution taxes. We lawyers at NRDC thought they were nuts, and feared that they would derail command-and-control measures like the Clean Air Act, so we opposed them. Looking back, I'd have to say this was the single biggest failure in environmental management—not getting the prices right."

A remarkable mea culpa; but in truth, the command-and-control approach was never as successful as its advocates claimed. For example, although it has cleaned up the air and water in rich countries, it has notably failed in dealing with waste management, hazardous emissions and fisheries depletion. Also, the gains achieved have come at a needlessly high price. That is because technology mandates and bureaucratic edicts stifle innovation and ignore local realities, such as varying costs of abatement. They also fail to use cost-benefit analysis to judge trade-offs.

Command-and-control methods will also be ill-suited to the problems of the future, which are getting trickier. One reason is that the obvious issues—like dirty air and water—have been tackled already. Another is increasing technological complexity: future problems are more likely to involve subtle linkages—like those involved in ozone depletion and global warming—that will require sophisticated responses. The most important factor may be society's ever-rising expectations; as countries grow wealthier, their people start clamouring for an ever-cleaner environment. But because the cheap and simple things have been done, that is proving increasingly expensive. Hence the greens' new interest in the market.

## Carrots, Not Just Sticks

In recent years, market-based greenery has taken off in several ways. With emissions trading, officials decide on a pollution target and then allocate tradable credits to companies based on that target. Those that find it expensive to cut emissions can buy credits from those that find it cheaper, so the target is achieved at the minimum cost and disruption.

The greatest green success story of the past decade is probably America's innovative scheme to cut emissions of sulphur dioxide ($SO_2$). Dan Dudek of Environmental Defence, a most unusual green group, and his market-minded colleagues persuaded the elder George Bush to agree to an amendment to the sacred Clean Air Act that would introduce an emissions-trading system to achieve sharp cuts in $SO_2$. At the time, this was hugely controversial: America's power industry insisted the cuts were prohibitively costly, while nearly every other green group decried the measure as a sham. In the event, ED has been vindicated. America's scheme has surpassed its initial objectives, and at far lower cost than expected. So great is the interest worldwide in trading that ED is now advising groups ranging from hard-nosed oilmen at BP to bureaucrats in China and Russia.

Europe, meanwhile, is forging ahead with another sort of market-based instrument: pollution taxes. The idea is to levy charges on goods and services so that their price reflects their "externalities"—jargon for how much harm they do to the environment and human health. Sweden introduced a sulphur tax a decade ago, and found that the sulphur content of fuels dropped 50% below legal requirements.

Though "tax" still remains a dirty word in America, other parts of the world are beginning to embrace green tax reform by shifting taxes from employment to pollution. Robert Williams of Princeton University has looked at energy use (especially the terrible effects on health of particulate pollution) and concluded that such externalities are comparable in size to the direct economic costs of producing that energy.

Externalities are only half the battle in fixing market distortions. The other half involves scrapping environmentally harmful subsidies. These range from prices below market levels for electricity and water to shameless cash handouts for industries such as coal. The boffins at the OECD reckon that stripping away harmful subsidies, along with introducing taxes on carbon-based fuels and chemicals use, would result in dramatically lower emissions by 2020 than current policies would be able to achieve. If the revenues raised were then used to reduce other taxes, the cost of these virtuous policies would be less than 1% of the OECD's economic output in 2020.

Such subsidies are nothing short of perverse, in the words of Norman Myers of Oxford University. They do double damage, by distorting markets and by encouraging behaviour that harms the environment. Development banks say such subsidies add up to $700 billion a year, but Mr Myers reckons the true sum is closer to $2 trillion a year. Moreover, the numbers do not fully reflect the harm done. For example, EU countries subsidise their fishing fleets to the tune of $1 billion a year, but that has encouraged enough overfishing to drive many North Atlantic fishing grounds to near-collapse.

Fishing is an example of the "tragedy of the commons", which pops up frequently in the environmental debate. A resource such as the ocean is common to many, but an individual "free rider" can benefit from plundering that commons or dumping waste into it, knowing that the costs of his actions will probably be distributed among many neighbours. In the case of shared fishing grounds, the absence of individual ownership drives each fisherman to snatch as many fish as he can—to the detriment of all.

# Of Rights and Wrongs

Assigning property rights can help, because providing secure rights (set at a sustainable level) aligns the interests of the individual with the wider good of preserving nature. This is what sceptical conservationists have observed in New Zealand and Iceland, where schemes for tradable quotas have helped revive fishing stocks. Similar rights-based approaches have led to revivals in stocks of African elephants in southern Africa, for example, where the authorities stress property rights and private conservation.

All this talk of property rights and markets makes many mainstream environmentalists nervous. Carl Pope, the boss of the Sierra Club, one of America's biggest green groups, does not reject market forces out of hand, but expresses deep scepticism about their scope. Pointing to the difficult problem of climate change, he asks: "Who has property rights over the commons?"

Even so, some greens have become converts. Achim Steiner of the IUCN reckons that the only way forward is rights-based conservation, allowing poor people "sustainable use" of their local environment. Paul Faeth of the World Resources Institute goes further. He says he is convinced that market forces could deliver that holy grail of environmentalism, sustainability—"but only if we get prices right."

# The Limits to Markets

Economic liberals argue that the market itself is the greatest price-discovery mechanism known to man. Allow it to function freely and without government meddling, goes the argument, and prices are discovered and internalised automatically. Jerry Taylor of the Cato Institute, a libertarian think-tank, insists that "The world today is already sustainable—except those parts where western capitalism doesn't exist." He notes that countries that have relied on central planning, such as the Soviet Union, China and India, have invariably misallocated investment, stifled innovation and fouled their environment far more than the prosperous market economies of the world have done.

All true. Even so, markets are currently not very good at valuing environmental goods. Noble attempts are under way to help them do better. For example, the Katoomba Group, a collection of financial and energy companies that have linked up with environmental outfits, is trying to speed the development of markets for some of forestry's ignored "co-benefits" such as carbon storage and watershed management, thereby producing new revenue flows for forest owners. This approach shows promise: water consumers ranging from officials in New York City to private hydro-electric operators in Costa Rica are now paying people upstream to manage their forests and agricultural land better. Paying for greenery upstream turns out to be cheaper than cleaning up water downstream after it has been fouled.

Economists too are getting into the game of helping capitalism "get prices right." The World Bank's Ian Johnson argues that conventional economic measures such as gross domestic product are not measuring wealth creation properly because they ignore the effects of environmental degradation. He points to the positive contribution to China's GDP from the logging industry, arguing that such a calculation completely ignores the billions of dollars-worth of damage from devastating floods caused by over-logging. He advocates a more comprehensive measure the Bank is working on, dubbed "genuine GDP", that tries (imperfectly, he accepts) to measure depletion of natural resources.

That could make a dramatic difference to how the welfare of the poor is assessed. Using conventional market measures, nearly the whole of the developing world save Africa has grown wealthier in the past couple of decades. But when the degradation of nature is properly accounted for, argues Mr Dasgupta at Cambridge, the countries of Africa and south Asia are actually much worse off today than they were a few decades ago—and even China, whose economic "miracle" has been much trumpeted, comes out barely ahead.

The explanation, he reckons, lies in a particularly perverse form of market distortion: "Countries that are exporting resource-based products (often among the poorest) may be subsidising the consumption of countries that are doing the importing (often among the richest)." As evidence, he points to the common practice in poor countries of encouraging resource extraction. Whether through licenses granted at below-market rates, heavily subsidised exports or corrupt officials tolerating illegal exploitation, he reckons the result is the same: "The cruel paradox we face may well be that contemporary economic development is unsustainable in poor countries because it is sustainable in rich countries."

One does not have to agree with Mr Dasgupta's conclusion to acknowledge that markets have their limits. That should not dissuade the world from attempting to get prices right—or at least to stop getting them so wrong. For grotesque subsidies, the direction of change should be obvious. In other areas, the market itself may not provide enough information to value nature adequately. This is true of threats to essential assets, such as nature's ability to absorb and "recycle" $CO_2$, that have no substitute at any price. That is when governments must step in, ensuring that an informed public debate takes place.

Robert Stavins of Harvard University argues that the thorny notion of sustainable development can be reduced to two simple ideas: efficiency and intergenerational equity. The first is about making the economic pie as large as possible; he reckons that economists are well equipped to handle it, and that market-based policies can be used to achieve it. On the second (the subject of the next section), he is convinced that markets must yield to public discourse and government policy: "Markets can be efficient, but nobody ever said they're fair. The question is, what do we owe the future?"

# Insuring a Brighter Future

## How to hedge against tomorrow's environmental risks

So what do we owe the future? A precise definition for sustainable development is likely to remain elusive but, as this survey has argued, the hazy outline of a useful one is emerging from the experience of the past decade.

For a start, we cannot hope to turn back the clock and return nature to a pristine state. Nor must we freeze nature in the state it is today, for that gift to the future would impose an unacceptable burden on the poorest alive today. Besides, we cannot forecast the tastes, demands or concerns of future generations. Recall that the overwhelming pollution problem a century ago was horse manure clogging up city streets: a century hence, many of today's problems will surely seem equally irrelevant. We should therefore think of our debt to the future as including not just natural resources but also technology, institutions and especially the capacity to innovate. Robert Solow got it mostly right a decade ago: the most important thing to leave future generations, he said, is the capacity to live as well as we do today.

However, as the past decade has made clear, there is a limit to that argument. If we really care about the "sustainable" part of sustainable development, we must be much more watchful about environmental problems with critical thresholds. Most local problems are reversible and hence no cause for alarm. Not all, however: the depletion of aquifers and the loss of topsoil could trigger irreversible changes that would leave future generations worse off. And global or long-term threats, where victims are far removed in time and space, are easy to brush aside.

In areas such as biodiversity, where there is little evidence of a sustainability problem, a voluntary approach is best. Those in the rich world who wish to preserve pandas, or hunt for miracle drugs in the rainforest, should pay for their predilections. However, where there are strong scientific indications of unsustainability, we must act on behalf of the future—even at the price of today's development. That may be expensive, so it is prudent to try to minimise those risks in the first place.

## A Riskier World

Human ingenuity and a bit of luck have helped mankind stay a few steps ahead of the forces degrading the environment this past century, the first full one in which the planet has been exposed to industrialisation. In the century ahead, the great race between development and degradation could well become a closer call.

On one hand, the demands of development seem sure to grow at a cracking pace in the next few decades as the Chinas, Indias and Brazils of this world grow wealthy enough to start enjoying not only the necessities but also some of the luxuries of life. On the other hand, we seem to be entering a period of huge technological advances in emerging fields such as biotechnology that could greatly increase resource productivity and more than offset the effect of growth on the environment. The trouble is, nobody knows for sure.

Since uncertainty will define the coming era, it makes sense to invest in ways that reduce that risk at relatively low cost. Governments must think seriously about the future implications of today's policies. Their best bet is to encourage the three powerful forces for sustainability outlined in this survey: the empowerment of local people to manage local resources and adapt to environmental change; the encouragement of science and technology, especially innovations that reduce the ecological footprint of consumption; and the greening of markets to get prices right.

To advocate these interventions is not to call for a return to the hubris of yesteryear's central planners. These measures would merely give individuals the power to make greener choices if they care to. In practice, argues Chris Heady of the OECD, this may still not add up to sustainability "because we might still decide to be greedy, and leave less for our children."

Happily, there are signs of an emerging bottom-up push for greenery. Even such icons of western consumerism as Unilever and Procter & Gamble now sing the virtues of "sustainable consumption." Unilever has vowed that by 2005 it will be buying fish only from sustainable sources, and P&G is coming up with innovative products such as detergents that require less water, heat and packaging. It would be naive to label such actions as expressions of "corporate social responsibility": in the long run, firms will embrace greenery only if they see profit in it. And that, in turn, will depend on choices made by individuals.

Such interventions should really be thought of as a kind of insurance that tilts the odds of winning that great race just a little in humanity's favour. Indeed, even some of the world's most conservative insurance firms increasingly see things this way. As losses from weather-related disasters have risen of late, the industry is getting more involved in policy debates on long-term environmental issues such as climate change.

Bruno Porro, chief risk officer at Swiss Re, argues that: "The world is entering a future in which risks are more concentrated and more complex. That is why we are pressing for policies that reduce those risks through preparation, adaptation and mitigation. That will be cheaper than covering tomorrow's losses after disaster strikes."

Jeffrey Sachs of Columbia University agrees: "When you think about the scale of risk that the world faces, it is clear that we grossly underinvest in knowledge . . . we have enough income to live very comfortably in the developed world and to prevent dire need in the developing world. So we should have the confidence to invest in longer-term issues like the environment. Let's help insure the sustainability of this wonderful situation."

He is right. After all, we have only one planet, now and in the future. We need to think harder about how to use it wisely.

# Acknowledgements

In addition to those cited in the text, the author would like to thank Robert Socolow, David Victor, Geoffrey Heal, and experts at Tsinghua University, Friends of the Earth, the European Commission, the World Business Council for Sustainable Development, the International Energy Agency, the OECD and the UN for sharing their ideas with him. A list of sources can be found on *The Economist's* website.

# Five Meta-Trends Changing the World

**Global, overarching forces such as modernization and widespread interconnectivity are converging to reshape our lives. But human adaptability—itself a "meta-trend"—will help keep our future from spinning out of control, assures THE FUTURIST's lifestyles editor.**

DAVID PEARCE SNYDER

Last year, I received an e-mail from a long-time Australian client requesting a brief list of the "meta-trends" having the greatest impact on global human psychology. What the client wanted to know was, which global trends would most powerfully affect human consciousness and behavior around the world?

The Greek root *meta* denotes a transformational or transcendent phenomenon, not simply a big, pervasive one. A meta-trend implies multidimensional or catalytic change, as opposed to a linear or sequential change.

What follows are five meta-trends I believe are profoundly changing the world. They are evolutionary, system-wide developments arising from the simultaneous occurrence of a number of individual demographic, economic, and technological trends. Instead of each being individual freestanding global trends, they are composites of trends.

## Trend 1—Cultural Modernization

Around the world over the past generation, the basic tenets of modern cultures—including equality, personal freedom, and self-fulfillment—have been eroding the domains of traditional cultures that value authority, filial obedience, and self-discipline. The children of traditional societies are growing up wearing Western clothes, eating Western food, listening to Western music, and (most importantly of all) thinking Western thoughts. Most Westerners—certainly most Americans—have been unaware of the personal intensities of this culture war because they are so far away from the "battle lines." Moreover, people in the West regard the basic institutions of modernization, including universal education, meritocracy, and civil law, as benchmarks of social progress, while the defenders of traditional cultures see them as threats to social order.

Demographers have identified several leading social indicators as key measures of the extent to which a nation's culture is modern. They cite the average level of education for men and for women, the percentage of the salaried workforce that is female, and the percentage of population that lives in urban areas. Other indicators include the percentage of the workforce that is salaried (as opposed to self-employed) and the percentage of GDP spent on institutionalized socioeconomic support services, including insurance, pensions, social security, civil law courts, worker's compensation, unemployment benefits, and welfare.

As each of these indicators rises in a society, the birthrate in that society goes down. The principal measurable consequence of cultural modernization is declining fertility. As the world's developing nations have become better educated, more urbanized, and more institutionalized during the past 20 years, their birthrates have fallen dramatically. In 1988, the United Nations forecast that the world's population would double to 12 billion by 2100. In 1992, their estimate dropped to 10 billion, and they currently expect global population to peak at 9.1 billion in 2100. After that, demographers expect the world's population will begin to slowly decline, as has already begun to happen in Europe and Japan.

> **Three signs that a culture is modern: its citizens' average level of education, the number of working women, and the percentage of the population that is urban. As these numbers increase, the birthrate in a society goes down, writes author David Pearce Snyder.**

The effects of cultural modernization on fertility are so powerful that they are reflected clearly in local vital statistics. In India, urban birthrates are similar to those in the United States, while rural birthrates remain unmanageably high. Cultural modernization is the linchpin of human sustainability on the planet.

The forces of cultural modernization, accelerated by economic globalization and the rapidly spreading wireless telecommunications info-structure, are likely to marginalize the world's traditional cultures well before the century is over. And because the wellsprings of modernization—secular industrial economies—are so unassailably powerful, terrorism is the only means by which the defenders of traditional culture can fight to preserve their values and way of life. In the near-term future, most observers believe that ongoing cultural conflict is likely to produce at least a few further extreme acts of terrorism, security measures not withstanding. But the eventual intensity and duration of the overt, violent phases of the ongoing global culture war are largely matters of conjecture. So, too, are the expert pronouncements of the probable long-term impacts of September 11, 2001, and terrorism on American priorities and behavior.

After the 2001 attacks, social commentators speculated extensively that those events would change America. Pundits posited that we would become more motivated by things of intrinsic value—children, family, friends, nature, personal self-fulfillment—and that we would see a sharp increase in people pursuing *pro bono* causes and public-service careers. A number of media critics predicted that popular entertainment such as television, movies, and games would feature much less gratuitous violence after September 11. None of that has happened. Nor have Americans become more attentive to international news coverage. Media surveys show that the average American reads less international news now than before September 11. Event-inspired changes in behavior are generally transitory. Even if current conflicts produce further extreme acts of terrorist violence, these seem unlikely to alter the way we live or make daily decisions. Studies in Israel reveal that its citizens have become habituated to terrorist attacks. The daily routine of life remains the norm, and random acts of terrorism remain just that: random events for which no precautions or mind-set can prepare us or significantly reduce our risk.

In summary, cultural modernization will continue to assault the world's traditional cultures, provoking widespread political unrest, psychological stress, and social tension. In developed nations, where the great majority embrace the tenets of modernization and where the threats from cultural conflict are manifested in occasional random acts of violence, the ongoing confrontation between tradition and modernization seems likely to produce security measures that are inconvenient, but will do little to alter our basic personal decision making, values, or day-to-day life. Developed nations are unlikely to make any serious attempts to restrain the spread of cultural modernization or its driving force, economic globalization.

# Trend 2—Economic Globalization

On paper, globalization poses the long-term potential to raise living standards and reduce the costs of goods and services for people everywhere. But the short-term marketplace consequences of free trade threaten many people and enterprises in both developed and developing nations with potentially insurmountable competition. For most people around the world, the threat from foreign competitors is regarded as much greater than the threat from foreign terrorists. Of course, risk and uncertainty in daily life is characteristically high in developing countries. In developed economies, however, where formal institutions sustain order and predictability, trade liberalization poses unfamiliar risks and uncertainties for many enterprises. It also appears to be affecting the collective psychology of both blue-collar and white-collar workers—especially males—who are increasingly unwilling to commit themselves to careers in fields that are likely to be subject to low-cost foreign competition.

Strikingly, surveys of young Americans show little sign of xenophobia in response to the millions of new immigrant workers with whom they are competing in the domestic job market. However, they feel hostile and helpless at the prospect of competing with Chinese factory workers and Indian programmers overseas. And, of course, economic history tells us that they are justifiably concerned. In those job markets that supply untariffed international industries, a "comparable global wage" for comparable types of work can be expected to emerge worldwide. This will raise workers' wages for freely traded goods and services in developing nations, while depressing wages for comparable work in mature industrial economies. To earn more than the comparable global wage, labor in developed nations will have to perform *incomparable* work, either in terms of their productivity or the superior characteristics of the goods and services that they produce. The assimilation of mature information technology throughout all production and education levels should make this possible, but developed economies have not yet begun to mass-produce a new generation of high-value-adding, middle-income jobs.

Meanwhile, in spite of the undeniable short-term economic discomfort that it causes, the trend toward continuing globalization has immense force behind it. Since World War II, imports have risen from 6% of world GDP to more than 22%, growing steadily throughout the Cold War, and even faster since 1990. The global dispersion of goods production and the uneven distribution of oil, gas, and critical minerals worldwide have combined to make international interdependence a fundamental economic reality, and corporate enterprises are building upon that reality. Delays in globalization, like the September 2003 World Trade Organization contretemps in Cancun, Mexico, will arise as remaining politically sensitive issues are resolved, including trade in farm products, professional and financial services, and the need for corporate social responsibility. While there will be enormous long-term economic benefits from globalization in both developed and developing nations, the short-term disruptions in local domestic employment will make free trade an ongoing political issue that will be manageable only so long as domestic economies continue to grow.

# Trend 3—Universal Connectivity

While information technology (IT) continues to inundate us with miraculous capabilities, it has given us, so far, only one new power that appears to have had a significant impact on our collective behavior: our improved ability to communicate with each other, anywhere, anytime. Behavioral researchers have found that cell phones have blurred or changed the boundaries between work and social life and between personal and public life. Cell phones have also increased users' propensity to "micromanage their lives, to be more spontaneous, and, therefore, to be late for everything," according to Leysia Palen, computer science professor at the University of Colorado at Boulder.

---

**Cell phones have blurred the lines between the public and the private. Nearly everyone is available anywhere, anytime—and in a decade cyberspace will be a town square, writes Snyder.**

---

Most recently, instant messaging—via both cell phones and online computers—has begun to have an even more powerful social impact than cell phones themselves. Instant messaging initially tells you whether the person you wish to call is "present" in cyberspace—that is, whether he or she is actually online at the moment. Those who are present can be messaged immediately, in much the same way as you might look out the window and call to a friend you see in the neighbor's yard. Instant messaging gives a physical reality to cyberspace. It adds a new dimension to life: A person can now be "near," "distant," or "in cyberspace." With video instant messaging—available now, and widely available in three years—the illusion will be complete. We will have achieved what Frances Cairncross, senior editor of *The Economist,* has called "the death of distance."

Universal connectivity will be accelerated by the integration of the telephone, cell phone, and other wireless telecom media with the Internet. By 2010, all long-distance phone calls, plus a third of all local calls, will be made via the Internet, while 80% to 90% of all Internet access will be made from Web-enabled phones, PDAs, and wireless laptops. Most important of all, in less than a decade, one-third of the world's population— 2 billion people—will have access to the Internet, largely via Web-enabled telephones. In a very real sense, the Internet will be the "Information Highway"—the infrastructure, or infostructure, for the computer age. The infostructure is already speeding the adoption of flexplace employment and reducing the volume of business travel, while making possible increased "distant collaboration," outsourcing, and offshoring.

---

**Corporate integrity and openness will grow steadily under pressure from watchdog groups and ordinary citizens demanding business transparency. The leader of tomorrow must adapt to this new openness or risk business disaster.**

---

As the first marketing medium with a truly global reach, the Internet will also be the crucible from which a global consumer culture will be forged, led by the first global youth peer culture. By 2010, we will truly be living in a global village, and cyberspace will be the town square.

# Trend 4—Transactional Transparency

Long before the massive corporate malfeasance at Enron, Tyco, and WorldCom, there was a rising global movement toward greater transparency in all private and public enterprises. Originally aimed at kleptocratic regimes in Africa and the former Soviet states, the movement has now become universal, with the establishment of more stringent international accounting standards and more comprehensive rules for corporate oversight and record keeping, plus a new UN treaty on curbing public-sector corruption. Because secrecy breeds corruption and incompetence, there is a growing worldwide consensus to expose the principal transactions and decisions of *all* enterprises to public scrutiny.

But in a world where most management schools have dropped all ethics courses and business professors routinely preach that government regulation thwarts the efficiency of the marketplace, corporate and government leaders around the world are lobbying hard against transparency mandates for the private sector. Their argument: Transparency would "tie their hands," "reveal secrets to their competition," and "keep them from making a fair return for their stockholders."

Most corporate management is resolutely committed to the notion that secrecy is a necessary concomitant of leadership. But pervasive, ubiquitous computing and comprehensive electronic documentation will ultimately make all things transparent, and this may leave many leaders and decision makers feeling uncomfortably exposed, especially if they were not provided a moral compass prior to adolescence. Hill and Knowlton, an international public-relations firm, recently surveyed 257 CEOs in the United States, Europe, and Asia regarding the impact of the Sarbanes-Oxley Act's reforms on corporate accountability and governance. While more than 80% of respondents felt that the reforms would significantly improve corporate integrity, 80% said they also believed the reforms would not increase ethical behavior by corporate leaders.

While most consumer and public-interest watchdog groups are demanding even more stringent regulation of big business, some corporate reformers argue that regulations are often counterproductive and always circumventable. They believe that only 100% transparency can assure both the integrity and competency of institutional actions. In the world's law courts—and in the court of public opinion—the case for transparency will increasingly be promoted by nongovernmental organizations (NGOs) who will take advantage of the global infostructure to document and publicize environmentally and socially abusive behaviors by both private and public enterprises. The ongoing battle between institutional and socioecological imperatives will become a central theme of Web newscasts, Netpress publications, and Weblogs that have already begun to supplant traditional media networks and newspaper chains among young adults worldwide. Many of these young people will sign up with NGOs to wage undercover war on perceived corporate criminals.

In a global marketplace where corporate reputation and brand integrity will be worth billions of dollars, businesses' response to this guerrilla scrutiny will be understandably hostile. In their recently released *Study of Corporate Citizenship,* Cone/Roper, a corporate consultant on social issues, found that a majority of consumers "are willing to use their individual power to punish those companies that do not share their values." Above all, our improving comprehension of humankind's innumerable interactions with the environment will make it increasingly clear that total transparency will be crucial to the security and sustainability of a modern global economy. But there will be skullduggery, bloodshed, and heroics before total transparency finally becomes international law—15 to 20 years from now.

# Trend 5—Social Adaptation

The forces of cultural modernization—education, urbanization, and institutional order—are producing social change in the developed world as well as in developing nations. During the twentieth century, it became increasingly apparent to the citizens of a growing number of modern industrial societies that neither the church nor the state was omnipotent and that their leaders were more or less ordinary people. This realization has led citizens of modern societies to assign less weight to the guidance of their institutions and their leaders and to become more self-regulating. U.S. voters increasingly describe themselves as independents, and the fastest-growing Christian congregations in America are nondenominational.

Since the dawn of recorded history, societies have adapted to their changing circumstances. Moreover, cultural modernization has freed the societies of mature industrial nations from many strictures of church and state, giving people much more freedom to be individually adaptive. And we can be reasonably certain that modern societies will be confronted with a variety of fundamental changes in circumstance during the next five, 10, or 15 years that will, in turn, provoke continuous widespread adaptive behavior, especially in America.

**Reaching retirement age no longer always means playing golf and spoiling the grandchildren. Seniors in good health who enjoy working probably won't retire, slowing the prophesied workforce drain, according to author David Pearce Snyder**

During the decade ahead, *infomation*—the automated collection, storage, and application of electronic data—will dramatically reduce paperwork. As outsourcing and off-shoring eliminate millions of U.S. middle-income jobs, couples are likely to work two lower-pay/lower-skill jobs to replace lost income. If our employers ask us to work from home to reduce the company's office rental costs, we will do so, especially if the arrangement permits us to avoid two hours of daily commuting or to care for our offspring or an aging parent. If a wife is able to earn more money than her spouse, U.S. males are increasingly likely to become househusbands and take care of the kids. If we are in good health at age 65, and still enjoy our work, we probably won't retire, even if that's what we've been planning to do all our adult lives. If adult children must move back home after graduating from college in order to pay down their tuition debts, most families adapt accordingly.

Each such lifestyle change reflects a personal choice in response to an individual set of circumstances. And, of course, much adaptive behavior is initially undertaken as a temporary measure, to be abandoned when circumstances return to normal. During World War II, millions of women voluntarily entered the industrial workplace in the United States and the United Kingdom, for example, but returned to the domestic sector as soon as the war ended and a prosperous normalcy was restored. But the Information Revolution and the aging of mature industrial societies are scarcely temporary phenomena, suggesting that at least some recent widespread innovations in lifestyle—including delayed retirements and "sandwich households"—are precursors of long-term or even permanent changes in society.

The current propensity to delay retirement in the United States began in the mid-1980s and accelerated in the mid-1990s. Multiple surveys confirm that delayed retirement is much more a result of increased longevity and reduced morbidity than it is the result of financial necessity. A recent AARP survey, for example, found that more than 75% of baby boomers plan to work into their 70s or 80s, regardless of their economic circumstances. If the baby boomers choose to age on the job, the widely prophesied mass exodus of retirees will not drain the workforce during the coming decade, and Social Security may be actuarially sound for the foreseeable future.

The Industrial Revolution in production technology certainly produced dramatic changes in society. Before the steam engine and electric power, 70% of us lived in rural areas; today 70% of us live in cities and suburbs. Before industrialization, most economic production was home- or family-based; today, economic production takes place in factories and offices. In preindustrial Europe and America, most households included two or three adult generations (plus children), while the great majority of households today are nuclear families with one adult generation and their children.

Current trends in the United States, however, suggest that the three great cultural consequences of industrialization—the urbanization of society, the institutionalization of work, and the atomization of the family—may all be reversing, as people adapt to their changing circumstances. The U.S. Census Bureau reports that, during the 1990s, Americans began to migrate out of cities and suburbs into exurban and rural areas for the first time in the twentieth century. Simultaneously, information work has begun to migrate out of offices and into households. Given the recent accelerated growth of telecommuting, self-employment, and contingent work, one-fourth to one-third of all gainful employment is likely to take place at home within 10 years. Meanwhile, growing numbers of baby boomers find themselves living with both their debt-burdened, underemployed adult children and their own increasingly dependent aging parents. The recent emergence of the "sandwich household" in America resonates powerfully with the multigenerational, extended families that commonly served as society's safety nets in preindustrial times.

# Leadership in Changing Times

The foregoing meta-trends are not the only watershed developments that will predictably reshape daily life in the decades ahead. An untold number of inertial realities inherent in the common human enterprise will inexorably change our collective circumstances—the options and imperatives that confront society and its institutions. Society's adaptation to these new realities will, in turn, create further changes in the institutional operating environment, among customers, competitors, and constituents. There is no reason to believe that the Information Revolution will change us any less than did the Industrial Revolution.

In times like these, the best advice comes from ancient truths that have withstood the test of time. The Greek philosopher-historian Heraclitus observed 2,500 years ago that "nothing about the future is inevitable except change." Two hundred years later, the mythic Chinese general Sun Tzu advised that "the wise leader exploits the inevitable." Their combined message is clear: "The wise leader exploits change."

DAVID PEARCE SNYDER is the lifestyles editor of THE FUTURIST and principal of The Snyder Family Enterprise, a futures consultancy located at 8628 Garfield Street, Bethesda, Maryland 20817. Telephone 301-530-5807; e-mail davidpearcesnyder@earthlink.net; Web site www.the-futurist.com.

Originally published in the July/August 2004 issue of *The Futurist,* pp. 22–27. Copyright © 2004 by World Future Society, 7910 Woodmont Avenue, Suite 450, Bethesda, MD 20814. Telephone: 301/656-8274; Fax: 301/951-0394; http://www.wfs.org. Used with permission from the World Future Society.

# Globalization's Effects on the Environment

Jo Kwong

In recent years, globalization has become a remarkably polarizing issue. In particular, discussions about globalization and its environmental impacts generate ferocious debate among policy analysts, environmental activists, economists and other opinion leaders. Is globalization a solution to serious economic and social problems of the world? Or is it a profit-motivated process that leads to oppression and exploitation of the world's less fortunate?

This article examines alternative perspectives about globalization and the environment. It offers an explanation for the conflicting visions that are frequently expressed and suggests elements of an institutional framework that can align the benefits of globalization with the objective of enhanced environmental protection.

Globalization, free of the emotional rhetoric, is simply about removing barriers so goods, services, people, and ideas, can freely move from place to place. At its most rudimentary level, globalization describes a process whereby people can make their own decisions about who their trading partners are and what opportunities they wish to pursue.

While this may seem fairly innocuous, globalization certainly raises many concerns. In developed nations, some people worry about globalization's impacts on culture, traditional ways of living, and indigenous control in less developed parts of the world. They wonder, "What's to stop profit-motivated companies from developing some of the pristine environments and fragile natural resources found in the developing world?" These critics of open trade fear that residents of developing nations will be the losers in more ways than one—stripped of their land's natural resources and hopelessly in debt to exploitative developed countries. This group takes a rather paternalistic view of the problems facing the world's poor.

Others—free marketers—believe that the developed world can produce positive benefits by exporting knowledge and technology to the developing world. By avoiding mistakes made in the developed world, it is argued that developing countries can advance in manners that sidestep some of the errors that occurred in others' development processes. Third-world poverty is cited as an important reason to foster greater economic growth in the developing world. To proponents of globalization, trade is seen as a way to lift the third world from poverty and enable local people to help themselves.

Moreover, there are divided views within the developing world. Some argue against so-called "eco-imperialism." "Why are others dictating whether or not we can develop our own resources? Who are these environmental activists that say billions of people in China shouldn't have cars because this will greatly accelerate global warming?" they ask. But others question, "Who are these corporations that come in and buy huge tracts of land in third-world interiors and develop large-scale forestry or oil developments, seemingly without concern about the impact on the local environment?"

In many ways, these alternative perspectives can be viewed as a "conflicts of visions" to steal a phrase from Thomas Sowell. Some people simply view the world fundamentally differently. In the globalization context, for example, one view values the protection of indigenous ways of life, even if that means living with greater poverty and fewer individual choices. Others believe economic efficiency is key—getting the most from our resources to provide the greatest amount of financial wealth and opportunity. Most likely, however, most people fall somewhere in between.

This discussion will offer an additional factor other than a "conflict of visions" that can help us understand the broad disparities in perspectives and understandings about the question, "Is globalization good for the environment?" In particular, it raises the possibility that perhaps we are not asking the right questions to address the set of concerns at hand.

In the 1990s, a number of economists sought to empirically answer the question of whether globalization helps or harms the environment. Some of the most often-cited findings are those from economists Gene Grossman and Alan Krueger. Grossman and Krueger investigated the relationship between the scale of economic activity and environmental quality for a broad set of environmental indicators. They found that environmental degradation and income have an inverted U-shaped relationship, with pollution increasing with income at low levels of income and decreasing with income at high levels of income. The turning point at which economic growth and pollution emissions switch from a positive to a negative relationship depends on the particular emissions and air quality measure tracked. For NOx, SOx and biological oxygen demand (BOD), the turning point appears to be around $5,000 per capita gross domestic product (GDP). This observation supports the view that countries can grow out of pollution problems with wealth.

These findings were followed by further studies that examined this "Environmental Kuznets Curve", as this inverted U-shaped curve was labeled, generating a new set of policy implications that supported the idea that trade can be good for the environment. If economic growth is good for the environment, policies that stimulate growth (trade liberalization, economic restructuring, and free markets) should also be good for the environment.

The most basic description of how this inverted curve can occur is to think about the types of activities that countries experience as they develop. At the most rudimentary level, people are burning cow dung and other readily available materials for heat and cooking sources. No controls are in place; the pollutants are released directly into the air. As economic activity increases and the economy reaches a point at which it can begin making investments, catalytic converters, furnaces, etc., pollution levels are reduced, and hence the inverted curve.

In "Poverty, Wealth and Waste," Barun Mitra compares patterns of waste distribution in India to those of the developed world. He addresses the myth that poor countries have lower levels of pollution:

> The painstaking efforts to recycle materials do not mean that a poor country like India is pollution-free. Indeed, the low quantity of waste generated in an economy with little capital and technological backwardness keeps the waste industry from graduating above small-scale local initiatives. And higher pollution occurs because there isn't the technology to capture highly dispersed waste such as sulfur dioxide from smokestacks or heavy metals that flow into wastewater.

A number of possible explanations for this observed relationship between pollution and income were advanced:

- As local economies grow and develop, they will inevitably change the way they use resources, creating different types of impacts upon the environment. A simple example is the pollution tradeoffs involved from our transition in transportation modes from horses to cars. Horses generated plenty of pollution in terms of manure, carcass disposal, etc. Cars, of course, generate an entirely different brand of pollution concerns. In other words, some environmental degradation along a country's development path is inevitable, especially during the take-off process of industrialization.
- Growth is associated with an increasing share of services and high-technology production, both of which tend to be more environment-friendly than production processes in earlier stages of industrialization.
- Knowledge and technology from the developed world can help ease this transition and lessen its duration, moving countries more quickly to the levels at which pollution will be decreasing. Free trade can promote a quicker diffusion of environment-friendly technologies and lead to a more efficient allocation of resources.
- The prosperity generated from economic activity will lead to more investments and higher standards of living that enable still greater investments in cleaner and newer technologies and processes. When a certain level of per capita income is reached, economic growth helps to undo the damage done in earlier years. As free trade expands, each 1 percent increase in per capita income tends to drive pollution concentrations down by 1.25 to 1.5 percent because of the movement to cleaner techniques of production.
- As individuals become richer they are willing to spend more on non-material goods, such as a cleaner environment. This point is made by Indur M. Goklany, in his description of earlier stages of development, "Society [initially] places a much higher priority on acquiring basic public health and other services such as sewage treatment, water supply, and electricity than on environmental quality, which initially worsens. But as the original priorities are met, environmental problems become higher priorities. More resources are devoted to solving those problems. Environmental degradation is arrested and then reversed."

These findings and explanations, unsurprisingly, generated an outpouring of negative response from environmental activists and anti-globalization proponents. "How can these economists be serious?" they, in effect, asked. "Do they really think it is wise to advocate policies that predictably increase pollution? Are we supposed to believe pollution will eventually decrease if we continue with the polluting activities? How absolutely ludicrous!"

Typical responses to the "growth is good" thesis include:

- Globalization will result in a "race to the bottom" as polluting companies relocate to countries with lax environmental standards.
- Trading with countries that do not have suitable environmental laws will lower environmental standards for all countries.
- Multinationals will exploit pristine environments in the developing world, reaping the resources for short-term growth, and then pulling out to repeat the process elsewhere—growth ruins the environment.
- Free trade provides a license to pollute—it is bad for the environment. Stronger environmental regulations at national and international levels are needed.

The Sierra Club summarized the widespread critiques to the Grossman and Krueger studies, drawing from research studies produced by the World Wide Fund for Nature and others. It argued that the findings were sufficiently over-generalized to dispense with any notion that they justify complacency about trade and the environment, pointing out several facts.

The empirical estimates of where "turning points" occur for different pollutants vary so widely as to cast doubt on the validity of any one set of results. For instance, where Grossman and Krueger found turning points for certain air pollutants at less than $5,000 per capita, others found turning points above $8,000 per capita.

For some air pollutants, Grossman and Krueger found that emissions levels don't follow an inverted U-curve, but following an S-curve that starts to rise again as incomes rise. For instance, they found that sulphur dioxide emissions start to rise when income increases above $14,000 per capita. The implication is that efficiency gains from improved technology at medium levels of per capita income are eventually overwhelmed by the growing size of the economy.

Since most of the world's population earns per capita incomes well below estimated turning points, global air pollution levels will continue to rise for nearly another century. By that time, emissions of some pollutants will be anywhere from two to four times higher than current levels.

Even for the limited number of pollutants that Grossman and Krueger study, they only demonstrate a correlation between changing per capita income and changing levels of environmental quality. They do not demonstrate a causal connection. The positive relationships they describe could actually be caused by noneconomic factors, such as the adoption of environmental legislation.

Both camps seem to have reasonable grounds for their views. Clearly there is a conflict of visions that is rooted in very different value systems. Can these two opposing perspectives be reconciled sufficiently to reach some type of consensus?

As noted earlier, many studies have re-examined the Environmental Kuznets curve since the publication of the Grossman and Krueger analysis in 1991, each attempting to prove or disprove the relationship between economic growth and environmental quality, or to isolate variables that may explain the observed relationships. In that same year, a fascinating monograph was published in London, called The Wealth of Nations and the Environment. Author Mikhail Bernstam set out to analyze the contention that economic growth negatively impacts the environment by examining how institutional structure impacts this relationship.

Bernstam examined and contrasted the impact of economic growth upon the environment in both capitalist and socialist countries. Interestingly, he found that the environmental Kuznets curve does in fact exist, but it does not apply to countries across the board. The Kuznets curve, he found, applies to market economies, but not to socialist ones. The difference, according to Bernstam, has its roots in the different structures of incentives and property rights of these two economic systems.

Under market economies with secure property rights and open trade, the pursuit of profits leads to the husbanding of resources. These capitalist economies use fewer resources to produce the equivalent level of output and hence do less damage to the environment. In contrast, in socialist countries, the managers of state enterprises operate under incentives that encourage them to maximize inputs, with little regard towards economic waste or damage to the environment.

More recently, a 2001 study by economists Werner Antweiler, Brian R. Copeland, and M. Scott Taylor asked, "Is Free Trade Good for the Environment?" They analyzed data on sulfur

dioxide over the period 1971 to 1996, a time when trade barriers were coming down and international trade was expanding. They found that countries that opened up to trade generated faster economic growth. Although economic growth produced more pollution, the greater wealth and higher incomes also generated a demand for a cleaner environment.

To separate these effects, the Antweiler model looked at the negative environmental consequences of increases in economic activity (the scale effect), the positive environmental consequences of increases in income that lead to cleaner production methods (the technique effect), and the impact of trade-induced changes in the composition of output upon pollution concentrations (the composition effect). When the scale, technique and composition effects estimates were combined, the Antweiler et al. model yielded the conclusion that free trade is good for the environment. For example, when analyzing sulfur dioxide, the authors estimate that for each 1 percent increase in per capita income in a nation, pollution *falls* by 1 percent.

The critical explanatory factor is that wealthier countries value environmental amenities more highly and enhance their production by employing environmentally friendly technologies. However, like Bernstam, these authors specified that it is important to distinguish between communist and noncommunist countries. Communist countries provided the exception to their rule about globalization's positive impacts upon the environment.

The studies, which consider the impact of institutional structures, make an important contribution to our understanding of the "economics vs. environment" debates. They suggest we consider other factors in our analysis of the effects of globalization. It is true that we often do find examples of disastrous environmental conditions, particularly when we look at socialist countries. But it is misleading to attribute the disasters to globalization. Instead, we need to examine the institutional arrangements in a particular country to see what role they play in economic development and environmental protection.

# Positive Globalization

As described earlier, at its most rudimentary level, globalization simply embodies a process of free and open trade, whereby people can make decisions about who their trading partners are and what opportunities they will choose to pursue.

But the cautions of the environmental activists are worthy of consideration. Free trade, in and of itself, will not guarantee positive outcomes. We also need guiding rules that essentially create the terms for fair and civil interaction.

In *Property Rights: A Practical Guide to Freedom & Prosperity,* Terry Anderson and Laura Huggins describe the importance of institutional rules. They use the example of children playing together and inventing games. In essence, the children work together to form rules that are fair. When they cannot agree on rules, chaos typically results and their play breaks down. The same is true for civil society. Institutional rules, in the form of constitutions, common law, and so on, provide the structure for human activity.

The critical role of institutions in shaping human behavior gained international attention in 1993 when Douglass C. North received the Nobel Prize in economics. North's groundbreaking research in economic history integrated economics, sociology, statistics and history to explain the role that institutions play in economic growth.

For several decades, North looked at the question, "Why do some countries become rich, while others remain poor?" In seeking answers to this query, he came to understand that institutions establish the formal and informal sets of rules that govern the behavior of human beings in a society. His research showed that, depending on their structure and enforcement, institutional arrangements can either foster or restrain economic development.

For the past nine years, the *Index of Economic Freedom,* jointly published by the Wall Street Journal and the Heritage Foundation (Washington, DC), has provided fascinating empirical evidence of the relationship of various institutions to economic prosperity. The study analyzes and ranks the economic freedom of 161 countries according to 10 institutional factors (trade policy, property rights, regulation, and black market, for example) in an effort to trace the path to economic prosperity.

The key finding of the research, supported year after year, is that countries with the most economic freedom enjoy higher rates of long-term economic growth and prosperity than those with less economic freedom. But, more relevant to this discussion, is the finding that economic freedom, which enables people to choose who and where their trading partners are, ultimately leads to more efficient resource use.

In another comparative index, *Economic Freedom of the World 2002,* published by the Fraser Institute in conjunction with public policy institutes around the world, Nobel laureate Milton Friedman describes the importance of private property and the rule of law as a basis for economic freedom. He spells out the three key ingredients key to establishing economic freedom as follows: "First of all, and most important, the rule of law, which extends to the protection of property. Second, widespread private ownership of the means of production. Third, freedom to enter or to leave industries, freedom of competition, freedom of trade. Those are essentially the basic requirements." These same factors also provide a framework for positive environmental development.

In the 1980s, a team of economists affiliated with the Property and Environmental Research Center (PERC) in Bozeman, MT, began developing a new paradigm for environmental policy. Their model, which eventually was coined "Free Market Environmentalism" described how incentives are the key to environmental stewardship. Not surprisingly, people who face little or no consequences for environmentally destructive actions face no incentive to protect the environment. Alternatively, people who are rewarded for good stewardship are much more likely to invest in environmental protection. The key, according to economists John Baden, Richard Stroup and Terry Anderson, are the very same three elements that Milton Friedman mentioned for economic prosperity: free and open markets, clearly established property rights, and rule of law.

*Free and open markets.* One of the most important benefits produced by a market economy is information, conveyed in the form of prices. Prices of natural and environmental resources provide clear signals about their availability. As a resource becomes scarcer, its price increases. And of course, the reverse is also true: When a resource becomes more abundant, the price decreases.

Many people fear that the profit motive leads to the depletion or degradation of environmental resources. As counterintuitive as it may sound, the profit motive actually works to the benefit of the environment.

Businesses face incentives to carefully consider the prices of the various natural resources that they use in their production processes. If a particular resource is in short supply, its price will be higher than others that are more readily available. It makes little sense for a producer to over utilize, or "waste," a high-priced resource.

High prices also encourage the search for, and development of, appropriate substitutes or alternatives. As companies search for ways to reduce costs, they naturally tend toward utilizing lower-priced, more abundant resources. Thus, the pursuit of profits is actually a driving force to conserve resources. In essence, under free market systems, entrepreneurs compete in developing low cost, efficient means to solve contemporary resource problems.

*Property rights.* Clearly established property rights generate another incentive for environmental stewardship. It makes no sense for private landowners, for example, to exploit and destroy their own property. Ownership creates a long-term perspective that leads to preserving and protecting property.

Careless destruction, however, does make sense for those who are only loosely held accountable for their actions. Politicians, bureaucrats, or others, who may be short-term managers, face the incentive to maximize immediate returns, even if this means long-term environmental damage. Even managers with longer tenures realize they can simply turn to the federal government for more funds to address the problems that short-sighted decision making may have created.

*Rule of law.* In many ways, the "rule of law" is the glue that holds market transactions and property rights together. Freedom to exchange is meaningless if individuals do not have secure rights to property, including the fruits of their labors. Failure of a country's legal system to provide for the security of property rights, enforcement of contracts, and the mutually agreeable settlement of disputes will undermine the operation of a market-exchange system. If individuals and businesses lack confidence that contracts will be enforced and the returns from their productive activity protected, their incentive to engage in innovative activities will be eroded.

With these elements in place, the economists' explanations prevail—globalization will enable local cultures to pick and choose the development and environmental paths that they wish to traverse. But without these institutional arrangements, the likelihood of negative consequences increases.

In countries that lack property rights and rule of law and that promote barriers to trade, an institutional structure develops that fosters destruction of the environment. For example, in Liberia, former President Charles Taylor rapidly sold off many of the nation's natural resources in order to fund his dictatorship. In the lawless structure of that country, Taylor was able to exploit the environment and his people. In a country that has clear property rights and rule of law, such corrupt options are closed off. Neither can corporations force a village, or a state, or a country to destroy its natural resources against the will of the people.

We see this illustrated in an ongoing controversy in Peru. In the 1990s, when then bankrupt Peru opened its statist economy to foreign investment, the nation drew almost $10 billion in mining capital. That sector now accounts for half of Peru's $8 billion in exports, and Peru has become one of the world's largest gold producers. Yet, the opening economy does not necessarily mean that multi nationals can run rough shod over the locals. It all depends on the institutional arrangements that are in place.

In the small town of Tambogrande, Peru, a Canadian mining company holds the rights to tap into $1 billion worth of copper and zinc beneath the town. To do so, however, requires demolishing many local homes. In a referendum held in 2002, the town residents voted to turn down the mining company's offer to build new homes in a different location. If the country's laws hold firm to the property rights of the villagers, the mining company will not be allowed to develop the copper mine without local consent. But, if the rule of law and respect of property rights are not upheld, then the foreign firm can force its will on the indigenous people.

Property rights provide a powerful incentive for people to carefully assess their options—in this case, whether the loss of their existing houses and the village is compensated for by the new homes they would be receiving. The nature of the property rights institutions indeed affects the range of outcomes. If the local government owned the rights to the housing, rather than individuals, we would expect an entirely different outcome. Local politicians likely would gain by acquiescing to the mining firm's proposal because the villagers, not the politicians, would incur the costs.

Unfortunately, in many developing countries, corruption and back door deal making, enabled by weak rule of law and property rights, proliferates. The result is that a few leaders come out ahead and the locals get short changed. Local protests are reportedly stalling at least 10 mining-investment projects in Peru that are worth $1.4 billion—and for good reason. The noted Peruvian economist, Hernando de Soto, author of the best-seller, *The Mystery of Capital,* comments that although the mines in some towns pay double the prevailing minimum wage, they do not compensate for "the loss of their sense of environmental and economic sovereignty." Consequently, the National Society of Mining, Petroleum & Energy is urging the government to adopt reforms that immediately give at least 20 percent of the royalties to on-site communities instead of sending all these funds to Lima. Manhattan Minerals, one of the companies interested in Tambogrande, thinks local communities should receive an even bigger cut, making these towns, in effect, feel more like shareholders. In other words, they need to give the locals an interest, or property right, in the operations.

In the southern Andean town of Lircay, Huanavelica, Peru's poorest state, residents are concerned that the mine will threaten adjacent agricultural lands. To show their anger, they have resorted to street demonstrations and setting fire to government installations. Their actions seem less extreme in light of previous experiences. For decades, state-owned mining created many environmental problems that residents are rightly worried about. This cultural legacy is a key factor for private mining companies as they hammer out new relationships and try to move forward.

Fortunately, their positive examples are evolving. The La Oroya copper smelter in the central Andes region was purchased by the Doe Run company—based in St. Louis. The Peruvian government gave the company 10 years to clean up the environmental mess that the government created. Doe Run has reportedly spent $40 million so far, including money for a program to reduce high blood-lead levels in area children.

Peru needs to continue to open its doors to foreign investment, or what some would call globalization, to lift its people out of poverty. It must establish institutions—rule of law, property rights and open markets that create a safe investment climate and allow corporations to prosper. Simultaneously, companies need to conduct business in a way that will benefit the local residents as well.

As another example of how incentives and disincentives can impact the environment, consider the case of India's automobile industry. Disincentives generated by the government's regulatory policies contributed to a stagnant, non-innovative industry which caused harm to the Indian economy and environment for decades.

Although Indian automobile manufacturers began producing cars in the 1930s, there was very little development and growth in that industry for over 50 years. Auto manufacturing was heavily regulated, licensed and protected. In addition, consumers faced high taxes and duties on imported automobiles and on gasoline. The upshot was that very little competition developed in India's automobile industry—autos with low fuel efficiency and high air emissions became the norm.

In recent years, however, the automobile sector has been slowly liberalizing, allowing some major multinational corporations to set up shop in India. As a result of the increased competition and relaxed barriers, more efficient and less-polluting automobiles are becoming available to Indian consumers. A free trade regime, from the outset, would have increased access to vehicles for consumers, lowered the cost of transportation, enabled the best technologies to be locally available, and improved air quality.

In other words, incentives matter. And the structure of institutions plays a key role in the nature of incentives that are in effect. Economists have raised interesting empirical questions by developing the environmental Kuznets curve, but, as the World Wide Fund for Nature (WWF) study and others suggest, there is no one curve that fits all pollutants for all places and times. Economist Bruce Yandle of Clemson University describes it this way, "There are families of relationships, and in many cases the inverted-U Environmental Kuznets Curve is the best way to approximate the link between environmental change and income growth."

Additionally, environmental activists are right in pointing out globalization's potentially negative impacts upon the environment. Income growth alone is insufficient to reduce environmental harms and may even increase these harms if a core set of institutional features are not in place.

As the Antweiler model indicates, economic growth creates the conditions for environmental protection by raising the demand for improved environmental quality and by providing the resources needed for protection. Whether environmental quality improvements materialize or not, or when, or how they develop, depends critically on government policies, social institutions, and the strength of markets. Better policies, such as removing distorting subsidies, introducing more secure property rights over resources, and using market-like mechanisms to connect the costs of pollution to prices paid for pollution-producing goods will lower peak environmental harm (flatten the underlying Environmental Kuznets Curve). These improved policies may also bring about an earlier environmental transition.

While it may seem to be an overwhelming challenge to accomplish the institutional reforms described above, the good news is that it is happening in some very unlikely parts of the world. Consider, for example, exciting changes that have recently been occurring in Rwanda, Africa.

Lawrence Reed, the founder and president of the Mackinac Center for Public Policy in Midland, Michigan, recently toured eastern Africa, home to the remaining wild mountain gorillas left in the world. Here, approximately 670 gorillas live on a string of lush, rain-forested volcanoes along the Rwanda border with Uganda and the Congo.

To Reed's surprise, native-owned and locally staffed companies conduct all gorilla safaris. Part of the fee goes to the government for salaries for national park employees and for programs that protect gorilla habitat. (These programs also are substantially supplemented by the efforts of private, non-profits that get support from around the world.) Two Rwandan entrepreneurs started the firm, Primate Safaris, three years ago. With six employees, they provide everything a gorilla safari enthusiast could hope for—a competent guide with a four-wheel drive vehicle, good meals and comfortable accommodations.

In fact, Reed's experience with Private Safaris was only the tip of the iceberg. Rwanda, he learned, is engaged in the continent's most ambitious privatization campaign. After experiencing the kind of stifling, socialist rule that consigned virtually all of Africa to grinding poverty for decades, this nation is now embracing the private sector with deliberate policy and enormous enthusiasm. Imagine Reed's surprise when, shortly after landing in Rwanda, he came across a sign at the airport outside the capital of Kigali which reads, "Privatization: A Loss? No Way." Further down the road, another sign says "Privatization fights laziness, privatization fights poverty, privatization fights smuggling, and privatization fights unemployment."

Several of the country's privatization efforts have had direct positive impacts on the environment. For example, in 1999, Shell Oil bought a portion of the assets of Petrorwanda (the bankrupt state oil company) and completely renovated 14 of the defunct firm's decrepit and environmentally hazardous gasoline stations.

An interesting development in Uganda suggests signs of similar institutional reforms. An English language, African-based band named "Afrigo" released a song entitled, "Today for Tomorrow," which celebrates the benefits of privatization. Here is a sample of the lyrics:

> Privatization, the surer route to economic emancipation/ Yeah, businessmen run businesses/government govern the nation/ You and I didn't create the situation/ Let's unite/ check the economy/ a better future for our children.

Apparently, citizens of Rwanda and Uganda are embracing private property rights and other economic and political changes to better their lives and those of the next generation. Environmental protection surely will fare better in this setting than in the failed socialist systems being replaced.

Is globalization good for the environment? Viewing globalization as the destroyer or savior of the environment misses the point. The problem is not globalization per se. A lack of key institutions, namely rule of law; property rights; free and open markets is the real villain in the tale. These institutions hold people accountable for their actions, and at the same time, reward them for positive behavior. They create conditions in which market competition rewards innovation and efficiency, and in which economic development and increased wealth can fuel improved environmental quality.

Globalization—free trade and multinational investments—can advance these institutional changes, leading to enhanced social and political stability. Concerns that multinational corporations might be engaged in a "race to the environmental bottom" seem unlikely in these circumstances. To the contrary, where these institutions are in place, the result can be a "race to the top," as jurisdictions compete to improve the quality of life for their constituents.

Globalization can be a means to accelerate learning about the importance of market institutions to economic growth. Environmental protection can be one of many important benefits resulting from such a transition. Getting back to my earlier comment about "a conflict of visions," I certainly hold a contrasting view from opponents of globalization. Critics believe globalization underlies many of the problems that plague the developing world. On the other hand, I see globalization as a basic part of the solution to these problems. Greater movement of goods, services, people and ideas can lead to economic prosperity, improved environmental protection, and a host of other social benefits.

# Suggested Further Readings

Antweiler, Werner, Brian R. Copeland, and M. Scott Taylor. 2001. "Is Free Trade Good for the Environment?" *American Economic Review.* Vol. 91(4), pp. 877–908.

Bernstam, Mikhail. 1991. "Is Free Trade Good for the Environment." Institute of Economic Affairs, p. 7.

Bhagwati, Jagdish. 2002. *Free Trade Today.* Princeton: Princeton University Press.

Goklany, Indur M. 1998. "The Environmental Transition to Air Quality." *Regulation.* Vol. 21(4), p. 36.

Grossman, G.M. and A.B. Krueger. 1995. "Economic Growth and the Environment." *Quarterly Journal of Economics.* Vol.110(2), pp. 353–377.

Yandle, Bruce, Maya Vijayaraghavan, and Madhusudan Bhattarai. 2002. "The Environmental Kuznets Curve: A Primer." *PERC Research Study 02-1,* p. 17.

**JO KWONG, PH.D.** is Director of Institute Relations, Atlas Economic Research Foundation in Fairfax, Virginia. This paper formed the basis for her address at Lindenwood University (St. Charles, Missouri) in the Economic Policy Lecture Series on March 29, 2004. The event was co-hosted by the Institute for Study of Economics and the Environment and the Division of Management.

# Do Global Attitudes and Behaviors Support Sustainable Development?

ANTHONY A. LEISEROWITZ, ROBERT W. KATES, AND THOMAS M. PARRIS

Many advocates of sustainable development recognize that a transition to global sustainability—meeting human needs and reducing hunger and poverty while maintaining the life-support systems of the planet—will require changes in human values, attitudes, and behaviors.[1] A previous article in *Environment* described some of the values used to define or support sustainable development as well as key goals, indicators, and practices.[2] Drawing on the few multinational and quasi-global-scale surveys that have been conducted,[3] this article synthesizes and reviews what is currently known about global attitudes and behavior that will either support or discourage a global sustainability transition.[4] (Table 1 provides details about these surveys.)

None of these surveys measured public attitudes toward "sustainable development" as a holistic concept. There is, however, a diverse range of empirical data related to many of the subcomponents of sustainable development: development and environment; the driving forces of population, affluence/poverty/consumerism, technology, and entitlement programs; and the gap between attitudes and behavior.

## Development

Concerns for environment and development merged in the early concept of sustainable development, but the meaning of these terms has evolved over time. For example, global economic development is widely viewed as a central priority of sustainable development, but development has come to mean human and social development as well.

### Economic Development

The desire for economic development is often assumed to be universal, transcending all cultural and national contexts. Although the surveys in Table 1 have no global-scale data on public attitudes toward economic development per se, this assumption appears to be supported by 91 percent of respondents from 35 developing countries, the United States, and Germany, who said that it is very important (75 percent) or somewhat important (16 percent) to live in a country where there is economic prosperity[5] What level of affluence is desired, how that economic prosperity is to be achieved, and how economic wealth should ideally be distributed within and between nations, however, are much more contentious questions. Unfortunately, there does not appear to be any global-scale survey research that has tried to identify public attitudes or preferences for particular levels or end-states of economic development (for example, infinite growth versus steady-state economies) and only limited or tangential data on the ideal distribution of wealth (see the section on affluence below).

Data from the World Values Survey suggest that economic development leads to greater perceived happiness as countries make the transition from subsistence to advanced industrial economies. But above a certain level of gross national product

## Table 1   Multinational Surveys

### One-time Surveys

| Name | Year(s) | Number of Countries |
|---|---|---|
| Pew Global Attitudes Project | 2002 | 43 |
| Eurobarometer | 2002 | 15 |
| International Social Science Program | 2000 | 25 |
| Health of the Planet | 1992 | 24 |

### Repeated Surveys

| Name | Year(s) | Number of Countries |
|---|---|---|
| GlobeScan International Environmental Monitor | 1997–2003 | 34 |
| World Values Survey | 1981–2002 | 79 |
| Demographic and Health Surveys | 1986–2002 | 17 |
| Organisation for Economic Co-operation and Development | 1990–2002 | 22 |

Note: Before November 2003, GlobeScan, Inc. was known as Environics International. Surveys before this time bear the older name.

Source: For more detail about these surveys and the countries sampled, see Appendix A in A. Leiserowitz, R. W. Kates, and T. M. Parris, *Sustainability Values, Attitudes and Behaviors: A Review of Multi-national and Global Trends,* CID Working Paper No. 113 (Cambridge, MA: Science, Environment and Development Group, Center for International Development, Harvard University, 2004), http://www.cid.harvard.edu/cidwp/113.htm.

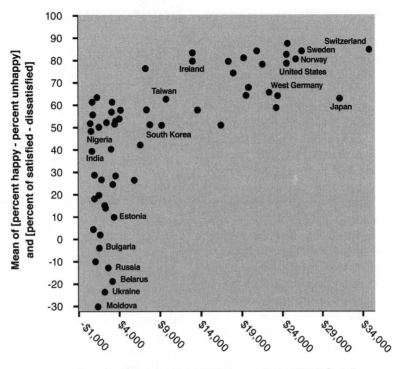

**Figure 1** Subjective Well-Being by Level of Economic Development.

Note: The subjective well-being index reflects the average of the percentage in each country who describe themselves as "very happy" or "happy" minus the percentage who describe themselves as "not very happy" or "unhappy"; and the percentage placing themselves in the 7–10 range, minus the percentage placing themselves in the 1–4 range, on a 10-point scale on which 1 indicates that one is strongly dissatisfied with one's life as a whole, and 10 indicates that one is highly satisfied with one's life as a whole.

Source: R. Inglehart, "Globalization and Postmodern Values," *Washington Quarterly* 23, no. 1 (1999): 215–228. Subjective well-being data from the 1990 and 1996 World Values Surveys. GNP per capita for 1993 data from *World Bank, World Development Report, 1995* (New York: Oxford University Press, 1995).

(GNP) per capita—approximately $14,000—the relationship between income level and subjective well-being disappears (see Figure 1). This implies that infinite economic growth does not lead to greater human happiness. Additionally, many of the unhappiest countries had, at the time of these surveys, recently experienced significant declines in living standards with the collapse of the Soviet Union. Yet GNP per capita remained higher in these ex-Soviet countries than in developing countries like India and Nigeria.[6] This suggests that relative trends in living standards influence happiness more than absolute levels of affluence, but the relationship between economic development and subjective well-being deserves more research attention.

## Human Development

Very limited data is available on public attitudes toward issues of human development, although it can be assumed that there is near-universal support for increased child survival rates, adult life expectancies, and educational opportunities. However, despite the remarkable increases in these indicators of human well-being

since World War II,[7] there appears to be a globally pervasive sense that human well-being has been deteriorating in recent years. In 2002, large majorities worldwide said that a variety of conditions had worsened over the previous five years, including the availability of well-paying jobs (58 percent); working conditions (59 percent); the spread of diseases (66 percent); the affordability of health care (60 percent); and the ability of old people to care for themselves in old age (59 percent). Likewise, thinking of their own countries, large majorities worldwide were concerned about the living conditions of the elderly (61 percent) and the sick and disabled (56 percent), while a plurality was concerned about the living conditions of the unemployed (42 percent).[8]

## Development Assistance

One important way to promote development is to extend help to poorer countries and people, either through national governments or nongovernmental organizations and charities. There is strong popular support but less official support for development assistance to poor countries. In 1970, the United Nations

General Assembly resolved that each economically advanced country would dedicate 0.7 percent of its gross national income (GNI) to official development assistance (ODA) by the middle of the 1970s—a target that has been reaffirmed in many subsequent international agreements.[9] As of 2004, only five countries had achieved this goal (Denmark, Norway, the Netherlands, Luxembourg, and Sweden). Portugal was close to the target at 0.63, yet all other countries ranged from a high of 0.42 percent (France) to lows of 0.16 and 0.15 percent (the United States and Italy respectively). Overall, the average ODA/GNI among the industrialized countries was only 0.25 percent—far below the UN target.[10]

By contrast, in 2002, more than 70 percent of respondents from 21 developed and developing countries said they would support paying 1 percent more in taxes to help the world's poor.[11] Likewise, surveys in the 13 countries of the Organisation for Economic Co-operation and Development's Development Assistance Committee (OECD-DAC) have found that public support for the principle of giving aid to developing countries (81 percent in 2003) has remained high and stable for more than 20 years.[12] Further, 45 percent said that their government's current (1999–2001) level of expenditure on foreign aid was too low, while only 10 percent said foreign aid was too high.[13] There is also little evidence that the public in OECD countries has developed "donor fatigue." Although surveys have found increasing public concerns about corruption, aid diversion, and inefficiency, these surveys also continue to show very high levels of public support for aid.

Public support for development aid is belied, however, by several factors. First, large majorities demonstrate little understanding of development aid, with most unable to identify their national aid agencies and greatly overestimating the percentage of their national budget devoted to development aid. For example, recent polls have found that Americans believed their government spent 24 percent (mean estimate) of the national budget on foreign assistance, while Europeans often estimated their governments spent 5 to 10 percent.[14] In reality, in 2004 the United States spent approximately 0.81 percent and the European Union member countries an average of approximately 0.75 percent of their national budgets on official development assistance, ranging from a low of 0.30 percent (Italy) to a high of 1.66 percent (Luxembourg).[15] Second, development aid is almost always ranked low on lists of national priorities, well below more salient concerns about (for example) unemployment, education, and health care. Third, "the overwhelming support for foreign aid is based upon the perception that it will be spent on remedying humanitarian crises," not used for other development-related issues like Third World debt, trade barriers, or increasing inequality between rich and poor countries—or for geopolitical reasons (for example, U.S. aid to Israel and Egypt).[16] Support for development assistance has thus been characterized as "a mile wide, but an inch deep" with large majorities supporting aid (in principle) and increasing budget allocations but few understanding what development aid encompasses or giving it a high priority.[17]

# Environment

Compared to the very limited or nonexistent data on attitudes toward economic and human development and the overall concept of sustainable development, research on global environmental attitudes is somewhat more substantial. Several surveys have measured attitudes regarding the intrinsic value of nature, global environmental concerns, the trade-offs between environmental protection and economic growth, government policies, and individual and household behaviors.

## Human-Nature Relationship

Most research has focused on anthropocentric concerns about environmental quality and natural resource use, with less attention to ecocentric concerns about the intrinsic value of nature. In 1967, the historian Lynn White Jr. published a now-famous and controversial article arguing that a Judeo-Christian ethic and attitude of domination, derived from Genesis, was an underlying historical and cultural cause of the modern environmental crisis.[18] Subsequent ecocentric, ecofeminist, and social ecology theorists have also argued that a domination ethic toward people, women, and nature runs deep in Western, patriarchal, and capitalist culture.[19] The 2000 World Values Survey, however, found that 76 percent of respondents across 27 countries said that human beings should "coexist with nature," while only 19 percent said they should "master nature" (see Figure 2). Overwhelming majorities of Europeans, Japanese, and North Americans said that human beings should coexist with nature, ranging from 85 percent in the United States to 96 percent in Japan. By contrast, only in Jordan, Vietnam, Tanzania, and the Philippines did more than 40 percent say that human beings should master nature.[20] In 2002, a national survey of the United States explored environmental values in more depth and found that Americans strongly agreed that nature has intrinsic value and that humans have moral duties and obligations to animals, plants, and non-living nature (such as rocks, water, and air). The survey found that Americans strongly disagreed that "humans have the right to alter nature to satisfy wants and desires" and that "humans are not part of nature" (see Figure 3).[21] This very limited data suggests that large majorities in the United States and worldwide now reject a domination ethic as the basis of the human-nature relationship, at least at an abstract level. This question, however, deserves much more cross-cultural empirical research.

## Environmental Concern

In 2000, a survey of 11 developed and 23 developing countries found that 83 percent of all respondents were concerned a fair amount (41 percent) to a great deal (42 percent) about environmental problems. Interestingly, more respondents from developing countries (47 percent) were "a great deal concerned" about the environment than from developed countries (33 percent), ranging from more than 60 percent in Peru, the Philippines, Nigeria, and India to less than 30 percent in the

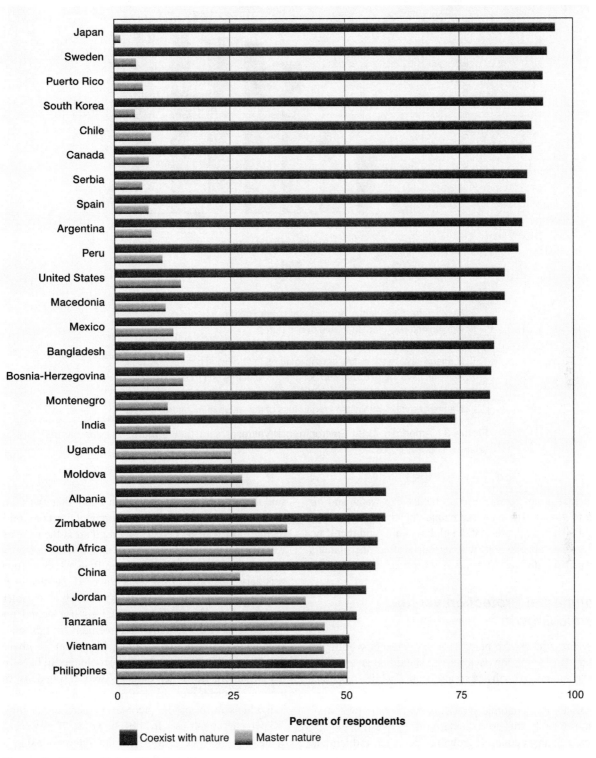

**Figure 2** Human-Nature Relationship.

Note: The question asked, "Which statement comes closest to your own views: human beings should master nature or humans should coexist with nature?"

Source: A. Leiserowitz, 2005. Data from world Values Survey, *The 1999–2002 Values Surveys Integrated Data File 1.0, CD-ROM in R. Inglehart, M. Basanez, J. Diez-Medrano, L. Halman, and R. Luijkx, eds., Human Beliefs and Values: A Cross-Cultural Sourcebook Based on the 1999–2002 Values Surveys, first edition* (Mexico City: Siglo XXI, 2004).

Netherlands, Germany, Japan, and Spain.[22] This survey also asked respondents to rate the seriousness of several environmental problems (see Figure 4). Large majorities worldwide selected the strongest response possible ("very serious") for seven of the eight problems measured. Overall, these results demonstrate very high levels of public concern about a wide range of

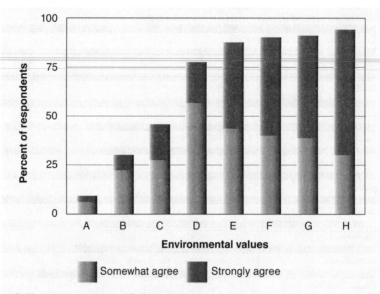

A: Humans are not part of nature.

B: Humans have the right to subdue and control nature.

C: Humankind was created to rule over nature.

D: Humans should adapt to nature rather than modify it to suit them.

E: Humans have moral duties and obligations to non-living nature.

F: Humans have moral duties and obligations to plants and trees.

G: Humans have moral duties and obligations to other animal species.

H: Nature has value within itself regardless of any value humans place on it.

**Figure 3** American (U.S.) Environmental Values.

Source: A. Leiserowitz, 2005.

environmental issues, from local problems like water and air pollution to global problems like ozone depletion and climate change.[23] Further, 52 percent of the global public said that if no action is taken, "species loss will seriously affect the planet's ability to sustain life" just 20 years from now.[24]

## Environmental Protection versus Economic Growth

In two recent studies, 52 percent of respondents worldwide agreed that "protecting the environment should be given priority" over "economic growth and creating jobs," while 74 percent of respondents in the G7 countries prioritized environmental protection over economic growth, even if some jobs were lost.[25] Unfortunately, this now-standard survey question pits the environment against economic growth as an either/or dilemma. Rarely do surveys allow respondents to choose an alternative answer, that environmental protection can generate economic growth and create jobs (for example, in new energy system development, tourism, and manufacturing).

## Attitudes toward Environmental Policies

In 1995, a large majority (62 percent) worldwide said they "would agree to an increase in taxes if the extra money were used to prevent environmental damage," while 33 percent

said they would oppose them.[26] In 2000, there was widespread global support for stronger environmental protection laws and regulations, with 69 percent saying that, at the time of the survey, their national laws and regulations did not go at all far enough.[27] The 1992 Health of the Planet survey found that a very large majority (78 percent) favored the idea of their own national government "contributing money to an international agency to work on solving global environmental problems." Attitudes toward international agreements in this survey, however, were less favorable. In 1992, 47 percent worldwide agreed that "our nation's environmental problems can be solved without any international agreements," with respondents from low-income countries more likely to strongly agree (23 percent) than individuals from middle-income (17 percent) or high-income (12 percent) countries.[28] In 2001, however, 79 percent of respondents from the G8 countries said that international negotiations and progress on climate change was either "not good enough" (39 percent) or "not acceptable" (40 percent) and needed faster action. Surprisingly, this latter 40 percent supported giving the United Nations "the power to impose legally-binding actions on national governments to protect the Earth's climate."[29]

## Environmental Behavior

Material consumption is one of the primary means by which environmental values and attitudes get translated into behavior.

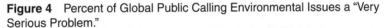

**Figure 4** Percent of Global Public Calling Environmental Issues a "Very Serious Problem."

Source: A. Leiserowitz, 2005. Data from Environics International (Globe Scan), *Environics International Environmental Monitor Survey Dataset* (Kingston, Canada: Environics international, 2000), http://jeff-lab.queensu.ca/poadata/info/iem/iemlist.shtml (accessed 5 October 2004).

(For attitudes toward consumption per se, see the following section on affluence, poverty, and consumerism.)

In 2002, Environics International (GlobeScan) found that 36 percent of respondents from 20 developed and developing countries stated that they had avoided a product or brand for environmental reasons, while 27 percent had refused packaging, and 25 percent had gathered environmental information.[30] Recycling was highly popular, with 6 in 10 people setting aside garbage for reuse, recycling, or safe disposal. These rates, however, reached 91 percent in North America versus only 36–38 percent in Latin America, Eastern Europe, and Central Asia,[31] which may be the result of structural barriers in these societies (for example, inadequate infrastructures, regulations, or markets). There is less survey data regarding international attitudes toward energy consumption, but among Europeans, large majorities said they had reduced or intended to reduce their use of heating, air conditioning, lighting, and domestic electrical appliances.[32]

In 1995, 46 percent of respondents worldwide reported having chosen products thought to be better for the environment, 50 percent of respondents said they had tried to reduce their own water consumption, and 48 percent reported that in the 12 months prior to the survey, they reused or recycled something rather than throwing it away. There was a clear distinction between richer and poorer societies: 67 percent of respondents from high-income countries reported that they had chosen "green" products, while only 30 percent had done so in low-income

countries. Likewise, 75 percent of respondents from high-income countries said that they had reused or recycled something, while only 27 percent in low-income countries said this.[33] However, the latter results contradict the observations of researchers who have noted that many people in developing countries reuse things as part of everyday life (for example, converting oil barrels into water containers) and that millions eke out an existence by reusing and recycling items from landfills and garbage dumps.[34] This disparity could be the result of inadequate survey representation of the very poor, who are the most likely to reuse and recycle as part of survival, or, alternatively, different cultural interpretations of the concepts "reuse" and "recycle."

In 2002, 44 percent of respondents in high-income countries were very willing to pay 10 percent more for an environmentally friendly car, compared to 41 percent from low-income countries and 29 percent from middle-income countries.[35] These findings clearly mark the emergence of a global market for more energy-efficient and less-polluting automobiles. However, while many people appear willing to spend more to buy an environmentally friendly car, most do not appear willing to pay more for gasoline to reduce air pollution. The same 2002 survey found that among high-income countries, only 28 percent of respondents were very willing to pay 10 percent more for gasoline if the money was used to reduce air pollution, compared to 23 percent in medium-income countries and 36 percent in low-income countries.[36] People appear to generally oppose higher gasoline prices, although public attitudes are probably affected, at least in part, by the prices extant at the time of a given survey, the rationale given for the tax, and how the income from the tax will be spent.

Despite the generally pro-environment attitudes and behaviors outlined above, the worldwide public is much less likely to engage in political action for the environment. In 1995, only 13 percent of worldwide respondents reported having donated to an environmental organization, attended a meeting, or signed a petition for the environment in the prior 12 months, with more doing so in high-income countries than in low-income countries.[37] Finally, in 2000, only 10 percent worldwide reported having written a letter or made a telephone call to express their concern about an environmental issue in the past year, 18 percent had based a vote on green issues, and 11 percent belonged to or supported an environmental group.[38]

# Drivers of Development and Environment

Many analyses of the human impact on life-support systems focus on three driving forces: population, affluence or income, and technology—the so-called I = PAT identity.[39] In other words, environmental impact is considered a function of these three drivers. In a similar example, carbon dioxide ($CO_2$) emissions from the energy sector are often considered a function of population, affluence (gross domestic product (GDP) per capita), energy intensity (units of energy per GDP), and technology ($CO_2$ emissions per unit of energy).[40] While useful, most analysts also recognize that these variables are not fundamental driving forces in and of themselves and are not independent from one another.[41] A similar approach has also been applied to human development (D = PAE), in which development is considered a function of population, affluence, and entitlements and equity.[42] What follows is a review of empirical trends in attitudes and behavior related to population, affluence, technology, and equity and entitlements.

# Population

Global population continues to grow, but the rate of growth continues to decline almost everywhere. Recurrent Demographic and Health Surveys (DHS) have found that the ideal number of children desired is declining worldwide. Globally, attitudes toward family planning and contraception are very positive, with 67 percent worldwide and large majorities in 38 out of 40 countries agreeing that birth control and family planning have been a change for the better.[43] Worldwide, these positive attitudes toward family planning are reflected in the behavior of more than 62 percent of married women of reproductive age who are currently using contraception. Within the developing world, the United Nations reports that from 1990 to 2000, contraceptive use among married women in Asia increased from 52 percent to 66 percent, in Latin American and the Caribbean from 57 percent to 69 percent, but in Africa from only 15 percent to 25 percent.[44] Notwithstanding these positive attitudes toward

contraception, in 1997, approximately 20 percent to 25 percent of births in the developing world were unwanted, indicating that access to or the use of contraceptives remains limited in some areas.[45]

DHS surveys have found that ideal family size remains significantly larger in western and middle Africa (5.2) than elsewhere in the developing world (2.9).[46] They also found that support for family planning is much lower in sub-Saharan Africa (44 percent) than in the rest of the developing world (74 percent).[47] Consistent with these attitudes, sub-Saharan Africa exhibits lower percentages of married women using birth control as well as lower rates of growth in contraceptive use than the rest of the developing world.[48]

# Affluence, Poverty, and Consumerism

Aggregate affluence and related consumption have risen dramatically worldwide with GDP per capita (purchasing-power parity, constant 1995 international dollars) more than doubling between 1975 and 2002.[49] However, the rising tide has not lifted all boats. Worldwide in 2001, more than 1.1 billion people lived on less than $1 per day, and 2.7 billion people lived on less than $2 per day—with little overall change from 1990. However, the World Bank projects these numbers to decline dramatically by 2015—to 622 million living on less than $1 per day and 1.9 billion living on less than $2 per day. There are also large regional differences, with sub-Saharan Africa the most notable exception: There, the number of people living on less than $1 per day rose from an estimated 227 million in 1990 to 313 million in 2001 and is projected to increase to 340 million by 2015.[50]

## Poverty

Poverty reduction is an essential objective of sustainable development.[51] In 1995, 65 percent of respondents worldwide said that more people were living in poverty than had been 10 years prior. Regarding the root causes of poverty, 63 percent blamed unfair treatment by society, while 26 percent blamed the laziness of the poor themselves. Majorities blamed poverty on the laziness and lack of willpower of the poor only in the United States (61 percent), Puerto Rico (72 percent), Japan (57 percent), China (59 percent), Taiwan (69 percent), and the Philippines (63 percent) (see Figure 5).[52] Worldwide, 68 percent said their own government was doing too little to help people in poverty within their own country, while only 4 percent said their government was doing too much. At the national level, only in the United States (33 percent) and the Philippines (21 percent) did significant proportions say their own government was doing too much to help people in poverty.[53]

## Consumerism

Different surveys paint a complicated and contradictory picture of attitudes toward consumption. On the one hand, majorities around the world agree that, at the societal level, material and status-related consumption are threats to human cultures and the

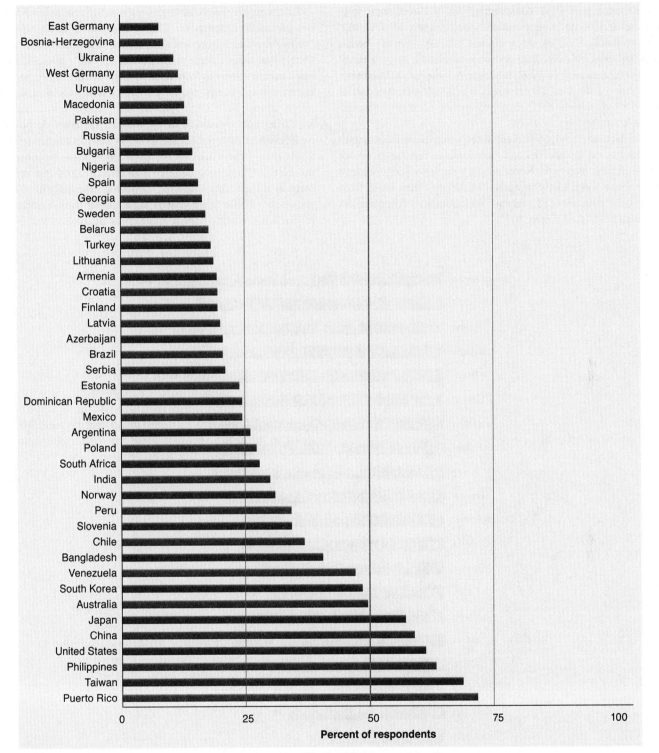

**Figure 5** Percent Blaming Poverty on the Laziness and Lack of Willpower of the Poor.

Source: A. Leiserowitz, 2005. Data from R. Inglehart, et al., *World Values Surveys and European Values Surveys, 1981–1984, 1990–1993, and 1995–1997* [computer file], Inter-university Consortium for Political and Social Research (ICPSR) version (Ann Arbor, MI: Institute for Social Research [producer], 2000; Ann Arbor, MI: ICPSR [distributor], 2000).

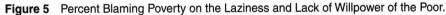

environment. Worldwide, 54 percent thought "less emphasis on money and material possessions" would be a good thing, while only 21 percent thought this would be a bad thing.[54] Further, large majorities agreed that gaining more time for leisure activities or family life is their biggest goal in life.[55]

More broadly, in 2002 a global study sponsored by the Pew Research Center for the People & the Press found that 45 percent worldwide saw consumerism and commercialism as a threat to their own culture. Interestingly, more respondents from high-income and upper middle-income countries (approximately

51 percent) perceived consumerism as a threat than low-middle- and low-income countries (approximately 43 percent).[56] Unfortunately, the Pew study did not ask respondents whether they believed consumerism and commercialism were a threat to the environment. In 1992, however, 41 percent said that consumption of the world's resources by industrialized countries contributed "a great deal" to environmental problems in developing countries."[57]

On the other hand, 65 percent of respondents said that spending money on themselves and their families represents one of life's greatest pleasures. Respondents from low-GDP countries were much more likely to agree (74 percent) than those from high-GDP countries (58 percent), which reflects differences in material needs (see Figure 6).[58]

Likewise, there may be large regional differences in attitudes toward status consumerism. Large majorities of Europeans and North Americans disagreed (78 percent and 76 percent respectively) that other people's admiration for one's possessions is important, while 54 to 59 percent of Latin American, Asian, and Eurasian respondents, and only 19 percent of Africans (Nigeria only), disagreed.[59] There are strong cultural norms against appearing materialistic in many Western societies, despite the high levels of material consumption in these countries relative to the rest of the world. At the same time, status or conspicuous consumption has long been posited as a significant driving force in at least some consumer behavior, especially in affluent societies.[60] While these studies are a useful start, much more research is needed to unpack and explain the roles of values and

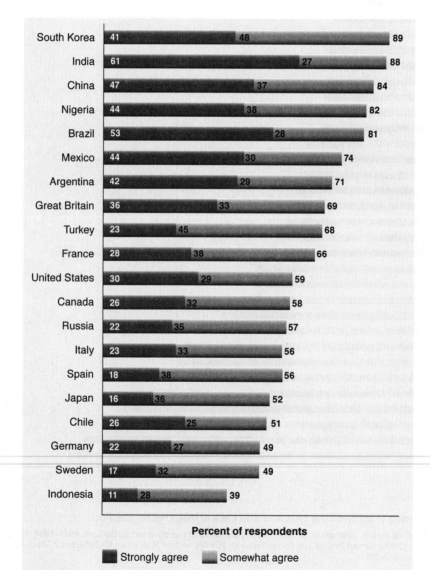

**Figure 6** Purchasing for Self and Family Gives Greatest Pleasure ("Strongly" and "Somewhat" Agree).

Note: The question was, "To spend money, to buy something new for myself or my family, is one of the greatest pleasures in my life."

Source: Environics International (GlobeScan), *Consumerism: A Special Report* (Toronto: Environics International, 2002). 6.

attitudes in material consumption in different socioeconomic circumstances.

# Science and Technology

Successful deployment of new and more efficient technologies is an important component of most sustainability strategies, even though it is often difficult to assess all the environmental, social, and public health consequences of these technologies in advance. Overall, the global public has very positive attitudes toward science and technology. The 1995 World Values Survey asked respondents, "In the long run, do you think the scientific advances we are making will help or harm mankind?" Worldwide, 56 percent of respondents thought science will help mankind, while 26 percent thought it will harm mankind. Further, 67 percent said an increased emphasis on technological development would be a good thing, while only 9 percent said it would be bad.[61] Likewise, in 2002, GlobeScan found large majorities worldwide believed that the benefits of modern technology outweigh the risks.[62] The support for technology, however, was significantly higher in countries with low GDPs (69 percent) than in high-GDP countries (56 percent), indicating more skepticism among people in technologically advanced societies. Further, this survey found dramatic differences in technological optimism between richer and poorer countries. Asked whether "new technologies will resolve most of our environmental challenges, requiring only minor changes in human thinking and individual behavior," 62 percent of respondents from low-GDP countries agreed, while 55 percent from high-GDP countries disagreed (see Figure 7).

.But what about specific technologies with sustainability implications? Do these also enjoy strong public support? What follows is a summary of global-scale data on attitudes toward renewable energy, nuclear power, the agricultural use of chemical pesticides, and biotechnology.

Europeans strongly preferred several renewable energy technologies (solar, wind, and biomass) over all other energy sources, including solid fuels (such as coal and peat), oil, natural gas, nuclear fission, nuclear fusion, and hydroelectric power. Also, Europeans believed that by the year 2050, these energy sources will be best for the environment (67 percent), be the least expensive (40 percent), and will provide the greatest amount of useful energy (27 percent).[63] Further, 37 percent of Europeans and approximately 33 percent of respondents in 16 developed and developing countries were willing to pay 10 percent more for electricity derived from renewable energy sources.[64]

Nuclear power, however, remains highly stigmatized throughout much of the developed world.[65] Among respondents from 18 countries (mostly developed), 62 percent considered nuclear power stations "very dangerous" to "extremely dangerous" for the environment.[66] Whatever its merits or demerits as an alternative energy source, public attitudes about nuclear power continue to constrain its political feasibility.

Regarding the use of chemical pesticides on food crops, a majority of people in poorer countries believed that the benefits are greater than the risks (54 percent), while respondents in high-GDP countries were more suspicious, with only 32 percent believing the benefits outweigh the risks.[67] Since 1998, however, support for the use of agricultural chemicals has dropped worldwide. Further, chemical pesticides are now one of the top food-related concerns expressed by respondents around the world.[68]

Additionally, the use of biotechnology in agriculture remains controversial worldwide, and views on the issue are divided between rich and poor countries. Across the G7 countries, 70 percent of respondents were opposed to scientifically altered fruits and vegetables because of health and environmental concerns,[69] while 62 percent of Europeans and 45 percent of Americans opposed the use of biotechnology in agriculture.[70] While majorities in poorer countries (65 percent) believed the benefits of using biotechnology on food crops are greater than

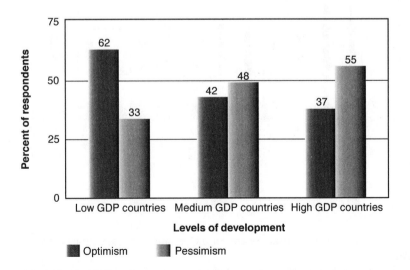

**Figure 7** Technological Optimism Regarding Environmental Problems.

Source: A. Leiserowitz, 2005. Data from Environics International (GlobeScan), *International Environmental Monitor* (Toronto: Environics International, 2002), 135.

the risks, majorities in high-GDP countries (51 percent) believed the risks outweigh the benefits.[71]

More broadly, public understanding of biotechnology is still limited, and slight variations in question wordings or framings can have significant impacts on support or opposition. For example, 56 percent worldwide thought that biotechnology will be good for society in the long term, yet 57 percent also agreed that "any attempt to modify the genes of plants or animals is ethically and morally wrong."[72] Particular applications of biotechnology also garnered widely different degrees of support. While 78 percent worldwide favored the use of biotechnology to develop new medicines, only 34 percent supported its use in the development of genetically modified food. Yet, when asked whether they supported the use of biotechnology to produce more nutritious crops, 61 percent agreed.[73]

## Income Equity and Entitlements

Equity and entitlements strongly determine the degree to which rising population and affluence affect human development, particularly for the poor. For example, as global population and affluence have grown, income inequality between rich and poor countries has also increased over time, with the notable exceptions of East and Southeast Asia—where incomes are on the rise on a par with (or even faster than) the wealthier nations of the world.[74] Inequality within countries has also grown in many rich and poor countries. Similarly, access to entitlements—the bundle of income, natural resources, familial and social connections, and societal assistance that are key determinants of hunger and poverty[75]—has recently declined with the emergence of market-oriented economies in Eastern and Central Europe,

Russia, and China; the rising costs of entitlement programs in the industrialized countries, including access to and quality of health care, education, housing, and employment; and structural adjustment programs in developing countries that were recommended by the International Monetary Fund. Critically, it appears there is no comparative data on global attitudes toward specific entitlements; however, there is much concern that living conditions for the elderly, unemployed, and the sick and injured are deteriorating, as cited above in the discussion on human development.

In 2002, large majorities said that the gap between rich and poor in their country had gotten wider over the previous 5 years. This was true across geographic regions and levels of economic development, with majorities ranging from 66 percent in Asia, 72 percent in North America, and 88 percent in Eastern Europe (excepting Ukraine) stating that the gap had gotten worse.[76] Nonetheless, 48 percent of respondents from 13 countries preferred a "competitive society, where wealth is distributed according to one's achievement," while 34 percent preferred an "egalitarian society, where the gap between rich and poor is small, regardless of achievement" (see Figure 8).[77]

More broadly, 47 percent of respondents from 72 countries preferred "larger income differences as incentives for individual effort," while 33 percent preferred that "incomes should be made more equal."[78] These results suggest that despite public perceptions of growing economic inequality, many accept it as an important incentive in a more individualistic and competitive economic system. These global results, however, are limited to just a few variables and gloss over many countries that strongly prefer more egalitarian distributions of wealth (such as India). Much more research is needed to understand how important the

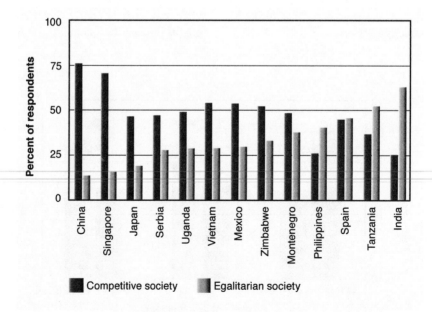

**Figure 8** Multinational Preferences for a Competitive Versus Egalitarian Society.

Source: A. Leiserowitz, 2005. Data from World Values Survey. *The 1999–2002 Values Surveys Integrated Data File 1.0,* CD-ROM in R. Inglehart, M. Basanez, J. Diez-Medrano, L. Halman, and R. Luijkx, eds., *Human Beliefs and Values: A Cross-Cultural Sourcebook Based on the 1999–2002 Values Surveys,* first edition (Mexico City: Siglo XX[[]]I, 2004).

principles of income equality and equal economic opportunity are considered globally, either as global goals or as means to achieve other sustainability goals.

# Does the Global Public Support Sustainable Development?

Surprisingly, the question of public support for sustainable development has never been asked directly, at least not globally. But two important themes emerge from the multinational data and analysis above. First, in general, the global public supports the main tenets of sustainable development. Second, however, there are many contradictions, including critical gaps between what people say and do—both as individuals and in aggregate. From these themes emerge a third finding: Diverse barriers stand between sustainability attitudes and action.

- *Large majorities worldwide appear to support environmental protection and economic and human development—the three pillars of sustainable development.* They express attitudes and have taken modest actions consonant with support for sustainable development, including support for environmental protection; economic growth; smaller populations; reduced poverty; improved technology; and care and concern for the poor, the marginal, the young, and the aged.

- *Amid the positive attitudes, however, are many contradictions.* Worldwide, all the components of the Human Development Index—life expectancy, adult literacy, and per capita income—have dramatically improved since World War II.[79] Despite the remarkable increases in human well-being, however, there appears to be a globally pervasive sense that human well-being has more recently been deteriorating. Meanwhile, levels of development assistance are consistently overestimated by lay publics, and the use of such aid is misunderstood, albeit strongly supported. Overall, there are very positive attitudes toward science and technology, but the most technologically sophisticated peoples are also the most pessimistic about the ability of technology to solve global problems. Likewise, attitudes toward biotechnology vary widely, depending on how the question is asked.

- Further, there are serious gaps between what people believe and what people do, both as individuals and as polities. Worldwide, the public strongly supports significantly larger levels of development assistance for poor countries, but national governments have yet to translate these attitudes into proportional action. Most people value the environment—for anthropocentric as well as ecocentric reasons—yet many ecological systems around the world continue to degrade, fragment, and lose resilience. Most favor smaller families, family planning, and contraception, but one-fifth to one-quarter of children born are not desired. Majorities are concerned with poverty and think more should be done to alleviate it, but important regions of the world

think the poor themselves are to blame, and a majority worldwide accepts large gaps between rich and poor. Most people think that less emphasis on material possessions would be a good thing and that more time for leisure and family should be primary goals, but spending money often provides one of life's greatest pleasures. While many would pay more for fuel-efficient cars, fuel economy has either stagnated or even declined in many countries. Despite widespread public support for renewable energy, it still accounts for only a tiny proportion of global energy production.

- *There are diverse barriers standing between pro-sustainability attitudes and individual and collective behaviors.*[80] These include at least three types of barriers. First are the direction, strength, and priority of particular attitudes. Some sustainability attitudes may be widespread but not strongly or consistently enough relative to other, contradictory attitudes. A second type of barrier between attitudes and behavior relates to individual capabilities. Individuals often lack the time, money, access, literacy, knowledge, skills, power, or perceived efficacy to translate attitudes into action. Finally, a third type of barrier is structural and includes laws; regulations; perverse subsidies; infrastructure; available technology; social norms and expectations; and the broader social, economic, and political context (such as the price of oil, interest rates, special interest groups, and the election cycle).

Thus, each particular sustainability behavior may confront a unique set of barriers between attitudes and behaviors. Further, even the same behavior (such as contraceptive use) may confront different barriers across society, space, and scale—with different attitudes or individual and structural barriers operating in developed versus developing countries, in secular versus religious societies, or at different levels of decisionmaking (for example, individuals versus legislatures). Explaining unsustainable behavior is therefore "dauntingly complex, both in its variety and in the causal influences on it."[81] Yet bridging the gaps between what people believe and what people do will be an essential part of the transition to sustainability.

# Promoting Sustainable Behavior

Our limited knowledge about global sustainability values, attitudes, and behaviors does suggest, however, that there are short and long-term strategies to promote sustainable behavior. We know that socially pervasive values and attitudes are often highly resistant to change. Thus, in the short term, leveraging the values and attitudes already dominant in particular cultures may be more practical than asking people to adopt new value orientations.[82] For example, economic values clearly influence and motivate many human behaviors, especially in the market and cash economies of the developed countries. Incorporating environmental and social "externalities" into prices or accounting for the monetary value of ecosystem services can thus encourage both individual and collective sustainable behavior.[83] Likewise, anthropocentric concerns about the impacts of environmental degradation and exploitative labor conditions on human health and social well-being remain strong motivators for action in both the developed and developing worlds.[84] Additionally, religious values are vital sources of meaning, motivation, and direction for much of the world, and many religions are actively re-evaluating and reinterpreting their traditions in support of sustainability.[85]

In the long term, however, more fundamental changes may be required, such as extending and accelerating the shift from materialist to post-materialist values, from anthropocentric to ecological worldviews, and a redefinition of "the good life."[86] These long term changes may be driven in part by impersonal forces, like changing economics (globalization) or technologies (for example, mass media and computer networks) or by broadly based social movements, like those that continue to challenge social attitudes about racism, environmental degradation, and human rights. Finally, sustainability science will play a critical role, at multiple scales and using multiple methodologies, as it works to identify and explain the key relationships between sustainability values, attitudes, and behaviors—and to apply this knowledge in support of sustainable development.

# Notes

1. For example, see U.S. National Research Council, Policy Division, Board on Sustainable Development, *Our Common Journey: A Transition toward Sustainability* (Washington, DC: National Academy Press, 1999); and P. Raskin et al., *Great Transition: The Promise and Lure of the Times Ahead* (Boston: Stockholm Environment Institute, 2002).

2. R. W. Kates, T. M. Parris, and A. Leiserowitz, "What Is Sustainable Development? Goals, Indicators, Values, and Practice," *Environment,* April 2005, 8–21.

3. For simplicity, the words "global" and "worldwide" are used throughout this article to refer to survey results. Please note, however, that there has never been a truly representative global survey with either representative samples from every country in the world or in which all human beings worldwide had an equal probability of being selected. Additionally, some developing country results are taken from predominantly urban samples and are thus not fully representative.

4. For more detail about these surveys and the countries sampled, see Appendix A in A. Leiserowitz, R. W. Kates, and T. M. Parris, *Sustainability Values, Attitudes and Behaviors: A Review of Multi-national and Global Trends* (No. CID Working Paper No. 113) (Cambridge, MA: Science, Environment and Development Group, Center for International Development, Harvard University, 2004), http://www.cid.harvard.edu/cidwp/113.htm.

5. Pew Research Center for the People & the Press, *Views of a Changing World* (Washington, DC: The Pew Research Center for the People & the Press, 2003), T72.

6. See R. Inglehart, "Globalization and Postmodern Values," *Washington Quarterly* 23, no. 1 (1999): 215–28.

7. Leiserowitz, Kates, and Parris, note 4 above, page 8.

8. Pew Research Center for the People & the Press, *The Pew Global Attitudes Project Dataset* (Washington, DC: The Pew Research Center for the People & the Press, 2004).

9. Gross national income (GNI) is "[t]he total market value of goods and services produced during a given period by labor and capital supplied by residents of a country, regardless of where the labor and capital are located. [GNI] differs from GDP primarily by including the capital income that residents earn from investments abroad and excluding the capital income that nonresidents earn from domestic investment." Official development assistance (ODA) is defined as "[t]hose flows to developing countries and multilateral institutions provided by official agencies, including state and local governments, or by their executive agencies, each transaction of which meets the following tests: (a) it is administered with the promotion of the economic development and welfare of developing countries as its main objective; and (b) it is concessional in character and conveys a grant element of at least 25 per cent." UN Millennium Project, *The O. 7% Target: An In-Depth Look,* http://www.unmillenniumproject.org/involved/action07.htm (accessed 24 August 2005). Official development assistance (ODA) does not include aid flows from private voluntary organizations (such as churches, universities, or foundations). For example, it is estimated that in 2000, the United States provided more than $4 billion in private grants for development assistance, versus nearly $10 billion in ODA. U.S. Agency for International Development (USAID), *Foreign Aid in the National Interest* (Washington. DC, 2002), 134.

10. Organisation for Economic Co-operation and Development (OECD), *Official Development Assistance Increases Further—But 2006 Targets Still a Challenge* (Paris: OECD, 2005), http://www.oecd.org/document/3/0,2340,en_2649_34447_34700611_1_1_1_1,00.html (accessed 30 July 2005).

11. Environics International (GlobeScan), *The World Economic Forum Poll: Global Public Opinion on Globalization* (Toronto: Environics International, 2002), http://www.globescan.com/brochures/WEF_Poll_Brief.pdf (accessed 5 October 2004), 3. Note that Environics International changed its name to Globe Scan Incorporated in November 2003.

12. OECD, *Public Opinion and the Fight Against Poverty* (Paris: OECD Development Centre, 2003), 17.

13. Ibid, page 19.

14. Program on International Policy Attitudes (PIPA), *Americans on Foreign Aid and World Hunger: A Study of U.S. Public Attitudes* (Washington, DC: PIPA, 2001), http://www.pipa.org/Online Reports/BFW (accessed 17 November 2004); and OECD, note 12 above, page 22.

15. See OECD Development Co-operation Directorate, *OECD-DAC Secretariat Simulation of DAC Members' Net ODA Volumes in 2006 and 2010,* http://www.oecd.org/dataoecd/57/30/35320618.pdf; and Central Intelligence Agency, The World Factbook, http://www.cia.gov/cia/publications/factbook/.

16. OECD, note 12 above, page 20.

17. I. Smillie and H. Helmich, eds., *Stakeholders: Government-NGO Partnerships for International Development* (London: Earthscan, 1999).

18. L. White Jr., "The Historical Roots of Our Ecologic Crisis," *Science,* l0 March 1967, 1203–07.

19. See C. Merchant, *The Death of Nature: Women, Ecology, and the Scientific Revolution* (1st ed.) (San Francisco: Harper & Row 1980); C. Merchant, *Radical Ecology: The Search for a Livable Worm* (New York: Routledge, 1992); and G. Sessions, *Deep Ecology for the Twenty-First Century* (1st ed.) (New York: Shambhala Press 1995).

20. World Values Survey, *The 1999–2002 Values Surveys Integrated Data File 1.0,* CD-ROM in R. Inglehart, M. Basanez, J. Diez-Medrano, L. Halman, and R. Luijkx, eds., *Human Beliefs and Values: A Cross-Cultural Sourcebook Based on the 1999–2002 Values Surveys,* first edition (Mexico City: Siglo XXI, 2004).

21. These results come from a representative national survey of American Climate change risk perceptions, policy preferences, and behaviors and broader environmental and cultural values. From November 2002 to February 2003, 673 adults (18 and older) completed a mail-out, mail-back questionnaire, for a response rate of 55 percent. The results are weighted to bring them in line with actual population proportions. See A. Leiserowitz, "American Risk Perceptions: Is Climate Change Dangerous?" *Risk Analysis,* in press; and A. Leiserowitz, "Climate Change Risk Perception and Policy Preferences: The Role of Affect, Imagery, and Values," *Climatic Change,* in press.

22. These results support the argument that concerns about the environment are not "a luxury affordable only by those who have enough economic security to pursue quality-of-life goals." See R. E. Dunlap, G. H. Gallup Jr., and A. M. Gallup, "Of Global Concern: Results of the Health of the Planet Survey," *Environment,* November 1993, 7–15, 33–39 (quote at 37); R. E. Dunlap, A. G. Mertig, "Global Concern for the Environment: Is Affluence a Prerequisite?" *Journal of Social Issues* 511, no. 4 (1995): 121–37; S. R. Brechin and W. Kempton, "Global Environmentalism: A Challenge to the Postmaterialism Thesis?" *Social Science Quarterly* 75, no. 2 (1994): 245–69.

23. Environics International (GlobeScan), *Environics International Environmental Monitor Survey Data-set* (Kingston, Canada: Environics International, 2000), http://jeff-lab.queensu.ca/poadata/info/iem/iemlist.shtml (accessed 5 October 2004). These multinational levels of concern and perceived seriousness of environmental problems remained roughly equivalent from 1992 to 2000, averaged across the countries sampled by the 1992 Health of the Planet and the Environics surveys, although some countries saw significant increases in perceived seriousness of environmental problems (India, the Netherlands, the Philippines, and South Korea), while others saw significant decreases (Turkey and Uruguay). See R. E. Dunlap, G. H. Gallup Jr., and A. M. Gallup, *Health of the Planet: Results of a 1992 International Environmental Opinion Survey of Citizens in 24 Nations* (Princeton, NJ: The George H. Gallup International Institute, 1993); and R. E. Dunlap, G. H. Gallup Jr., and A. M. Gallup, "Of Global Concern: Results of the Health of the Planet Survey," *Environment,* November 1993, 7–15, 33–39.

24. GlobeScan, *Results of First-Ever Global Poll on Humanity's Relationship with Nature* (Toronto: GlobeScan Incorporated, 2004), http://www.globescan.com/news_archives/IUCN_PR.html (accessed 30 July 2005).

25. World Values Survey, note 20 above; and Pew Research Center for the People & the Press, *What the World Thinks in 2002* (Washington, DC: The Pew Research Center for the People & the Press, 2002), T-9. The G7 includes Canada, France, Germany, Great Britain, Italy, Japan and the United States. It expanded to the G8 with the addition of Russia in 1998.

26. R. Inglehart, et al., *World Values Surveys and European Values Surveys, 1981–1984, 1990–1993, and 1995–1997* [computer file], Inter-university Consortium for Political and Social Research (ICPSR) version (Ann Arbor, MI: Institute for Social Research [producer], 2000; Ann Arbor, MI: ICPSR [distributor], 2000).

27. Environics International (GlobeScan), note 23 above.

28. Dunlap, Gallup Jr., and Gallup, *Health of the Planet: Results of a 1992 International Environmental Opinion Survey of Citizens in 24 Nations,* note 23 above.

29. Environics International (GlobeScan), *New Poll Shows G8 Citizens Want Legally-Binding Climate Accord* (Toronto: Environics International, 2001), http://www.globescan.com/news_archives/IEM_climatechange.pdf (accessed 30 July 2005).

30. Environics International (GlobeScan), *International Environmental Monitor* (Toronto: Environics International, 2002), 44.

31. Ibid., page 49.

32. The European Opinion Research Group, *Eurobarometer: Energy: Issues, Options and Technologies, Science and Society,* EUR 20624 (Brussels: European Commission, 2002), 96–99.

33. Inglehart, note 26 above.

34. C. M. Rogerson, "The Waste Sector and Informal Entrepreneurship in Developing World Cities," *Urban Forum* 12, no. 2 (2001): 247–59.

35. Environics International (GlobeScan), note 30 above, page 63. These results are based on the sub-sample of those who own or have regular use of a car.

36. Environics International (GlobeScan), note 30 above, page 65.

37. Inglehart, note 26 above.

38. Environics International (GlobeScan), note 23 above.

39. P. A. Ehrlich and J. P. Holdren, review of *The Closing Circle,* by Barry Commoner, *Environment,* April 1972, 24, 26–39.

40. Y. Kaya, "Impact of Carbon Dioxide Emission Control on GNP Growth: Interpretation of Proposed Scenarios," paper presented at the Intergovernmental Panel on Climate Change (IPCC) Energy and Industry Subgroup, Response Strategies Working Group, Paris, France, 1990; and R. York, E. Rosa, and T. Dietz, "STIRPAT, IPAT and ImPACT: Analytic Tools for Unpacking the Driving Forces of Environmental Impacts," *Ecological Economics* 46, no. 3 (2003): 351.

41. IPCC, *Emissions Scenarios* (Cambridge: Cambridge University Press, 2000); and E. F. Lambin, et al., "The Causes of Land-Use and Land-Cover Change: Moving Beyond the Myths," *Global Environmental Change: Human and Policy Dimensions* 11, no. 4 (2001):

42. T. M. Parris and R. W. Kates, "Characterizing a Sustainability Transition: Goals, Targets, Trends, and Driving Forces," *Proceedings of the National Academy of Sciences of the United States of America* 100, no. 14 (2003): 6.

43. Pew Research Center for the People & the Press, note 8 above, page T17.

44. United Nations, *Majority of World's Couples Are Using Contraception* (New York: United Nations Population Division, 2001).

45. J. Bongaarts, "Trends in Unwanted Childbearing in the Developing World," *Studies in Family Planning* 28, no. 4 (1997): 267–77.

46. Demographic and Health Surveys (DHS), *STATCompiler* (Calverton, MD: Measure DHS, 2004), http://www.measuredhs.com/ (accessed 5 October, 2004).

47. Ibid.

48. U.S. Bureau of the Census, *World Population Profile: 1998,* WP/98 (Washington, DC, 1999), 45.

49. World Bank, *World Development Indicators CD-ROM 2004* [computer file] (Washington, DC: International Bank for Reconstruction and Development (IBRD) [producer], 2004).

50. World Bank, *Global Economic Prospects 2005: Trade, Regionalism, and Development* [computer file] (Washington, DC: IBRD [producer] 2005).

51. For more information on poverty reduction strategies, see T. Banuri, review of *Investing in Development: A Practical Plan to Achieve the Millennium Goals,* by UN Millennium Project, *Environment,* November 2005 (this issue), 37.

52. Inglehart, note 26 above.

53. Inglehart, note 26 above.

54. Inglehart, note 26 above.

55. Environics International (GlobeScan), *Consumerism: A Special Report* (Toronto: Environics International, 2002), 6.

56. Pew Research Center for the People & the Press, note 25 above.

57. Dunlap, Gallup Jr., and Gallup, *Health of the Planet: Results of a 1992 International Environmental Opinion Survey of Citizens in 24 Nations,* note 23 above, page 57.

58. Environics International (GlobeScan), note 55 above, pages 3–4.

59. Environics International (GlobeScan), note 55 above, pages 3–4.

60. T. Veblen, *The Theory of the Leisure Class: An Economic Study of Institutions* (New York: Macmillan 1899).

61. Inglehart, note 26 above.

62. Environics International (GlobeScan), note 30 above, page 133.

63. The European Opinion Research Group, note 32 above, page 70.

64. Environics International (GlobeScan), note 23 above.

65. For example, see J. Flynn, P. Slovic, and H. Kunreuther, *Risk, Media and Stigma: Understanding Public Challenges to Modern Science and Technology* (London: Earthscan, 2001).

66. International Social Science Program, *Environment II,* (No. 3440) (Cologne: Zentralarchiv für Empirische Sozialforschung, Universitaet zu Koeln (Central Archive for Empirical Social Research, University of Cologne), 2000), 114.

67. Environics International (GlobeScan), note 30 above, page 139.

68. Environics International (GlobeScan), note 30 above, page 141.

69. Pew Research Center for the People & the Press, note 25 above, page T20.

70. Chicago Council on Foreign Relations (CCFR), *Worldviews 2002* (Chicago: CCFR, 2002), 26.

71. Environics International (GlobeScan), note 30 above, page 163.

72. Environics International (GlobeScan), note 30 above, page 156–57.

73. Environics International (GlobeScan), note 30 above, page 57.

74. W. J. Baumol, R. R. Nelson, and E. N. Wolff, *Convergence of Productivity: Cross-National Studies and Historical Evidence* (New York: Oxford University Press, 1994).

75. A. K. Sen, *Poverty and Famines: An Essay on Entitlement and Deprivation* (Oxford: Oxford University Press, 1981).

76. Pew Research Center for the People & the Press, note 5 above, page 37.

77. World Values Survey, note 20 above.

78. World Values Survey, note 20 above.

79. The human development index (HDI) measures a country's average achievements in three basic aspects of human development: longevity, knowledge, and a decent standard of living. Longevity is measured by life expectancy at birth; knowledge is measured with the adult literacy rate and the combined primary, secondary, and tertiary gross enrollment ratio; and standard of living is measured by gross domestic product per capita (purchase-power parity US$). The UN Development Programme (UNDP) has used the HDI for its annual reports since 1993. UNDP, *Questions About the Human Development Index (HDI),* http://www.undp.org/hdr2003/faq.html#21 (accessed 25 August 2005).

80. See, for example, J. Blake, "Overcoming the 'Value-Action Gap' in Environmental Policy: Tensions Between National Policy and Local Experience," *Local Environment* 4, no. 3 (1999): 257–78; A. Kollmuss and J. Agyeman, "Mind the Gap: Why Do People Act Environmentally and What Are the Barriers to Pro-Environmental Behavior?" *Environmental Education Research* 8, no. 3 (2002): 239–60; and E C. Stem, "Toward a Coherent Theory of Environmentally Significant Behavior," *Journal of Social Issues* 56, no. 3 (2000): 407–24.

81. Stern, ibid., page 421.

82. See, for example, P. W. Schultz and L. Zelezny, "Reframing Environmental Messages to Be Congruent with American Values," *Human Ecology Review* 10, no. 2 (2003): 126–36.

83. Millennium Ecosystem Assessment, *Ecosystems and Human Well-Being: Synthesis* (Washington, DC: Island Press, 2005).

84. Dunlap, Gallup Jr., and Gallup, *Health of the Planet: Results of a 1992 International Environmental Opinion Survey of Citizens in 24 Nations,* note 23 above, page 36.

85. See *The Harvard Forum on Religion and Ecology,* http://environment.harvard.edu/religion/main.html; R. S. Gottlieb, *This Sacred Earth: Religion, Nature, Environment* (New York: Routledge, 1996); and G. Gardner, *Worldwatch Paper # 164: Invoking the Spirit: Religion and Spirituality in the Quest for a Sustainable World* (Washington, DC: Worldwatch Institute, 2002).

86. R. Inglehart, *Modernization and Postmodernization: Cultural, Economic and Political Change in 43 Societies* (Princeton: Princeton University Press, 1997); T. O'Riordan, "Frameworks for Choice: Core Beliefs and the Environment," *Environment,* October 1995, 4–9, 25–29; and E Raskin and Global Scenario Group, *Great Transition: The Promise and Lure of the Times Ahead* (Boston: Stockholm Environment Institute, 2002).

---

**ANTHONY A. LEISEROWITZ** is a research scientist at Decision Research and an adjunct professor of environmental studies at the University of Oregon, Eugene. He is also a principal investigator at the Center for Research on Environmental Decisions at Columbia University. Leiserowitz may be reached at (541) 485-2400 or by email at ecotone@uoregon.edu. **ROBERT W. KATES** is an independent scholar based in Trenton, Maine, and a professor emeritus at Brown University, where he served as director of the Feinstein World Hunger Program. He is also a former vice-chair of the Board of Sustainable Development of the U.S National Academy's National Research Council. In 1991, Kates was awarded the National Medal of Science for his work on hunger, environment, and natural hazards. He is an executive editor of *Environment* and may be contacted at rkates@acadia.net. **THOMAS M. PARRIS** is a research scientist at and director of the New England office of ISCIENCES, LLC. He is a contributing editor of *Environment.* Parris may be reached at parris@isciences.com. The authors retain copyright.

Article originally appeared in *Environment,* Vol. 47, no. 9, November 2005, pp. 23–38. Published by Heldref Publications, Washington, DC. Copyright © 2005 by Anthony A. Leiserowitz, Robert W. Kates, and Thomas M. Parris. Reprinted by permission of the authors. (Figure 1 © 1999 by Ronald Inglehart).

# The Climax of Humanity

**Demographically and economically, our era is unique in human history. Depending on how we manage the next few decades, we could usher in environmental sustainability—or collapse**

GEORGE MUSSER

The 21st century feels like a letdown. We were promised flying cars, space colonies and 15-hour workweeks. Robots were supposed to do our chores, except when they were organizing rebellions; children were supposed to learn about disease from history books; portable fusion reactors were supposed to be on sale at the Home Depot. Even dystopian visions of the future predicted leaps of technology and social organization that leave our era in the dust.

Looking beyond the blinking lights and whirring gizmos, though, the new century is shaping up as one of the most amazing periods in human history. Three great transitions set in motion by the Industrial Revolution are reaching their culmination. After several centuries of faster-than-exponential growth, the world's population is stabilizing. Judging from current trends, it will plateau at around nine billion people toward the middle of this century. Meanwhile extreme poverty is receding both as a percentage of population and in absolute numbers. If China and India continue to follow in the economic footsteps of Japan and South Korea, by 2050 the average Chinese will be as rich as the average Swiss is today; the average Indian, as rich as today's Israeli. As humanity grows in size and wealth, however, it increasingly presses against the limits of the planet. Already we pump out carbon dioxide three times as fast as the oceans and land can absorb it; midcentury is when climatologists think global warming will really begin to bite. At the rate things are going, the world's forests and fisheries will be exhausted even sooner.

These three concurrent, intertwined transitions—demographic, economic, environmental—are what historians of the future will remember when they look back on our age. They are transforming everything from geopolitics to the structure of families. And they pose problems on a scale that humans have little experience with. As Harvard University biologist E. O. Wilson puts it, we are about to pass through "the bottleneck," a period of maximum stress on natural resources and human ingenuity.

The trends are evident in everyday life. Many of us have had the experience of getting lost in our hometowns because they have grown so much. But growth is slowing as families shrink.

## POPULATION GROWTH IS SLOWING ...

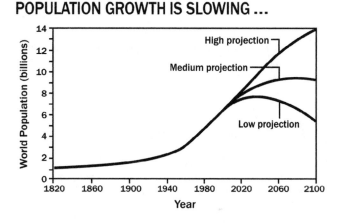

**Figure 1** Three World-Changing Transitions.

Ever more children grow up not just without siblings but also without aunts, uncles or cousins. (Some people find that sad, but the only other way to have a stable population is for death rates to rise.) Chinese goods line Wal-Mart shelves, Indians handle customer-service calls, and, in turn, ever more Asians buy Western products. Spring flowers bloom a week earlier than they did 50 years ago because of global warming, and restaurants serve different types of fish than they used to because species that were once common have been fished out.

Looking at the present era in historical context helps to put the world's myriad problems in perspective. Many of those problems stem, directly or indirectly, from growth. As growth tapers off, humanity will have a chance to close the books on them. A bottleneck may be tough to squeeze through, but once you do, the worst is behind you.

The transitions we are undergoing define the scope of the challenges. Scientists can estimate, at least roughly, how many people will inhabit Earth, what they are going to need and want, what resources are available, and when it is all going to happen. By the latter half of this century, humanity could enter an equilibrium in which economic growth, currently driven by the combination of more productivity, more people and more

# Action Plan for the 21st Century

**1. Understand the changes.** Obvious though it may seem, this first step is so often neglected. It can be hard to look past the daily headlines to understand the core trends we are experiencing. Demographer Joel E. Cohen paints the broad picture of a larger, slower-growing, more urbanized and older population. The detailed projections are uncertain, but what is important is the general issues that they raise.

**2. Achieve Millennium Development Goals.** This month the United Nations General Assembly is reviewing the mixed progress toward these quantitative goals for reducing poverty and inequality. Economist Jeffrey D. Sachs, head of the U.N. Millennium Project, argues for a concerted aid effort. Besides advancing human well-being, it would ease environmental problems that are linked with poverty, such as air pollution and deforestation.

**3. Preserve crucial habitats.** Extinction is irreversible, so avoiding it is a top priority. Obscure creatures are not the only victims; economically valuable species, such as sturgeon and wild grain varieties, are also in trouble. Ecologists Stuart L. Pimm and Clinton Jenkins argue that rounding out nature reserves will cost money but bring multiple benefits. Even in narrow economic terms, countries are often better off saving old-growth forest than converting it to farms or ranches.

**4. Wean off fossil fuels.** The atmosphere can hold only so much carbon dioxide before the climate goes haywire. Reducing emissions requires extensive changes to how we produce and use energy, but Amory B. Lovins, one of the country's most innovative thinkers on the subject, argues that the task is not nearly as daunting or costly as you might think. Accelerating the existing trend toward higher efficiency could do the trick.

**5. Provide cheap irrigation to poor farmers.** How can we feed all those new mouths without trashing the soil, exhausting aquifers and damming every last river? Development specialist Paul Polak argues that small-scale appropriate technology, such as manual pumps and drip irrigation, can boost yields, stretch out limited water supplies and start farmers on the path to prosperity

**6. Beef up health systems.** In rich countries and rapidly developing ones such as China and India, more people now get sick from chronic conditions, such as heart disease and mental illness, than from infections. In poorer countries, malaria, tuberculosis and other bugs remain the big burden. Epidemiologist Barry R. Bloom argues that in both cases, the top priority is better prevention, ranging from vaccines and mosquito nets to antismoking campaigns.

**7. Brace for slower growth.** Political and financial institutions will have to retool as the economy approaches global constraints. Economist Herman E. Daly argues for new ways to collect taxes, set interest rates, and regulate pollution and resource extraction. In an accompanying commentary, economist Partha Dasgupta agrees with much of what Daly says but suggests that rich-country economies are already more sustainable than many people assume.

**8. Prioritize more rationally.** Right now priorities are set largely by who shouts the loudest or plays golf with the right people. As staff writer W. Wayt Gibbs describes, economists and environmental scientists have been working on better approaches. With costs and benefits properly priced in, markets can act as giant distributed computers that weigh trade-offs. But they can fail, for example, when costs are concentrated and benefits are diffuse.

---

resources, will flow entirely from productivity—which would take much of the edge off conflicts between the economy and the environment. Old challenges will give way to new ones. This process is already evident in countries at the leading edge of the transitions. The Social Security debate in the U.S., like worries about pensions in Europe and Japan, is the sound of a society planning for life after growth.

In the public's eyes, demographers have a checkered reputation. Thirty years ago, wasn't overpopulation the big concern? Paul Ehrlich's book *The Population Bomb* was a best seller. The film *Soylent Green,* starring Charlton Heston, dramatized a future in which people would be stacked like cordwood and fed little squares that looked like tofu but weren't. Lately, though, underpopulation has become the cause célèbre, heralded by neoconservatives such as Nicholas Eberstadt. Their concern is epitomized by another Heston movie: *The Omega Man,* in which humanity dwindles to nothingness. So which will it be: Too many people or too few?

Mainstream demographers have not swung back and forth nearly as much as these extreme depictions might suggest. Families in the developing world have shrunk faster than expected, but the forecasts described in *Scientific American*'s 1974 special issue on population have largely stood the test

of time. In fact, the *Soylent Green* and *Omega Man* scenarios each contain an element of truth. Humanity is still growing enormously in absolute terms, and past success at avoiding Malthusian nightmares is no guarantee of future performance. The decline in growth rates is a worry, though. Historically, most stable or shrinking societies have been down at the heel.

Partisans of one scenario shrug off the challenges of the other, expressing "confidence" that they can be handled without actually doing much to ensure that they are. Once you blow away the fog of ideology, the outlines of a comprehensive action plan begin to emerge. It is hardly the only way forward, but it can serve as a starting point for discussion.

A recurring theme of this plan is that business is not necessarily the enemy of nature, or vice versa. Traditionally the economy and environment have not even been described in like terms. The most-watched economic statistics, such as gross domestic product (GDP), do not measure resource depletion; they are essentially measures of cash flow rather than balance sheets of assets and liabilities. If you clear-cut a forest, GDP jumps even though you have wiped out an asset that could have brought in a steady stream of income.

More broadly, the prices we pay for goods and services seldom include the associated environmental costs. Someone else

## ... PROSPERITY IS SPREADING ...

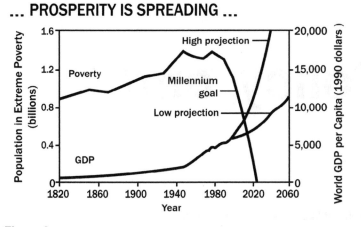

## ... BUT CO₂ EMISSIONS ARE TROUBLING

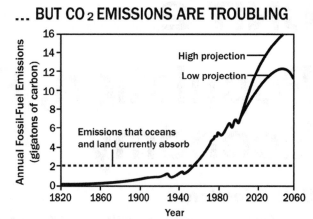

**Figure 2**

picks up the tab—and that someone is usually us, in another guise. By one estimate, the average American taxpayer forks out $2,000 a year to subsidize farming, driving, mining and other activities with a heavy environmental footprint. The distorted market gives consumers and producers little incentive to clean up. Environmentalists inadvertently reinforce this tendency when they focus on the priceless attractions of nature, which are deeply meaningful but difficult to weigh against more pressing concerns. The Endangered Species Act has provided iconic examples of advocates talking past one another. Greens blamed the plight of spotted owls on loggers; the loggers blamed unemployment on self-indulgent ornithology. In fact, both were victims of unsustainable forestry.

In recent years, economists and environmental scientists have come together to hang a price tag on nature's benefits. Far from demeaning nature, this exercise reveals how much we depend on it. *The Millennium Ecosystem Assessment,* published earlier this year, identified services—from pollination to water filtration—that humans would have to provide for themselves, at great cost, if nature did not. Of the 24 broad categories of services, the team found that 15 are being used faster than they regenerate.

When the environment is properly accounted for, what is good for nature is often what is good for the economy and even for individual business sectors. Fishers, for example, maximize their profits when they harvest fisheries at a sustainable level; beyond that point, both yields and profits decline as more people chase ever fewer fish. To be sure, life is not always so convenient. Society must sometimes make real trade-offs. But it is only beginning to explore the win-win options.

If decision makers can get the framework right, the future of humanity will be secured by thousands of mundane decisions: how many babies people have, where they graze their cattle, how they insulate their houses. It is usually in mundane matters that the most profound advances are made. What makes a community rich is not the computers and the DVDs, which you can find nowadays even in humble villages. It is the sewage pipes, the soft beds, the sense of physical and economic security. By helping to bring these benefits of modernity to all, science and technology will have done something more spectacular than building space colonies.

---

**GEORGE MUSSER** is a staff writer and editor.

# UNIT 2

# Population, Policy, and Economy

## Unit Selections

## Key Points to Consider

- Why should policy makers in the more developed countries of the world become more aware of the true dimensions of the world's food problem? How can increased awareness of food scarcity and misallocation lead to solutions for both food production and environmental protection?

- What is the relationship between environmental conditions and national security? Are there non-military solutions that might be more effective in achieving global security than are military ones?

- How and why has thinking about organic farming changed from the idea that farming without agricultural chemicals is too expensive for the world's poorest farmers to the concept that organic farming may, in fact, help to solve food supply problems?

- What is the relationship between resource scarcity and political stability in the world's primary oil-producing region: North Africa and the Middle East? Which resource will be more important for future regional stability: oil or water?

- How is global climate change producing problems for farmers? Are there certain areas of the world where increasing uncertainty about temperature and precipitation are likely to have greater impact?

- What strategies have been used by marketers to sell environmentally-friendly or "green" products? Why have those strategies been largely unsuccessful, and how might "green marketing" be made more effective to shift consumer attitudes toward products and services that provide satisfied customers and a safer environment?

## Student Web Site

www.mhcls.com/online

## Internet References

Further information regarding these Web sites may be found in this book's preface or online.

**The Hunger Project**
*http://www.thp.org*

**Poverty Mapping**
*http://www.povertymap.net*

**World Health Organization**
*http://www.who.int*

**World Population and Demographic Data**
*http://geography.about.com/cs/worldpopulation*

**WWW Virtual Library: Demography & Population Studies**
*http://demography.anu.edu.au/VirtualLibrary*

One of the greatest setbacks on the road to the development of more stable and sensible population policies came about as a result of inaccurate population growth projections were made in the late 1960s and early 1970s. The world was in for a population explosion—the experts told us back then—and food riots would take place in the streets of even the richer countries by the 1980s. But shortly after the publication of such heralded works as *The Population Bomb* (Paul Ehrlich, 1975) and *Limits to Growth* (D. H. Meadows et al., 1974), the growth rate of the world's population began to decline slightly. There was no cause and effect relationship at work here. The decline in growth was simply the demographic transition at work, a process in which declining population growth tends to accompany increasing levels of economic development. Since the alarming predictions did not come to pass, the world began to relax a little. However, two facts still remain: population growth in biological systems must be limited by available resources, and the availability of Earth's resources is finite.

That population growth cannot continue indefinitely is a mathematical certainty. But it is also a certainty that contemporary notions of a continually expanding economy must give way before the realities of a finite resource base. Consider the following: In developing countries, high and growing rural population densities have forced the use of increasingly marginal farmland once considered to be too steep, too dry, too wet, too sterile, or too far from markets for efficient agricultural use. Farming this land damages soil and watershed systems, creates deforestation problems, and adds relatively little to total food production. In the more developed world, farmers also have been driven—usually by market forces—to farm more marginal lands and to rely more on environmentally harmful farming methods utilizing high levels of agricultural chemicals (such as pesticides and artificial fertilizers). These chemicals create hazards for all life and rob the soil of its natural ability to renew itself. The increased demand for economic expansion has also created an increase in the use of precious groundwater reserves for irrigation purposes, depleting those reserves beyond their natural capacity to recharge and creating the potential for once-fertile farmland and grazing land to be transformed into desert. The continued demand for higher production levels also contributes to a soil erosion problem that has reached alarming proportions in all agricultural areas of the world, whether high or low on the scale of economic development. The need to increase the food supply and its consequent effects on the agricultural environment are not the only results of continued population growth. For industrialists, the larger market creates an almost irresistible temptation to accelerate production, requiring the use of more marginal resources and resulting in the destruction of more fragile ecological systems, particularly in the tropics. For consumers, the increased demand for products means increased competition for scarce resources, driving up the cost of those resources until only the wealthiest can afford what our grandfathers would have viewed as an adequate standard of living.

The articles selected for this second unit all relate, in one way or another, to the theory and reality of population growth and its relationship to public policy and economic growth. In the first selection, "Population and Consumption: What We Know, What We Need to Know," geographer and MacArthur Fellow Robert Kates argues that the present set of environmental problems are tied to both the expanding human population in strict numerical terms and the tendency of that growing population to demand more per capita shares of the world's dwindling resources. Kates also recognizes the role of increasing levels of technology as a factor in increasing levels of environmental disturbance. Kates' conclusion is that, while economic development is a global good, increasingly developed societies must learn to curb their tastes for material things if the capacity of the environment to supply both necessary and unnecessary materials is not to be outstripped.

In the following selection—"A New Security Paradigm," Gregory D. Foster of the National Defense University, picks up on some of Kates' arguments, extending them from the economic to the political. Foster notes that while there has been a tendency to equate "national security" with military action against rogue states or terrorism, far more important in the long run may be defense measures that are not military, directed toward environmental security—an area where environmental conditions and concerns over national security converge. Such global environmental problems as climate change, Foster contends, "take us much farther along the path to ultimate causes than terrorism ever could." If nations can view pre-emptive strikes against supposed weapons of mass destruction as being in their best interests, then they should take a similar view of pre-emptive strikes against the conditions that produced an enhanced greenhouse world.

The next two articles in this section move from the theoretical to the practical and from global to regional scales in addressing issues of population, resources, and environment. In "Can Organic Farming Feed Us All?" *Worldwatch Institute* research scientist Brian Halweil suggests that the accepted view of "organic farming as too expensive

to be effective in feeding those areas where food is in short supply" is simply not borne out by the facts. New organic farming is dependent upon new technology—but it is not dependent upon the agricultural chemicals that have become the norm in many of the world's agricultural regions. Studies in developing countries have demonstrated that farmers are capable of increasing yields by mixing traditional organic farming methods with very small inputs of chemicals. In "Where Oil and Water Do Mix: Environmental Scarcity and Future Conflict in the Middle East and North Africa," Jason J. Morrissette and Douglas A. Borer of the University of Georgia and the Naval Postgraduate School respectively, contend that control of water resources has been a principal challenge for human societies since the emergence of agricultural civilizations. In the world's greatest arid region, stretching from the Atlantic coast of Morocco to Iran, contemporary conflicts tend to revolve around control of petroleum and natural gas resources. The reasons why "water wars" have been reduced in the arid world are rooted largely in linkages between water use and the global economy. But the economic factors that have reduced the importance of water shortages cannot extend indefinitely into the future: water is ultimately a nonrenewable resource and its use must be sustainable. The basic environmental and demographic trends of this region suggest that future conflicts will revolve not around oil but around water.

The unit's fifth selection discusses one of agriculture's more serious environmental problems at a global scale. In "The Irony of Climate," *Worldwatch Institute* researcher Brian Halweil suggests the presence of irony in the fact that while natural global climate change thousands of years ago may have pushed humans into food production (agriculture) instead of hunting and gathering, the human society made possible by agriculture is now creating anthropogenic or human-induced climate change. Indeed, the nature of modern agriculture itself, driven by increasing fossil-fuel consumption and use of agricultural chemicals derived from petroleum, is a contributing factor in global warming. As local disturbances in climate alter agricultural patterns, moving food over long distances can accelerate the impact of food production upon climate—the ultimate feedback loop in terms of population and environmental carrying capacity. Sharply contrasting in tone but with an equal goal of improving upon global environmental quality, the unit's concluding article deals with something that environmentalists seldom think about: marketing. In "Avoiding Green Marketing Myopia" marketing experts Jacquelyn Ottman, Edwin Stafford, and Cathy Hartman suggest that while the development of environmentally sustainable products and services would benefit everyone, the traditional marketing approach to selling "green" products and services is incorrect and unproductive. Rather than using marketing strategies designed only to appeal to the "deepest green" niche of consumers, those who want to sell environmentally sound products need to demonstrate to main-stream consumers that green products can save them money, give them a better product, and be good for the environment as well. Green marketing needs to satisfy two objectives: improved environment quality and increased customer satisfaction.

All the authors of selections in this unit make it clear that the global environment is being stressed by population growth as well as environmental and economic policies that result in more environmental pressure and degradation. While it should be evident that we can no longer afford to permit the unplanned and unchecked growth of the planet's dominant species, it should also be apparent that the unchecked growth of economic systems without some kind of environmental accounting systems is just as dangerous.

# Population and Consumption

## What We Know, What We Need to Know

ROBERT W. KATES

Thirty years ago, as Earth Day dawned, three wise men recognized three proximate causes of environmental degradation yet spent half a decade or more arguing their relative importance. In this classic environmentalist feud between Barry Commoner on one side and Paul Ehrlich and John Holdren on the other, all three recognized that growth in population, affluence, and technology were jointly responsible for environmental problems, but they strongly differed about their relative importance. Commoner asserted that technology and the economic system that produced it were primarily responsible.[1] Ehrlich and Holdren asserted the importance of all three drivers: population, affluence, and technology. But given Ehrlich's writings on population,[2] the differences were often, albeit incorrectly, described as an argument over whether population or technology was responsible for the environmental crisis.

Now, 30 years later, a general consensus among scientists posits that growth in population, affluence, and technology are jointly responsible for environmental problems. This has become enshrined in a useful, albeit overly simplified, identity known as IPAT, first published by Ehrlich and Holdren in *Environment* in 1972[3] in response to the more limited version by Commoner that had appeared earlier in *Environment* and in his famous book *The Closing Circle*.[4] In this identity, various forms of environmental or resource impacts (I) equals population (P) times affluence (A) (usually income per capita) times the impacts per unit of income as determined by technology (T) and the institutions that use it. Academic debate has now shifted from the greater or lesser importance of each of these driving forces of environmental degradation or resource depletion to debate about their interaction and the ultimate forces that drive them.

However, in the wider global realm, the debate about who or what is responsible for environmental degradation lives on. Today, many Earth Days later, international debates over such major concerns as biodiversity, climate change, or sustainable development address the population and the affluence terms of Holdrens' and Ehrlich's identity, specifically focusing on the character of consumption that affluence permits. The concern with technology is more complicated because it is now widely recognized that while technology can be a problem, it can be a solution as well. The development and use of more environmentally benign and friendly technologies in industrialized countries have slowed the growth of many of the most pernicious forms of pollution that originally drew Commoner's attention and still dominate Earth Day concerns.

A recent report from the National Research Council captures one view of the current public debate, and it begins as follows:

*For over two decades, the same frustrating exchange has been repeated countless times in international policy circles. A government official or scientist from a wealthy country would make the following argument: The world is threatened with environmental disaster because of the depletion of natural resources (or climate change or the loss of biodiversity), and it cannot continue for long to support its rapidly growing population. To preserve the environment for future generations, we need to move quickly to control global population growth, and we must concentrate the effort on the world's poorer countries, where the vast majority of population growth is occurring.*

Government officials and scientists from low-income countries would typically respond:

*If the world is facing environmental disaster, it is not the fault of the poor, who use few resources. The fault must lie with the world's wealthy countries, where people consume the great bulk of the world's natural resources and energy and cause the great bulk of its environmental degradation. We need to curtail overconsumption in the rich countries which use far more than their fair share, both to preserve the environment and to allow the poorest people on earth to achieve an acceptable standard of living.*[5]

It would be helpful, as in all such classic disputes, to begin by laying out what is known about the relative responsibilities of both population and consumption for the environmental crisis, and what might need to be known to address them. However, there is a profound asymmetry that must fuel the frustration of the developing countries' politicians and scientists: namely, how much people know about population and how little they know about consumption. Thus, this article begins by examining these differences in knowledge and action and concludes with the alternative actions needed to go from more to enough in both population and consumption.[6]

# Population

What population is and how it grows is well understood even if all the forces driving it are not. Population begins with people and their key events of birth, death, and location. At the margins, there is some debate over when life begins and ends or whether residence is temporary or permanent, but little debate in between. Thus, change in the world's population or any place is the simple arithmetic of adding births, subtracting deaths, adding immigrants, and subtracting outmigrants. While whole subfields of demography are devoted to the arcane details of these additions and subtractions, the error in estimates of population for almost all places is probably within 20 percent and for countries with modern statistical services, under 3 percent—better estimates than for any other living things and for most other environmental concerns.

Current world population is more than six billion people, growing at a rate of 1.3 percent per year. The peak annual growth rate in all history—about 2.1 percent—occurred in the early 1960s, and the peak population increase of around 87 million per year occurred in the late 1980s. About 80 percent or 4.8 billion people live in the less developed areas of the world, with 1.2 billion living in industrialized countries. Population is now projected by the United Nations (UN) to be 8.9 billion in 2050, according to its medium fertility assumption, the one usually considered most likely, or as high as 10.6 billion or as low as 7.3 billion.[7]

A general description of how birth rates and death rates are changing over time is a process called the demographic transition.[8] It was first studied in the context of Europe, where in the space of two centuries, societies went from a condition of high births and high deaths to the current situation of low births and low deaths. In such a transition, deaths decline more rapidly than births, and in that gap, population grows rapidly but eventually stabilizes as the birth decline matches or even exceeds the death decline. Although the general description of the transition is widely accepted, much is debated about its cause and details.

The world is now in the midst of a global transition that, unlike the European transition, is much more rapid. Both births and deaths have dropped faster than experts expected and history foreshadowed. It took 100 years for deaths to drop in Europe compared to the drop in 30 years in the developing world. Three is the current global average births per woman of reproductive age. This number is more than halfway between the average of five children born to each woman at the post World War II peak of population growth and the average of 2.1 births required to achieve eventual zero population growth.[9] The death transition is more advanced, with life expectancy currently at 64 years. This represents three-quarters of the transition between a life expectancy of 40 years to one of 75 years. The current rates of decline in births outpace the estimates of the demographers, the UN having reduced its latest medium expectation of global population in 2050 to 8.9 billion, a reduction of almost 10 percent from its projection in 1994.

Demographers debate the causes of this rapid birth decline. But even with such differences, it is possible to break down the projected growth of the next century and to identify policies that would reduce projected populations even further. John Bongaarts of the Population Council has decomposed the projected developing country growth into three parts and, with his colleague Judith Bruce, has envisioned policies that would encourage further and more rapid decline.[10] The first part is unwanted fertility, making available the methods and materials for contraception to the 120 million married women (and the many more unmarried women) in developing countries who in survey research say they either want fewer children or want to space them better. A basic strategy for doing so links voluntary family planning with other reproductive and child health services.

Yet in many parts of the world, the desired number of children is too high for a stabilized population. Bongaarts would reduce this desire for large families by changing the costs and benefits of childrearing so that more parents would recognize the value of smaller families while simultaneously increasing their investment in children. A basic strategy for doing so accelerates three trends that have been shown to lead to lower desired family size: the survival of children, their education, and improvement in the economic, social, and legal status for girls and women.

However, even if fertility could immediately be brought down to the replacement level of two surviving children per woman, population growth would continue for many years in most developing countries because so many more young people of reproductive age exist. So Bongaarts would slow this momentum of population growth by increasing the age of childbearing, primarily by improving secondary education opportunity for girls and by addressing such neglected issues as adolescent sexuality and reproductive behavior.

How much further could population be reduced? Bongaarts provides the outer limits. The population of the developing world (using older projections) was expected to reach 10.2 billion by 2100. In theory, Bongaarts found that meeting the unmet need for contraception could reduce this total by about 2 billion. Bringing down desired family size to replacement fertility would reduce the population a billion more, with the remaining growth—from 4.5 billion today to 7.3 billion in 2100—due to population momentum. In practice, however, a recent U.S. National Academy of Sciences report concluded that a 10 percent reduction is both realistic and attainable and could lead to a lessening in projected population numbers by 2050 of upwards of a billion fewer people.[11]

# Consumption

In contrast to population, where people and their births and deaths are relatively well-defined biological events, there is no consensus as to what consumption includes. Paul Stern of the National Research Council has described the different ways physics, economics, ecology, and sociology view consumption.[12] For physicists, matter and energy cannot be consumed, so consumption is conceived as transformations of matter and energy with increased entropy. For economists, consumption is spending on consumer goods and services and thus distinguished from their production and distribution. For ecologists, consumption is obtaining energy and nutrients by eating something else, mostly green plants or other consumers of green plants. And for some sociologists, consumption is a status symbol—keeping up with the Joneses—when individuals and households use their incomes to increase their social status through certain kinds of purchases. These differences are summarized in the box below.

In 1977, the councils of the Royal Society of London and the U.S. National Academy of Sciences issued a joint statement on consumption, having previously done so on population. They chose a variant of the physicist's definition:

> *Consumption is the human transformation of materials and energy. Consumption is of concern to the extent that it makes the transformed materials or energy less available for future use, or negatively impacts biophysical systems in such a way as to threaten human health, welfare, or other things people value.*[13]

On the one hand, this society/academy view is more holistic and fundamental than the other definitions; on the other hand, it is more focused, turning attention to the environmentally damaging. This article uses it as a working definition with one modification, the addition of information to energy and matter, thus completing the triad of the biophysical and ecological basics that support life.

In contrast to population, only limited data and concepts on the transformation of energy, materials, and information exist.[14] There is relatively good global knowledge of energy transformations due in part to the common units of conversion between different technologies. Between 1950 and today, global energy production and use increased more than fourfold.[15] For material transformations, there are no aggregate data in common units on a global basis, only for some specific classes of materials including materials for energy production, construction, industrial minerals and metals, agricultural crops, and water.[16] Calculations of material use by volume, mass, or value lead to different trends.

Trend data for per capita use of physical structure materials (construction and industrial minerals, metals, and forestry products) in the United States are relatively complete. They show an inverted S shaped (logistic) growth pattern: modest doubling between 1900 and the depression of the 1930s (from two to four metric tons), followed by a steep quintupling with economic recovery until the early 1970s (from two to eleven tons), followed by a leveling off since then with fluctuations related to economic downturns (see Figure 1).[17] An aggregate analysis of all current material production and consumption in the United States averages more than 60 kilos per person per day (excluding water). Most of this material flow is split

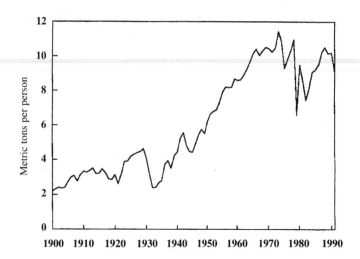

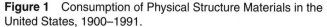

**Figure 1** Consumption of Physical Structure Materials in the United States, 1900–1991.

Source: I. Wernick, "Consuming Materials: The American Way," *Technological Forecasting and Social Change,* 53 (1996): 114.

between energy and related products (38 percent) and minerals for construction (37 percent), with the remainder as industrial minerals (5 percent), metals (2 percent), products of fields (12 percent), and forest (5 percent).[18]

A massive effort is under way to catalog biological (genetic) information and to sequence the genomes of microbes, worms, plants, mice, and people. In contrast to the molecular detail, the number and diversity of organisms is unknown, but a conservative estimate places the number of species on the order of 10 million, of which only one-tenth have been described.[19] Although there is much interest and many anecdotes, neither concepts nor data are available on most cultural information. For example, the number of languages in the world continues to decline while the number of messages expands exponentially.

Trends and projections in agriculture, energy, and economy can serve as surrogates for more detailed data on energy and material transformation.[20] From 1950 to the early 1990s, world population more than doubled (2.2 times), food as measured by grain production almost tripled (2.7 times), energy more than quadrupled (4.4 times), and the economy quintupled (5.1 times). This 43-year record is similar to a current 55-year projection (1995–2050) that assumes the continuation of current trends or, as some would note, "business as usual." In this 55-year projection, growth in half again of population (1.6 times) finds almost a doubling of agriculture (1.8 times), more than

## What Is Consumption?

**Physicist:** "What happens when you transform matter/energy"
**Ecologist:** "What big fish do to little fish"
**Economist:** "What consumers do with their money"
**Sociologist:** "What you do to keep up with the Joneses"

twice as much energy used (2.4 times), and a quadrupling of the economy (4.3 times).[21]

Thus, both history and future scenarios predict growth rates of consumption well beyond population. An attractive similarity exists between a demographic transition that moves over time from high births and high deaths to low births and low deaths with an energy, materials, and information transition. In this transition, societies will use increasing amounts of energy and materials as consumption increases, but over time the energy and materials input per unit of consumption decrease and information substitutes for more material and energy inputs.

Some encouraging signs surface for such a transition in both energy and materials, and these have been variously labeled as decarbonization and dematerialization.[22] For more than a century, the amount of carbon per unit of energy produced has been decreasing. Over a shorter period, the amount of energy used to produce a unit of production has also steadily declined. There is also evidence for dematerialization, using fewer materials for a unit of production, but only for industrialized countries and for some specific materials. Overall, improvements in technology and substitution of information for energy and materials will continue to increase energy efficiency (including decarbonization) and dematerialization per unit of product or service. Thus, over time, less energy and materials will be needed to make specific things. At the same time, the demand for products and services continues to increase, and the overall consumption of energy and most materials more than offsets these efficiency and productivity gains.

# What to Do about Consumption

While quantitative analysis of consumption is just beginning, three questions suggest a direction for reducing environmentally damaging and resource-depleting consumption. The first asks: *When is more too much for the life-support systems of the natural world and the social infrastructure of human society?* Not all the projected growth in consumption may be resource-depleting—"less available for future use"—or environmentally damaging in a way that "negatively impacts biophysical systems to threaten human health, welfare, or other things people value."[23] Yet almost any human-induced transformations turn out to be either or both resource-depleting or damaging to some valued environmental component. For example, a few years ago, a series of eight energy controversies in Maine were related to coal, nuclear, natural gas, hydroelectric, biomass, and wind generating sources, as well as to various energy policies. In all the controversies, competing sides, often more than two, emphasized environmental benefits to support their choice and attributed environmental damage to the other alternatives.

Despite this complexity, it is possible to rank energy sources by the varied and multiple risks they pose and, for those concerned, to choose which risks they wish to minimize and which they are more willing to accept. There is now almost 30 years of experience with the theory and methods of risk assessment and 10 years of experience with the identification and setting of environmental priorities. While there is still no readily accepted methodology for separating resource-depleting or environmentally damaging consumption from general consumption or for identifying harmful transformations from those that are benign, one can separate consumption into more or less damaging and depleting classes and *shift* consumption to the less harmful class. It is possible to *substitute* less damaging and depleting energy and materials for more damaging ones. There is growing experience with encouraging substitution and its difficulties: renewables for nonrenewables, toxics with fewer toxics, ozone-depleting chemicals for more benign substitutes, natural gas for coal, and so forth.

The second question, *Can we do more with less?*, addresses the supply side of consumption. Beyond substitution, shrinking the energy and material transformations required per unit of consumption is probably the most effective current means for reducing environmentally damaging consumption. In the 1997 book, *Stuff: The Secret Lives of Everyday Things,* John Ryan and Alan Durning of Northwest Environment Watch trace the complex origins, materials, production, and transport of such everyday things as coffee, newspapers, cars, and computers and highlight the complexity of reengineering such products and reorganizing their production and distribution.[24]

Yet there is growing experience with the three Rs of consumption shrinkage: reduce, recycle, reuse. These have now been strengthened by a growing science, technology, and practice of industrial ecology that seeks to learn from nature's ecology to reuse everything. These efforts will only increase the existing favorable trends in the efficiency of energy and material usage. Such a potential led the Intergovernmental Panel on Climate Change to conclude that it was possible, using current best practice technology, to reduce energy use by 30 percent in the short run and 50–60 percent in the long run.[25] Perhaps most important in the long run, but possibly least studied, is the potential for and value of substituting information for energy and materials. Energy and materials per unit of consumption are going down, in part because more and more consumption consists of information.

The third question addresses the demand side of consumption —*When is more enough?*[26] Is it possible to reduce consumption by more satisfaction with what people already have, by *satiation,* no more needing more because there is enough, and by *sublimation,* having more satisfaction with less to achieve some greater good? This is the least explored area of consumption and the most difficult. There are, of course, many signs of *satiation* for some goods. For example, people in the industrialized world no longer buy additional refrigerators (except in newly formed households) but only replace them. Moreover, the quality of refrigerators has so improved that a 20-year or more life span is commonplace. The financial pages include frequent stories of the plight of this industry or corporation whose markets are saturated and whose products no longer show the annual growth equated with profits and progress. Such enterprises are frequently viewed as failures of marketing or entrepreneurship rather than successes in meeting human needs sufficiently and efficiently. Is it possible to reverse such views, to create a standard of satiation, a satisfaction in a need well met?

Can people have more satisfaction with what they already have by using it more intensely and having the time to do so? Economist Juliet Schor tells of some overworked Americans who would willingly exchange time for money, time to spend with family and using what they already have, but who are constrained by an uncooperative employment structure.[27] Proposed U.S. legislation would permit the trading of overtime for such compensatory time off, a step in this direction. *Sublimation,* according to the dictionary, is the diversion of energy from an immediate goal to a higher social, moral, or aesthetic purpose. Can people be more satisfied with less satisfaction derived from the diversion of immediate consumption for the satisfaction of a smaller ecological footprint?[28] An emergent research field grapples with how to encourage consumer behavior that will lead to change in environmentally damaging consumption.[29]

A small but growing "simplicity" movement tries to fashion new images of "living the good life."[30] Such movements may never much reduce the burdens of consumption, but they facilitate by example and experiment other less-demanding alternatives. Peter Menzel's remarkable photo essay of the material goods of some 30 households from around the world is powerful testimony to the great variety and inequality of possessions amidst the existence of alternative life styles.[31] Can a standard of "more is enough" be linked to an ethic of "enough for all"? One of the great discoveries of childhood is that eating lunch does not feed the starving children of some far-off place. But increasingly, in sharing the global commons, people flirt with mechanisms that hint at such—a rationing system for the remaining chlorofluorocarbons, trading systems for reducing emissions, rewards for preserving species, or allowances for using available resources.

A recent compilation of essays, *Consuming Desires: Consumption, Culture, and the Pursuit of Happiness,*[32] explores many of these essential issues. These elegant essays by 14 well-known writers and academics ask the fundamental question of why more never seems to be enough and why satiation and sublimation are so difficult in a culture of consumption. Indeed, how is the culture of consumption different for mainstream America, women, inner-city children, South Asian immigrants, or newly industrializing countries?

# Why We Know and Don't Know

In an imagined dialog between rich and poor countries, with each side listening carefully to the other, they might ask themselves just what they actually know about population and consumption. Struck with the asymmetry described above, they might then ask: "Why do we know so much more about population than consumption?"

The answer would be that population is simpler, easier to study, and a consensus exists about terms, trends, even policies. Consumption is harder, with no consensus as to what it is, and with few studies except in the fields of marketing and advertising. But the consensus that exists about population comes from substantial research and study, much of it funded by governments and groups in rich countries, whose asymmetric concern readily identifies the troubling fertility behavior of others and only reluctantly considers their own consumption behavior. So while consumption is harder, it is surely studied less (see Table 1).

The asymmetry of concern is not very flattering to people in developing countries. Anglo-Saxon tradition has a long history of dominant thought holding the poor responsible for their condition—they have too many children—and an even longer tradition of urban civilization feeling besieged by the barbarians at their gates. But whatever the origins of the asymmetry, its persistence does no one a service. Indeed, the stylized debate of population versus consumption reflects neither popular understanding nor scientific insight. Yet lurking somewhere beneath the surface concerns lies a deeper fear.

Consumption is more threatening, and despite the North–South rhetoric, it is threatening to all. In both rich and poor countries alike, making and selling things to each other, including unnecessary things, is the essence of the economic system. No longer challenged by socialism, global capitalism seems inherently based on growth—growth of both consumers and their consumption. To study consumption in this light is to risk concluding that a transition to sustainability might require profound changes in the making and selling of things and in the opportunities that this provides. To draw such conclusions, in the absence of convincing alternative visions, is fearful and to be avoided.

# What We Need to Know and Do

In conclusion, returning to the 30-year-old IPAT identity—a variant of which might be called the Population/Consumption (PC) version—and restating that identity in terms of population and consumption, it would be: $I = P^*C/P^*I/C$, where I equals environmental degradation and/or resource depletion; P equals the number of people or households; and C equals the transformation of energy, materials, and information (see Figure 2).

With such an identity as a template, and with the goal of reducing environmentally degrading and resource-depleting influences, there are at least seven major directions for research and policy. To reduce the level of impacts per unit of consumption, it is necessary to separate out more damaging consumption and shift to less harmful forms, *shrink* the amounts of environmentally damaging energy and materials per unit of consumption,

## Table 1 A Comparison of Population and Consumption

| Population | Consumption |
| --- | --- |
| Simpler, easier to study | More complex |
| Well-funded research | Unfunded, except marketing |
| Consensus terms, trends | Uncertain terms, trends |
| Consensus policies | Threatening policies |

Source: Robert W. Kates.

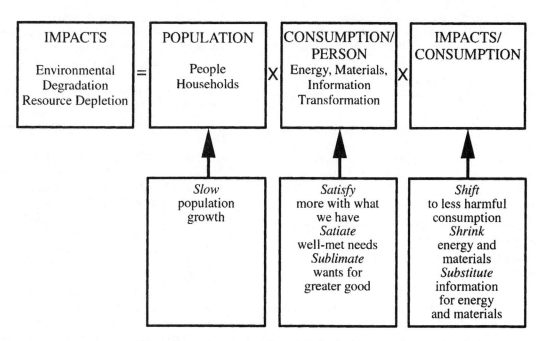

**Figure 2** IPAT (Population/Consumption Version): A Template for Action.
Source: Robert W. Kates.

and *substitute* information for energy and materials. To reduce consumption per person or household, it is necessary to *satisfy* more with what is already had, *satiate* well-met consumption needs, and *sublimate* wants for a greater good. Finally, it is possible to *slow* population growth and then to *stabilize* population numbers as indicated above.

However, as with all versions of the IPAT identity, population and consumption in the PC version are only proximate driving forces, and the ultimate forces that drive consumption, the consuming desires, are poorly understood, as are many of the major interventions needed to reduce these proximate driving forces. People know most about slowing population growth, more about shrinking and substituting environmentally damaging consumption, much about shifting to less damaging consumption, and least about satisfaction, satiation, and sublimation. Thus the determinants of consumption and its alternative patterns have been identified as a key understudied topic for an emerging sustainability science by the recent U.S. National Academy of Science study.[33]

But people and society do not need to know more in order to act. They can readily begin to separate out the most serious problems of consumption, shrink its energy and material throughputs, substitute information for energy and materials, create a standard for satiation, sublimate the possession of things for that of the global commons, as well as slow and stabilize population. To go from more to enough is more than enough to do for 30 more Earth Days.

# Notes

1. B. Commoner, M. Corr, and P. Stamler, "The Causes of Pollution," *Environment,* April 1971, 2–19.

2. P. Ehrlich, *The Population Bomb* (New York: Ballantine, 1966).

3. P. Ehrlich and J. Holdren, "Review of The Closing Circle," *Environment,* April 1972, 24–39.

4. B. Commoner, *The Closing Circle* (New York: Knopf, 1971).

5. P. Stern, T. Dietz, V. Ruttan, R. H. Socolow, and J. L. Sweeney, eds., *Environmentally Significant Consumption: Research Direction* (Washington, D.C.: National Academy Press, 1997), 1.

6. This article draws in part upon a presentation for the 1997 De Lange-Woodlands Conference, an expanded version of which will appear as: R. W. Kates, "Population and Consumption: From More to Enough," in *In Sustainable Development: The Challenge of Transition,* J. Schmandt and C. H. Wards, eds. (Cambridge, U.K.: Cambridge University Press, forthcoming), 79–99.

7. United Nations, Population Division, *World Population Prospects: The 1998 Revision* (New York: United Nations, 1999).

8. K. Davis, "Population and Resources: Fact and Interpretation," K. Davis and M. S. Bernstam, eds., in *Resources, Environment and Population: Present Knowledge, Future Options,* supplement to *Population and Development Review,* 1990: 1–21.

9. Population Reference Bureau, *1997 World Population Data Sheet of the Population Reference Bureau* (Washington, D.C.: Population Reference Bureau, 1997).

10. J. Bongaarts, "Population Policy Options in the Developing World," *Science,* 263: (1994), 771–776; and J. Bongaarts and J. Bruce, "What Can Be Done to Address Population Growth?" (unpublished background paper for The Rockefeller Foundation, 1997).

11. National Research Council, Board on Sustainable Development, *Our Common Journey: A Transition Toward Sustainability* (Washington, D.C.: National Academy Press, 1999).

12. See Stern, et al., note 5 above.

13. Royal Society of London and the U.S. National Academy of Sciences, "Towards Sustainable Consumption," reprinted in *Population and Development Review,* 1977, 23 (3): 683–686.

14. For the available data and concepts, I have drawn heavily from J. H. Ausubel and H. D. Langford, eds., *Technological Trajectories and the Human Environment.* (Washington, D.C.: National Academy Press, 1997).

15. L. R. Brown, H. Kane, and D. M. Roodman, *Vital Signs 1994: The Trends That Are Shaping Our Future* (New York: W. W. Norton and Co., 1994).

16. World Resources Institute, United Nations Environment Programme, United Nations Development Programme, World Bank, *World Resources, 1996–97* (New York: Oxford University Press, 1996); and A. Gruebler, *Technology and Global Change* (Cambridge, Mass.: Cambridge University Press, 1998).

17. I. Wernick, "Consuming Materials: The American Way," *Technological Forecasting and Social Change,* 53 (1996): 111–122.

18. I. Wernick and J. H. Ausubel, "National Materials Flow and the Environment," *Annual Review of Energy and Environment,* 20 (1995): 463–492.

19. S. Pimm, G. Russell, J. Gittelman, and T. Brooks, "The Future of Biodiversity," *Science,* 269 (1995): 347–350.

20. Historic data from L. R. Brown, H. Kane, and D. M. Roodman, note 15 above.

21. One of several projections from P. Raskin, G. Gallopin, P. Gutman, A. Hammond, and R. Swart, *Bending the Curve: Toward Global Sustainability,* a report of the Global Scenario Group, Polestar Series, report no. 8 (Boston: Stockholm Environmental Institute, 1995).

22. N. Nakicénovíc, "Freeing Energy from Carbon," in *Technological Trajectories and the Human Environment,* eds., J. H. Ausubel and H. D. Langford. (Washington, D.C.: National Academy Press, 1997); I. Wernick, R. Herman, S. Govind, and J. H. Ausubel, "Materialization and Dematerialization: Measures and Trends," in J. H. Ausubel and H. D. Langford, eds., *Technological Trajectories and the Human Environment* (Washington, D.C.: National Academy Press, 1997), 135–156; and see A. Gruebler, note 16 above.

23. Royal Society of London and the U.S. National Academy of Science, note 13 above.

24. J. Ryan and A. Durning, *Stuff: The Secret Lives of Everyday Things* (Seattle, Wash.: Northwest Environment Watch, 1997).

25. R. T. Watson, M. C. Zinyowera, and R. H. Moss, eds., *Climate Change 1995: Impacts, Adaptations, and Mitigation of Climate Change—Scientific-Technical Analyses* (Cambridge, U.K.: Cambridge University Press, 1996).

26. A sampling of similar queries includes: A. Durning, *How Much Is Enough?* (New York: W. W. Norton and Co., 1992); Center for a New American Dream, *Enough!: A Quarterly Report on Consumption, Quality of Life and the Environment* (Burlington, Vt.: The Center for a New American Dream, 1997); and N. Myers, "Consumption in Relation to Population, Environment, and Development," *The Environmentalist,* 17 (1997): 33–44.

27. J. Schor, *The Overworked American* (New York: Basic Books, 1991).

28. A. Durning, *How Much Is Enough?: The Consumer Society and the Future of the Earth* (New York: W. W. Norton and Co., 1992); Center for a New American Dream, note 26 above; and M. Wackernagel and W. Ress, *Our Ecological Footprint: Reducing Human Impact on the Earth* (Philadelphia. Pa.: New Society Publishers, 1996).

29. W. Jager, M. van Asselt, J. Rotmans, C. Vlek, and P. Costerman Boodt, *Consumer Behavior: A Modeling Perspective in the Contest of Integrated Assessment of Global Change,* RIVM report no. 461502017 (Bilthoven, the Netherlands: National Institute for Public Health and the Environment, 1997); and P. Vellinga, S. de Bryn, R. Heintz, and P. Molder, eds., *Industrial Transformation: An Inventory of Research.* IHDP-IT no. 8 (Amsterdam, the Netherlands: Institute for Environmental Studies, 1997).

30. H. Nearing and S. Nearing. *The Good Life: Helen and Scott Nearing's Sixty Years of Self-Sufficient Living* (New York: Schocken, 1990); and D. Elgin, *Voluntary Simplicity: Toward a Way of Life That Is Outwardly Simple, Inwardly Rich* (New York: William Morrow, 1993).

31. P. Menzel, *Material World: A Global Family Portrait* (San Francisco: Sierra Club Books, 1994).

32. R. Rosenblatt, ed., *Consuming Desires: Consumption, Culture, and the Pursuit of Happiness* (Washington, D.C.: Island Press, 1999).

33. National Research Council, Board on Sustainable Development, *Our Common Journey: A Transition Toward Sustainability* (Washington, D.C.: National Academy Press, 1999).

---

**ROBERT W. KATES** is an independent scholar in Trenton, Maine; a geographer; university professor emeritus at Brown University; and an executive editor of *Environment.* The research for "Population and Consumption: What We Know, What We Need to Know" was undertaken as a contribution to the recent National Academies/National Research Council report, *Our Common Journey: A Transition Toward Sustainability.* The author retains the copyright to this article. Kates can be reached at RR1, Box 169B, Trenton, ME 04605.

From *Environment,* April 2000, pp. 10–19. Copyright © 2000 by Robert Kates. Reprinted by permission of the author. Published by Heldref Publications, 1319 Eighteenth St., NW, Washington, DC 20036-1802. www.heldref.org

# A New Security Paradigm

**It's easy to equate "national security" or "global security" with military defense against rogue states and terrorism, but a leading U.S. military expert says that view is far too narrow—and could lead to catastrophe if not changed.**

GREGORY D. FOSTER

Whatever else the year 2004 might be noted for by future historians—the U.S. political wars, the genocide in Darfur, the strategic debacle in Iraq—it may well turn out to have been a seminal year for the field of environmental security—the intellectual, operational, and policy space where environmental conditions and security concerns converge.

So too, one hopes, might it have been the year when U.S. policymakers and the American public began to awaken, however belatedly, to the need for an entirely new approach to security and for the fundamental strategic transformation necessary to achieve such security.

The highlight of the year in this regard was, for two essentially countervailing reasons, the award of the Nobel Peace Prize to Kenyan environmental activist Wangari Maathai. On one hand, by broadening the definitional bounds of peace, the award gave new legitimacy to those who would embrace unconventional conceptions of security, especially involving the environment. "This is the first time environment sets the agenda for the Nobel Peace Prize, and we have added a new dimension to peace," said committee chairman Ole Danbolt Mjoes in announcing the award. "Peace on earth depends on our ability to secure our living environment."

On the other hand, no less noteworthy were critics of the award, whose expressions of disparagement typified and reaffirmed the stultifying hold of traditionalist thinking on the conduct of international relations. Espen Barth Eide, former Norwegian deputy foreign minister, observed: "The one thing the Nobel Committee does is define the topic of this epoch in the field of peace and security. If they widen it too much, they risk undermining the core function of the Peace Prize; you end up saying everything that is good is peace." Traditionalists everywhere, including most of the U.S. policy establishment, no doubt took succor from such self-righteous indignation and resolved to perpetuate the received truths of the past that have made real peace so elusive and illusory to date.

Beyond the Peace Prize, two other events ten months apart served as defining bookends for what could turn out to have been the undeclared Year of Environmental Security. The first was an attention-grabbing article, "The Pentagon's Weather Nightmare," that appeared in the February 9, 2004 issue of *Fortune* magazine. Describing a report two futurists—Peter Schwartz and Doug Randall of Global Business Network—had recently prepared for the Defense Department on the national security implications of abrupt climate change, the article generated a flurry of intense but short-lived excitement and speculation on whether, why, and to what extent the Pentagon was finally taking climate change seriously.

The second bookend event came at the end of the year with the issuance of the final report of the internationally distinguished, 16-member High-Level Panel on Threats, Challenges and Change that UN Secretary-General Kofi Annan had appointed in November 2003 to examine the major threats and challenges the world faces in the broad field of peace and security.

These two particular events, potentially significant enough in their own right, should be viewed in the larger context of several other magnifying events that occurred over the course of the year.

For starters, Sir David King, chief science adviser to British prime minister Tony Blair, raised eyebrows and hackles with a controversial article in the January 9, 2004 issue of *Science* magazine. King cited climate change as "the most severe problem that we are facing today—more serious even than the threat of terrorism," and accused the U.S. government of "failing to take up the challenge of global warming." In a subsequent speech to the American Association for the Advancement of Science, he added: "Climate change is real. Millions will increasingly be exposed to hunger, drought, flooding and debilitating diseases such as malaria. Inaction due to questions over the science [a thinly veiled reference to Bush administration foot-dragging] is no longer defensible."

In March, former UN chief weapons inspector Hans Blix added further fuel to the fire in a BBC television interview with David Frost: "I think we still overestimate the danger of terror. There are other things that are of equal, if not greater, magnitude, like the environmental global risks." This statement reinforced an equally pointed one Blix had made a year earlier: "To me the question of the environment is more ominous than that of peace and war. . . . I'm more worried about global warming than I am of any major military conflict."

In May, the blockbuster 20th Century Fox disaster movie *The Day After Tomorrow,* portraying the cataclysmic global consequences of accelerated climate change, was released to theaters nationwide (with European release scheduled for October). Some, such as Sir David King and former vice president Al Gore, promoted or endorsed the movie, clearly not because of its admittedly unrealistic compression of time and exaggeration of catastrophic effects, but because of its potential for awakening and sensitizing the public to the plausibility and seriousness of abrupt climate change. Others fiercely criticized the movie for trivializing such a vital issue. Anti-doomsayer Gregg Easterbrook, senior editor of *The New Republic,* assailed the "cheapo, third-rate disaster movie" for its "imbecile-caliber" science: "By presenting global warming in a laughably unrealistic way, the movie will only succeed in making audiences think that climate change is a big joke, when in fact the real science case for greenhouse-gas reform gets stronger all the time."

In a major September address in London, Tony Blair, faced with continuing criticism from his opposition, called climate change "the world's greatest environmental challenge . . . a challenge so far-reaching in its impact and irreversible in its destructive power, that it alters radically human existence." "Apart from a diminishing handful of skeptics," he said, "there is a virtual worldwide scientific consensus on the scope of the problem."

Then in October, the United Nations Environment Programme's Division of Early Warning and Assessment issued a thought-provoking new report, *Understanding Environment, Conflict, and Cooperation,* that resulted from the deliberations of participants in a new initiative to leverage environmental activities, policies, and actions for promoting international conflict prevention, peace, and cooperation. The subject matter of the report is not new, but the question it implies is: whether new life can be breathed into what was, throughout most of the 1990s, a lively debate over whether and how the environment and security are related and interact. Since the Kyoto negotiations of 1997, that debate has been largely moribund, to the detriment of both U.S. policy and strategic discourse more generally.

# Revivifying Environmental Security

Even if the events recounted above had not occurred this past year, the findings and recommendations of the UN High-Level Threat Panel and the introduction into the public imagination of abrupt climate change as a matter of prospective national security concern would stand as forceful stimuli for policy practitioners, scholars, and the general public to accord environmental security more serious and immediate attention.

This article goes to press before the actual release of the High-Level Panel's final report; but publicly available preliminary work by the United Nations Foundation's United Nations & Global Security Initiative, in cooperation with the Environmental Change & Security Project of the Woodrow Wilson Center, prefigures how the Panel's thinking is likely to be guided on environmental matters. This introductory passage from a discussion summary presented to the Panel is indicative of that thrust:

> *Environmental changes can threaten global, national, and human security. Environmental issues include land degradation, climate change, water quality and quantity, and the management and distribution of natural-resource assets (such as oil, forests, and minerals). These factors can contribute directly to conflict, or can be linked to conflict, by exacerbating other causes such as poverty, migration, small arms, and infectious diseases. For example, experts predict that climate change will trigger enormous physical and social changes like water shortages, natural disasters, decreased agricultural productivity, increased rates and scope of infectious diseases, and shifts in human migration; these changes could significantly impact international security by leading to competition for natural resources, destabilizing weak states, and increasing humanitarian crises. However, managing environmental issues and natural resources can also build confidence and contribute to peace through cooperation across lines of tension.*

Add to this Secretary-General Annan's own words in announcing the High-Level Panel to the UN General Assembly in September 2003, and it seems clear that the Panel will endorse the environment-security linkage and acknowledge that environmental degradation, resource scarcity, and climate change are threats or challenges that face the world and demand collective response:

> *All of us know there are new threats that must be faced—or, perhaps, old threats in new and dangerous combinations: new forms of terrorism, and the proliferation of weapons of mass destruction. But, while some consider these threats as self-evidently the main challenge to world peace and security, others feel more immediately menaced by small arms employed in civil conflict, or by so-called "soft threats" such as the persistence of extreme poverty, the disparity of income between and within societies, and the spread of infectious diseases, or climate change and environmental degradation.*

The February 2004 *Fortune* article was a dispassionate but revealing summary of a Pentagon-commissioned study that, though unclassified, ordinarily wouldn't have received much—if any—public exposure. Substantively, the article did two things. First, judging from the volume and intensity of follow-up commentary it generated, it clearly raised expectations—positive and negative—about the content and ramifications of the Pentagon report. Was the military actually interested in climate change? Why? Enough to do something about it? To what end and with what effect (especially on the military's principal mission)?

Second, the article—and the report it reported—upped the ante in the continuing debate over climate change. In addressing *abrupt* climate change, it accentuated an emerging thesis that gives urgency to what otherwise is considered (by some, perhaps many) to be a long-term, gradual phenomenon that, if real, can be passed off, without present political or economic regret, for future generations to deal with. And in tying abrupt climate change to national security, the article and report give added—ultimate—importance to the subject. National security is, after all, the public-policy holy of holies—the iconic totem that takes precedence over all else. National security is about endangerment and survival, the thinking goes. So if something can be shown to have national security implications (however defined), then perhaps it too is about such things; perhaps it too, therefore, warrants serious attention and the commitment of resources.

For people familiar with the U.S. military's normal modes of communication, the release of the Schwartz-Randall report to *Fortune* was unusual enough to cause speculation about whether the man who commissioned it, Andrew W. Marshall—the Pentagon's director of net assessment for the past 30 years—may have been signaling concerns that went well beyond the report's scientific message: first, that the institution he works for is intractably parochial and resistant to change; second, that the Pentagon is particularly inbred and close-minded about matters as esoteric and ideologically encumbered as the environment; third, that since imaginative futurists had prepared the report, it could more easily be dismissed as speculative fantasy by bureaucratic pragmatists who prefer to think they are grounded in reality; fourth, that however long he (Marshall) may have served in the Pentagon, he has little clout in influencing the military to actually take action based on his office's analytical products; fifth, that going public therefore offers more hope for forcing internal Pentagon awakening (if not change) in response to external pressure from arguably less parochial outside parties such as Congress and the media; and sixth, that perhaps the most potent force for movement on this particular front is the business community, which has the most to both gain and lose from climate change—especially when the political regime in power opts for dogmatic inaction in deference to the cosmic invisible hand of the marketplace.

The importance of this episode, as well as its relevance for the future, lies in both the message and the method of the Schwartz-Randall report itself. The implicit message is that even worse than climate change is the not unrealistic possibility of *abrupt* climate change. For those who had not heard of it, the article made clear that abrupt climate change is not just global warming speeded up, but a wholly different kind of event triggered by the baseline climate change we already know. In brief, the global warming now taking place could conceivably lead to a halting of the ocean currents that now keep Europe temperate—global *warming* thus ironically leading to regionally much colder conditions and widespread accelerations of the catastrophic effects

already commonly associated with "normal" climate change: floods, droughts, windstorms, wave events, wildfires, disease epidemics, species loss, famine, and more. The explicit message is that the concatenation of such effects could then lead to additional, national security consequences—most notably military confrontations between states over access to scarce food, water, and energy supplies, or what the authors describe as a "world of warring states."

Paradoxically, portraying what is relevant to national security as essentially that which invites or involves military force is perhaps necessary to grab the attention of purported experts on the subject, but it thereby also betrays the shallowness and narrowness of the canonical security paradigm most of us have unthinkingly bought into. This state of affairs is reinforced by the methodology of the Schwartz-Randall report, which seeks not to predict whether, when, or how abrupt climate change and its attendant effects would occur, but merely to present a plausible scenario of what might happen if and when it does. In the authors' words, "The duration of this event could be decades, centuries, or millennia and it could begin this year or many years in the future." Despite this caveat, the theme of abrupt climate change as a national security concern may be sufficiently eye-opening and provocative that, in conjunction with the other motivating forces of the year just past, it could take public consciousness of environmental security to a new level.

# Rethinking Security, Reassessing the Threat

However many people there may now be who recognize that environmental conditions precipitate or contribute to other conditions—violent conflict, civil unrest, instability, regime or state failure—regularly associated with security as usually defined, they are vastly outnumbered by those who either openly oppose the environment-security linkage or ignore it as irrelevant or inconsequential.

These oppositionists come from two different but overlapping camps: ideological conservatives and the mainstream traditionalists who dominate the national security community. This distinction is crucial because the latter—the technocratic mandarins inside and outside government—have the final exegetical say about what security is and what therefore is allowed to be a legitimate part of the security dialogue.

Oppositionists treat the environment as a purely *ideological* issue, and climate change as the most ideological of all—accordingly as dismissible as feminism, the homosexual agenda, or any other reflection of "political correctness." This despite the fact that, in a purer ontological sense, the environment is an inherently strategic matter, and climate change the most strategic of all. The environment is everywhere. It respects no borders, physical or otherwise. In its reach, its effects, and its consequences, it is truly global. And, fully understood, it brings into question all of our prevailing notions of sovereignty, territorial integrity, and even aggression and intervention. Nonetheless, just as to a hammer everything looks like a nail, to an ideologue everything looks ideological—to be accepted or rejected on the basis not of reason but of internalized dogma.

One of the major issues that has most divided those who debate the environment-security relationship is how broadly or narrowly to define security. Oppositionists invariably take the narrow road—basically equating security with defense, just as they similarly equate power with force. To them, security has axiomatic meaning that derives from its historical roots. Ironically, on this particular count the oppositionists are abetted by a shadow contingent of like-minded liberal environmentalists, who believe that linking the environment to security is dangerous because it will inevitably militarize the former and relinquish vital resources needed for environmental protection to an already bloated, profligate military establishment. In their fear of militarizing the environment, they risk getting into bed with betes noir who are fully committed to militarizing our entire strategic posture.

The counterpoise to this narrow construction of security begins with the recognition that security is, at root, a psychological and sociological phenomenon that starts—and ends—with the individual. To be secure is, literally, to be free—from harm and danger, threat and intimidation, doubt and fear, need and want. In the hierarchy of human needs, security is one of the most basic impulses—exceeded in its primacy only by the even more basic physiological needs for food, water, shelter, and the like, each of which is dependent on environmental well-being. Such primal needs translate into the natural rights that all human beings deserve to enjoy and that governments, as we have learned from America's founders, are instituted to secure.

Individual or human security, then, is the necessary precondition for national security, not merely its residual by-product. Accordingly, *assured security* stands as the primary overarching strategic aim a democratic society such as ours must seek to attain. In this supernal sense, security is something much more robust than defense. It encompasses the totality of conditions enumerated in America's security credo, the Preamble to the Constitution—not just the common defense, but no less importantly, national unity (a more perfect union), justice, domestic tranquility, the general welfare, and liberty. Only where all of these conditions exist in adequate measure is there true security. Where even one—liberty, say, or the general welfare—is sacrificed or compromised for another—say the common defense—the result is some degree of insecurity. Thus, in the final analysis, everything is related to security; everything is related to *national* security.

However broadly or narrowly security is defined, whatever endangers it or places it at risk is a threat; and whatever constitutes or qualifies as a threat is crucial because, in the idealized protocol of traditional national security planning, threats are the ostensible starting point for determining the requirements that produce capabilities and programs for countering these threats. (In reality, of course, capabilities and programs acquire their own bureaucratic life and thus are more likely to determine than to derive from threats.)

Oppositionists generally accept as legitimate threats only those parties or phenomena that, beyond their perceived potential for harm, are considered capable of or the product of malevolent

intent. *Intentionality* is the key legitimizing element. Terrorism fills this bill, just as state-based adversaries traditionally have. Weapons of mass destruction seem to qualify because, though inert entities in themselves, in human hands they can be ominous instruments of harm. Climate change and assorted forms of environmental degradation, though, typically don't pass muster as credible threats, no matter how much death and destruction they can wreak. Instead, they are implicitly written off as pure acts of nature, assuming metaphysical proportions that place them beyond human control and therefore outside the bounds of either preventive or retributive concern.

Such blinkered threat assessment is entirely characteristic of the policy establishment. To cite just a few notable examples:

- The 2002 White House national security strategy, in 34 pages of text, mentions the word environment in only one short paragraph about U.S. trade negotiations.

- In his February 2004 "Worldwide Threat Briefing" to Congress, Director of Central Intelligence George Tenet devoted five pages of testimony each to terrorism, Iraq, and proliferation, three paragraphs to global narcotics, a paragraph each to population trends, infectious disease, and humanitarian food insecurity, but nothing at all to environmental matters.

- The much bally-hooed, future-oriented Hart-Rudman Commission, whose members extolled their own prescience for adumbrating 9/11-type terrorist attacks on the United States, gave only the most cursory treatment to 21st-century environmental challenges in its initial September 1999 report. Arguing innocuously that pollution can be—and implicitly will be—counteracted by economic growth and the spread of remediation technologies, the commission essentially dismissed the subject with this (dare we say, ideological) statement: "There is fierce disagreement over several major environmental issues. Many are certain that global warming will produce major social traumas within 25 years, but the scientific evidence does not yet support such a conclusion. Nor is it clear that recent weather patterns result from anthropogenic activity as opposed to natural fluctuations."

- Somewhat in contrast, the National Intelligence Council's *Global Trends 2015* report, issued in December 2000 (before the following year's 9/11 attacks), identified natural resources and the environment as one of the most important "drivers and trends that will shape the world of 2015." Focusing principally on food, water, and energy security developments, the experts who collaborated on the report acknowledged the persistence and growth of global environmental problems in the years ahead, a growing consensus on the need to deal with such problems, and the prospect that "global warming will challenge the international community."

Typifying the thinking of policymakers and other members of the national security mandarinate, such assessments also seem more representative than not of general public sentiment. A particularly revealing indication of this is the most recent Chicago

Council on Foreign Relations study of U.S. public opinion on international issues, *Global Views 2004*. Asked to identify the most critical threats to U.S. vital interests, the public ranks global warming a distant seventh (37% of respondents), behind the likes of international terrorism (75%), chemical and biological weapons (66%), unfriendly countries becoming nuclear powers (64%), immigration into the United States (58%), and other developments. Another recent (February 2004) poll by Gallup found that environmental concerns don't even make the public's top-eleven list of possible threats to U.S. vital interests— international terrorism and the spread of weapons of mass destruction far outpacing all other prospective threats.

That environmental matters should be of such little overall public concern is a reflection of how limited and unstrategic our thinking about security actually is. Perhaps if we were to pay more attention to the documented effects of particular conditions and events, rather than to the nebulous, abstract notion of intentionality implicitly embedded in our prevailing standards of threat-worthiness, we could see the world differently—and more accurately.

Look, for example, at comparative fatalities from the highly credible threat of terrorism and the highly dubious threat of natural disasters. Since 1968, there have been 19,114 incidents of terrorism world wide, resulting in a total of 23,961 deaths and 62,502 associated injuries. However disturbing these figures may be, they pale in comparison to those resulting from natural disasters.

The average *annual* death toll over the past century due to drought, famine, floods, windstorms, temperature extremes, wave surges, and wildfires has been 243,577. Thus, even if we ignore earthquakes, volcanic eruptions, and disease epidemics, and don't count injuries or other harmful effects (such as homelessness), three times as many people die each year on average in natural disasters that could be linked to—and exacerbated by—climate change as have been killed and injured together in 37 years of terrorist incidents. And lest the use of a century-long average seem skewed, consider that just since 1990, there have been more than 207,000 fatalities from the foregoing types of disasters in South Asia alone, more than 23,000 in Central America and Mexico, and tens of thousands more in other parts of the world.

These figures are startling in their empirical exactitude, more so if one accepts estimates that average annual economic losses to such disasters were on the order of $660 billion in the 1990s. They lead us to consider a final argument that ideological conservatives invoke to discredit environmental and climate concerns—the need for sounder, more defensible science—and the associated argument national security mandarins use to deny or ignore the environment-security linkage—the lack of unequivocal evidence that environmental conditions actually cause diminished security in the form of violence.

Both arguments are excuses for denial and inaction; and both are suffused with hypocrisy. Those who demand conclusive *proof* that environmental conditions *cause* violence set a disingenuously unattainable legitimizing standard that permits them to perpetuate their own established preference for dealing with visible, immediate, politically remunerative symptoms.

Terrorism is a cardinal example of this—singularly symptomatic, never causative, except at some advanced, derivative level, where violence produces further violence.

Those who call for science as the only proper basis for public policy—at least climate policy—pretend to be motivated by a rigorous quest for objective (nonpolitical, non-ideological) truth. Yet they shamelessly accept or reject truth claims, labeling them "scientific" or not, based on whether those claims support or contradict their pre-established ideological beliefs. President Bush, for example, has repeatedly stated that climate policy must be based on better science (that is not yet available). But when asked about embryonic stem-cell research in this past year's second presidential debate, he stated that science is important, but it must be balanced by ethics. So, when the issue is stem-cell research—or perhaps abortion or homosexuality or capital punishment—ethics can take precedence over non-cooperative science; but when the issue is climate change or the environment more generally, not so. Maybe the earth really is flat.

Of more immediate relevance to this discussion is the practice common to many who call for better climate science. Paradoxically, they are perfectly content to unquestioningly accept and espouse demonstrably unscientific *assertions* from the military—especially concerning the degradation of military readiness that allegedly results from the so-called "encroachment" of environmental restrictions (e.g., species protection) on military installations. This despite the fact that the General Accounting Office has strongly criticized the military for failing to document whether and how much encroachment has actually degraded readiness.

Senator James Inhofe (R-OK) and the Senate Republican Policy Committee both exemplify this particular hypocrisy. Inhofe has said that "catastrophic global warming is a hoax"— "alarmism not based on objective science" even as he has said that "readiness problems . . . are caused by an ever-growing maze of environmental procedures and regulations in which we are losing the ability to prepare our patriot children, our war fighters, for war." Similarly the Senate Republican Policy Committee claims that "what scientists do agree on [with regard to climate change] is not policy-relevant, and on policy-relevant issues, there is little scientific agreement," while also asserting: "Among the most burdensome [examples of encroachment] are environmental laws and lawsuits that hinder or even ban military training and testing—thereby impairing readiness. . . . The evidence of detrimental impact is ample."

# Searching for a Strategic Response

What the foregoing contradictions suggest, among other things, is that the prevailing paradigm of security, according primacy as it does to the military and the use of force, long ago hijacked us intellectually and continues to hold us hostage; and, moreover, that in the absence of countervailing strategic thought of any consequence, ideology inevitably rushes in to fill the intellectual void, as it has in the case of environmental security, thereby forcing out rationality and blinding us to the future. The only remedy for this state of affairs, the only hope that the environment, climate change in particular, and, for that matter, other unconventional threats and challenges might be taken seriously as matters of serious security concern, is for fundamental strategic transformation to take place.

It seems insultingly obvious that strategic threats demand strategic response. But let us grasp the magnitude of this statement, for in the media age in which we live, there is virtually nothing—however obscure, however remote that is without almost instantaneous strategic consequence: Let us further understand why being strategic is therefore so intrinsically important. First, it is a moral obligation of government—to take the long view, to grasp the big picture, to anticipate and prevent, to appreciate the hidden, residual consequences of action or inaction, to recognize and capitalize on the interrelatedness of all things otherwise seemingly discrete and unrelated.

Second, being strategic inoculates us against crisis. Where crisis occurs, be it a terrorist incident or a natural disaster, strategic thinking has failed—with the unwanted result that decisionmaking must be artificially compressed and forced, and resources diverted from their intended purposes. Thus does *crisis prevention* stand alongside assured security as an overarching strategic aim of democratic society.

Third, being strategic provides the intellectual basis for both the strategic leadership expected of a superpower and the enduring, broad-based consensus necessary to galvanize a diverse, pluralistic society in common cause in the face of uncertainty, complexity, and ambiguity.

Four strategic imperatives should guide our future. The first let us call *targeted causation management*—focusing our thinking and our actions on identifying and eradicating the underlying causes of insecurity, thereby curing the disease rather than treating the symptoms. Environmental degradation and climate change take us much farther along the path to ultimate causes than terrorism ever could, especially if we acknowledge that the social, political, economic, and military conditions we prefer to deal with and attribute violence to may mask disaffection and unrest more deeply attributable to an environmentally degraded quality of life.

A second strategic imperative, *institutionalized anticipatory response,* calls for institutionalizing—giving permanence and legitimacy to the capacity and inclination for preventive action. This would enhance the prospects that conditions and events can be dealt with when they are manageable, before they mutate out of control and demand forceful response. Examples could range from a Manhattan Project-like effort to develop alternative energy sources and technologies, to greater inter-jurisdictional intelligence sharing, to massive disaster-resistant infrastructure development in the developing world.

A third strategic imperative is *appropriate situational tailoring*—dealing with conditions and events on their own geographic, cultural, and political terms rather than, as we are wont to do, inviting failure by imposing our preferred capabilities and approaches on the situations at hand. In a purely institutional sense, such tailoring might take the form, for example,

of new multilateral collective security regimes in each region of the world, with major environmental preparedness and enforcement arms.

The fourth strategic imperative is *comprehensive operational integration*—achieving fuller organizational, doctrinal, procedural, and technological integration across military-nonmilitary, governmental-non-governmental, and national-international lines. In a conceptual policy sense, this might assume the form of an overarching strategic architecture for unifying the activities of five organizational and cultural pillars—sustainable development, sustainable energy, sustainable business, sustainable consumption, and sustainable security. In a purely structural sense, the recognition that reorganization may be required to give birth and life to needed rethinking might produce such measures as the addition of a new Cabinet-level secretary of energy and environmental affairs to formal National Security Council membership, the creation of a UN under secretary-general for environmental affairs (or environmental security), or the expansion of the United Nations Environment Programme into an organization with operational capabilities and enforcement authority. In any event, all such measures would have to be underwritten by a firm commitment to more thoroughgoing transparency and multilateralism.

Finally, let us turn to the military. On the one hand, military action represents the least strategic option available for addressing environmental security (or virtually anything else for that matter). At least this is true so long as the military continues to be configured and oriented as it is and always has been—that is, for warfighting. On the other hand, the military is so central to our governing conception of security that true strategic transformation can take place only if it includes, or perhaps is preceded by, far-reaching military transformation—making real what until now has been only tiresome rhetoric from the Pentagon.

If the military has shown itself serious to date about environmental matters—even to the extent of crediting itself with being an excellent steward of nature—it is entirely a reflection of a distinctly engineering and management orientation dedicated principally to installation cleanup and remediation. Environmental security—the stuff of operations and intelligence, rather than of engineering and logistics—has been largely alien to the military ethos and identity. One need only consider the military's efforts under President Clinton to seek and gain selected exemptions from the Kyoto Protocol, or its tireless (if not entirely successful) attempts under President Bush to seek exemption from an array of environmental laws alleged to degrade readiness.

Two overriding considerations must guide military transformation. The first is the realization that what we ought to want is a military that is not just militarily effective—an instrument of force that serves the state—but that is *strategically effective*—an instrument of power that serves the larger aims of society and even humanity. The second overriding consideration is the concomitant realization that the military must be, and be seen to be, not a warfighting machine so much as a self-contained, self-sufficient enterprise that is capable of being projected over long distances for sustained periods of time to effectively manage all stages of a full range of complex emergencies.

Such considerations, taken to heart, ideally would produce a completely revamped military organized, manned, equipped, and trained primarily for nation-building, peacekeeping, humanitarian assistance, and disaster response, and only residually for warfighting. Such a military not only would possess the requisite capabilities for fulfilling the strategic imperatives enumerated above; it also would project the all-important imagery of a force truly committed to the pursuit of peace rather than to the enduringly illogical proposition that peace can be purchased by practicing war.

If we are to think and act strategically, which we must, we do well to recall the declaration from the Gayanashagowa, the Great Law of Peace of the Six Nations Iroquois Confederacy: "In our every deliberation we must consider the impact of our decisions on the next seven generations." And in applying this strategic precept to the matter at hand, which we must, we do no less well to take up the challenge issued recently by former Soviet President Mikhail Gorbachev. Interviewed some months ago, he was asked what he thought of the American doctrine of preemption. To which he responded:

> *Those who talk about leadership of the world all the time ought to exercise it. Rather than develop strategic doctrines of military preemption—as we've seen in Iraq, where no weapons of mass destruction have yet been found—let's act where the intelligence is clear: on climate change and other issues such as water, where today 2 billion people in the world don't have access to clean water. Let's talk instead about preempting global warming and the looming water crisis.*

Indeed. Words for the self-proclaimed world's only superpower to act on.

---

GREGORY D. FOSTER is a professor at the Industrial College of the Armed Forces, National Defense University, Washington, D.C., where he previously has served as George C. Marshall Professor and J. Carlton Ward Distinguished Professor and Director of Research. The views presented here are strictly his own.

From *World Watch Magazine*, January/February 2005, pp. 36–46. Copyright © 2005 by Worldwatch Institute. Reprinted by permission. www.worldwatch.org

# Can Organic Farming Feed Us All?

Brian Halweil

The only people who think organic farming can feed the world are delusional hippies, hysterical moms, and self-righteous organic farmers. Right?

Actually, no. A fair number of agribusiness executives, agricultural and ecological scientists, and international agriculture experts believe that a large-scale shift to organic farming would not only increase the world's food supply, but might be the only way to eradicate hunger.

This probably comes as a surprise. After all, organic farmers scorn the pesticides, synthetic fertilizers, and other tools that have become synonymous with high-yield agriculture. Instead, organic farmers depend on raising animals for manure, growing beans, clover, or other nitrogen-fixing legumes, or making compost and other sources of fertilizer that cannot be manufactured in a chemical plant but are instead grown—which consumes land, water, and other resources. (In contrast, producing synthetic fertilizers consumes massive amounts of petroleum.) Since organic farmers can't use synthetic pesticides, one can imagine that their fields suffer from a scourge of crop-munching bugs, fruit-rotting blights, and plant-choking weeds. And because organic farmers depend on rotating crops to help control pest problems, the same field won't grow corn or wheat or some other staple as often.

As a result, the argument goes, a world dependent on organic farming would have to farm more land than it does today—even if it meant less pollution, fewer abused farm animals, and fewer carcinogenic residues on our vegetables. "We aren't going to feed 6 billion people with organic fertilizer," said Nobel Prize-winning plant breeder Norman Borlaug at a 2002 conference. "If we tried to do it, we would level most of our forest and many of those lands would be productive only for a short period of time." Cambridge chemist John Emsley put it more bluntly: "The greatest catastrophe that the human race could face this century is not global warming but a global conversion to 'organic farming'—an estimated 2 billion people would perish."

In recent years, organic farming has attracted new scrutiny, not just from critics who fear that a large-scale shift in its direction would cause billions to starve, but also from farmers and development agencies who actually suspect that such a shift could better satisfy hungry populations. Unfortunately, no one had ever systematically analyzed whether in fact a widespread shift to organic farming would run up against a shortage of nutrients and a lack of yields—until recently. The results are striking.

## High-Tech, Low-Impact

There are actually myriad studies from around the world showing that organic farms can produce about as much, and in some settings much more, than conventional farms. Where there is a yield gap, it tends to be widest in wealthy nations, where farmers use copious amounts of synthetic fertilizers and pesticides in a perennial attempt to maximize yields. It is true that farmers converting to organic production often encounter lower yields in the first few years, as the soil and surrounding biodiversity recover from years of assault with chemicals. And it may take several seasons for farmers to refine the new approach.

But the long-standing argument that organic farming would yield just one-third or one-half of conventional farming was based on biased assumptions and lack of data. For example, the often-cited statistic that switching to organic farming in the United States would only yield one-quarter of the food currently produced there is based on a U.S. Department of Agriculture study showing that all the manure in the United States could only meet one-quarter of the nation's fertilizer needs—even though organic farmers depend on much more than just manure.

More up-to-date research refutes these arguments. For example, a recent study by scientists at the Research Institute for Organic Agriculture in Switzerland showed that organic farms were only 20 percent less productive than conventional plots over a 21-year period. Looking at more than 200 studies in North America and Europe, Per Pinstrup Andersen (a Cornell professor and winner of the World Food Prize) and colleagues recently concluded that organic yields were about 80 percent of conventional yields. And many studies show an even narrower gap. Reviewing 154 growing seasons' worth of data on various crops grown on rain-fed and irrigated land in the United States, University of California–Davis agricultural scientist Bill Liebhardt found that organic corn yields were 94 percent of conventional yields, organic wheat yields were 97 percent, and organic soybean yields were 94 percent. Organic tomatoes showed no yield difference.

More importantly, in the world's poorer nations where most of the world's hungry live, the yield gaps completely disappear. University of Essex researchers Jules Pretty and Rachel Hine looked at over 200 agricultural projects in the developing world that converted to organic and ecological approaches, and found that for all the projects—involving 9 million farms on nearly 30 million hectares—yields increased an average of 93 percent. A seven-year study from Maikaal District in central India involving 1,000 farmers cultivating 3,200 hectares found that average yields for cotton, wheat, chili, and soy were as much as 20 percent higher on the organic farms than on nearby conventionally managed ones. Farmers and agricultural scientists attributed the higher yields in this dry region to the emphasis on cover crops, compost, manure, and other practices that increased organic matter (which helps retain water) in the soils. A study from Kenya found that while organic farmers in "high-potential areas" (those with above average rainfall and high soil quality) had lower maize yields than nonorganic farmers, organic farmers in areas with poorer resource endowments consistently outyielded conventional growers. (In both regions, organic farmers had higher net profits, return on capital, and return on labor.)

Contrary to critics who jibe that it's going back to farming like our grandfathers did or that most of Africa already farms organically and it can't do the job, organic farming is a sophisticated combination of old wisdom and modern ecological innovations that help harness the yield-boosting effects of nutrient cycles, beneficial insects, and crop synergies. It's heavily dependent on technology—just not the technology that comes out of a chemical plant.

## High-Calorie Farms

So could we make do without the chemical plants? Inspired by a field trip to a nearby organic farm where the farmer reported that he raised an amazing 27 tons of vegetables on six-tenths of a hectare in a relatively short growing season, a team of scientists from the University of Michigan tried to estimate how much food could be raised following a global shift to organic farming. The team combed through the literature for any and all studies comparing crop yields on organic farms with those on nonorganic farms. Based on 293 examples, they came up with a global dataset of yield ratios for the world's major crops for the developed and the developing world. As expected, organic farming yielded less than conventional farming in the developed world for most food categories, while studies from the developing world showed organic farming boosting yields. The team then ran two models. The first was conservative in the sense that it applied the yield ratio for the developed world to the entire planet, i.e., they assumed that every farm regardless of location would get only the lower developed-country yields. The second applied the yield ratio for the developed world to wealthy nations and the yield ratio for the developing world to those countries.

"We were all surprised by what we found," said Catherine Badgley, a Michigan paleoecologist who was one of the lead researchers. The first model yielded 2,641 kilocalories ("calories") per person per day, just under the world's current

## Enough Nitrogen to Go Around?

In addition to looking at raw yields, the University of Michigan scientists also examined the common concern that there aren't enough available sources of non-synthetic nitrogen—compost, manure, and plant residues—in the world to support large-scale organic farming. For instance, in his book *Enriching the Earth: Fritz Haber, Carl Bosch, and the Transformation of World Food Production,* Vaclav Smil argues that roughly two-thirds of the world's food harvest depends on the Haber-Bosch process, the technique developed in the early 20th century to synthesize ammonia fertilizer from fossil fuels. (Smil admits that he largely ignored the contribution of nitrogen-fixing crops and assumed that some of them, like soybeans, are net users of nitrogen, although he himself points out that on average half of all the fertilizer applied globally is wasted and not taken up by plants.) Most critics of organic farming as a means to feed the world focus on how much manure—and how much related pasture land and how many head of livestock—would be needed to fertilize the world's organic farms. "The issue of nitrogen is different in different regions," says Don Lotter, an agricultural consultant who has published widely on organic farming and nutrient requirements. "But lots more nitrogen comes in as green manure than animal manure."

Looking at 77 studies from the temperate areas and tropics, the Michigan team found that greater use of nitrogen-fixing crops in the world's major agricultural regions could result in 58 million metric tons more nitrogen than the amount of synthetic nitrogen currently used every year. Research at the Rodale Institute in Pennsylvania showed that red clover used as a winter cover in an oat/wheat–corn–soy rotation, with no additional fertilizer inputs, achieved yields comparable to those in conventional control fields. Even in arid and semi-arid tropical regions like East Africa, where water availability is limited between periods of crop production, drought-resistant green manures such as pigeon peas or groundnuts could be used to fix nitrogen. In Washington state, organic wheat growers have matched their non-organic neighbor's wheat yields using the same field pea rotation for nitrogen. In Kenya, farmers using leguminous tree crops have doubled or tripled corn yields as well as suppressing certain stubborn weeds and generating additional animal fodder.

The Michigan results imply that no additional land area is required to obtain enough biologically available nitrogen, even without including the potential for intercropping (several crops grown in the same field at the same time), rotation of livestock with annual crops, and inoculation of soil with *Azobacter, Azospirillum,* and other free-living nitrogen-fixing bacteria.

production of 2,786 calories but significantly higher than the average caloric requirement for a healthy person of between 2,200 and 2,500. The second model yielded 4,381 calories per person per day, 75 percent greater than current availability—and a quantity that could theoretically sustain a much larger human population than is currently supported on the world's farmland. (It also laid to rest another concern about organic agriculture; see sidebar.)

The team's interest in this subject was partly inspired by the concern that a large-scale shift to organic farming would require clearing additional wild areas to compensate for lower yields—an obvious worry for scientists like Badgley, who studies present and past biodiversity. The only problem with the argument, she said, is that much of the world's biodiversity exists in close proximity to farmland, and that's not likely to change anytime soon. "If we simply try to maintain biodiversity in islands around the world, we will lose most of it," she said. "It's very important to make areas between those islands friendly to biodiversity. The idea of those areas being pesticide-drenched fields is just going to be a disaster for biodiversity, especially in the tropics. The world would be able to sustain high levels of biodiversity much better if we could change agriculture on a large scale."

Badgley's team went out of the way to make its assumptions as conservative as possible: most of the studies they used looked at the yields of a single crop, even though many organic farms grow more than one crop in a field at the same time, yielding more total food even if the yield of any given crop may be lower. Skeptics may doubt the team's conclusions—as ecologists, they are likely to be sympathetic to organic farming—but a second recent study of the potential of a global shift to organic farming, led by Niels Halberg of the Danish Institute of Agricultural Sciences, came to very similar conclusions, even though the authors were economists, agronomists, and international development experts.

Like the Michigan team, Halberg's group made an assumption about the differences in yields with organic farming for a range of crops and then plugged those numbers into a model developed by the World Bank's International Food Policy Research Institute (IFPRI). This model is considered the definitive algorithm for predicting food output, farm income, and the number of hungry people throughout the world. Given the growing interest in organic farming among consumers, government officials, and agricultural scientists, the researchers wanted to assess whether a large-scale conversion to organic farming in Europe and North America (the world's primary food exporting regions) would reduce yields, increase world food prices, or worsen hunger in poorer nations that depend on imports, particularly those people living in the Third World's swelling megacities. Although the group found that total food production declined in Europe and North America, the model didn't show a substantial impact on world food prices. And because the model assumed, like the Michigan study, that organic farming would boost yields in Africa, Asia, and Latin America, the most optimistic scenario even had hunger-plagued sub-Saharan Africa exporting food surpluses.

"Modern non-certified organic farming is a potentially sustainable approach to agricultural development in areas with low yields due to poor access to inputs or low yield potential because it involves lower economic risk than comparative interventions based on purchased inputs and may increase farm level resilience against climatic fluctuations," Halberg's team concluded. In other words, studies from the field show that the yield increases from shifting to organic farming are highest and most consistent in exactly those poor, dry, remote areas where hunger is most severe. "Organic agriculture could be an important part of increased food security in sub-Saharan Africa," says Halberg.

That is, if other problems can be overcome. "A lot of research is to try to kill prejudices," Halberg says—like the notion that organic farming is only a luxury, and one that poorer nations cannot afford. "I'd like to kill this once and for all. The two sides are simply too far from each other and they ignore the realities of the global food system." Even if a shift toward organic farming boosted yields in hungry African and Asian nations, the model found that nearly a billion people remained hungry, because any surpluses were simply exported to areas that could best afford it.

## Wrong Question?

These conclusions about yields won't come as a surprise to many organic farmers. They have seen with their own eyes and felt with their own hands how productive they can be. But some supporters of organic farming shy away from even asking whether it can feed the world, simply because they don't think it's the most useful question. There is good reason to believe that a global conversion to organic farming would not proceed as seamlessly as plugging some yield ratios into a spreadsheet.

To begin with, organic farming isn't as easy as farming with chemicals. Instead of choosing a pesticide to prevent a pest outbreak, for example, a particular organic farmer might consider altering his crop rotation, planting a crop that will repel the pest or one that will attract its predators—decisions that require some experimentation and long-term planning. Moreover, the IFPRI study suggested that a large-scale conversion to organic farming might require that most dairy and beef production eventually "be better integrated in cereal and other cash crop rotations" to optimize use of the manure. Bringing cows back to one or two farms to build up soil fertility may seem like a no-brainer, but doing it wholesale would be a challenge—and dumping ammonia on depleted soils still makes for a quicker fix.

Again, these are just theoretical assumptions, since a global shift to organic farming could take decades. But farmers are ingenious and industrious people and they tend to cope with whatever problems are at hand. Eliminate nitrogen fertilizer and many farmers will probably graze cows on their fields to compensate. Eliminate fungicides and farmers will look for fungus-resistant crop varieties. As more and more farmers begin to farm organically, everyone will get better at it. Agricultural research centers, universities, and agriculture ministries will throw their resources into this type of farming—in sharp contrast to their current neglect of organic agriculture, which partly stems from the assumption that organic farmers will never play a major role in the global food supply.

So the problems of adopting organic techniques do not seem insurmountable. But those problems may not deserve most of our attention; even if a mass conversion over, say, the next two decades, dramatically increased food production, there's little guarantee it would eradicate hunger. The global food system can be a complex and unpredictable beast. It's hard to anticipate how China's rise as a major importer of soybeans for its feedlots, for instance, might affect food supplies elsewhere. (It's likely to drive up food prices.) Or how elimination of agricultural subsidies in wealthy nations might affect poorer countries. (It's likely to boost farm incomes and reduce hunger.) And would less meat eating around the world free up food for the hungry? (It would, but could the hungry afford it?) In other words, "Can organic farming feed the world?" is probably not even the right question, since feeding the world depends more on politics and economics than any technological innovations.

"'Can organic farming feed the world' is indeed a bogus question," says Gene Kahn, a long-time organic farmer who founded Cascadian Farms organic foods and is now vice president of sustainable development for General Mills. "The real question is, can we feed the world? Period. Can we fix the disparities in human nutrition?" Kahn notes that the marginal difference in today's organic yields and the yields of conventional agriculture wouldn't matter if food surpluses were redistributed.

But organic farming will yield other benefits that are too numerous to name. Studies have shown, for example, that the "external" costs of organic farming—erosion, chemical pollution to drinking water, death of birds and other wildlife—are just one-third those of conventional farming. Surveys from every continent show that organic farms support many more species of birds, wild plants, insects, and other wildlife than conventional farms. And tests by several governments have shown that organic foods carry just a tiny fraction of the pesticide residues of the nonorganic alternatives, while completely banning growth hormones, antibiotics, and many additives allowed in many conventional foods. There is even some evidence that crops grown organically have considerably higher levels of health-promoting antioxidants.

There are social benefits as well. Because organic farming doesn't depend on expensive inputs, it might help shift the balance towards smaller farmers in hungry nations. A 2002 report from the UN Food and Agriculture Organization noted that "organic systems can double or triple the productivity of traditional systems" in developing nations but suggested that yield comparisons offer a "limited, narrow, and often misleading picture" since farmers in these countries often adopt organic farming techniques to save water, save money, and reduce the variability of yields in extreme conditions. A more recent study by the International Fund for Agricultural Development found that the higher labor requirements often mean that "organic agriculture can prove particularly effective in bringing redistribution of resources in areas where the labour force is underemployed. This can help contribute to rural stability."

## Food Versus Fuel

Sometimes, when humans try to solve one problem, they end up creating another. The global food supply is already under serious strain: more than 800 million people go hungry every day, the world's population continues to expand, and a growing number of people in the developing world are changing to a more Western, meat-intensive diet that requires more grain and water per calorie than traditional diets do. Now comes another potential stressor: concern about climate change means that more nations are interested in converting crops into biofuels as an alternative to fossil fuels. But could this transition remove land from food production and further intensify problems of world hunger?

For several reasons, some analysts say no, at least not in the near future. First, they emphasize that nearly 40 percent of global cereal crops are fed to livestock, not humans, and that global prices of grains and oil seeds do not always affect the cost of food for the hungry, who generally cannot participate in formal markets anyway.

Second, at least to date, hunger has been due primarily to inadequate income and distribution rather than absolute food scarcity. In this regard, a biofuels economy may actually help to reduce hunger and poverty. A recent UN Food and Agriculture Organization report argued that increased use of biofuels could diversify agricultural and forestry activities, attract investment in new small and medium-sized enterprises, and increase investment in agricultural production, thereby increasing the incomes of the world's poorest people.

Third, biofuel refineries in the future will depend less on food crops and increasingly on organic wastes and residues. Producing biofuels from corn stalks, rice hulls, sawdust, or waste paper is unlikely to affect food production directly. And there are drought-resistant grasses, fast-growing trees, and other energy crops that will grow on marginal lands unsuitable for raising food.

Nonetheless, with growing human appetites for both food and fuel, biofuels' long-run potential may be limited by the priority given to food production if bioenergy systems are not harmonized with food systems. The most optimistic assessments of the long-term potential of biofuels have assumed that agricultural yields will continue to improve and that world population growth and food consumption will stabilize. But the assumption about population may prove to be wrong. And yields, organic or otherwise, may not improve enough if agriculture in the future is threatened by declining water tables or poor soil maintenance.

## Middle Earth

These benefits will come even without a complete conversion to a sort of organic Utopia. In fact, some experts think that a more hopeful, and reasonable, way forward is a sort of middle ground, where more and more farmers adopt the principles of organic farming even if they don't follow the approach religiously. In this scenario, both poor farmers and the environment come out way ahead. "Organic agriculture is *not* going to do the trick," says Roland Bunch, an agricultural extensionist who has worked for decades in Africa and the Americas and is now with COSECHA (Association of Consultants for a Sustainable, Ecological, and People-Centered Agriculture) in Honduras. Bunch knows first-hand that organic agriculture can produce more than conventional farming among poorer farmers. But he also knows that these farmers cannot get the premium prices paid for organic produce elsewhere, and that they are often unable, and unwilling, to shoulder some of the costs and risks associated with going completely organic.

Instead, Bunch points to "a middle path," of eco-agriculture, or low-input agriculture that uses many of the principles of organic farming and depends on just a small fraction of the chemicals. "These systems can immediately produce two or three times what smallholder farmers are presently producing" Bunch says. "And furthermore, it is attractive to smallholder farmers because it is less costly per unit produced." In addition to the immediate gains in food production, Bunch suggests that the benefits for the environment of this middle path will be far greater than going "totally organic," because "something like five to ten times as many smallholder farmers will adopt it per unit of extension and training expense, because it behooves them economically. They aren't taking food out of their kids' mouths. If five farmers eliminate half their use of chemicals, the effect on the environment will be two and one-half times as great as if one farmer goes totally organic."

And farmers who focus on building their soils, increasing biodiversity, or bringing livestock into their rotation aren't precluded from occasionally turning to biotech crops or synthetic nitrogen or any other yield-enhancing innovations in the future, particularly in places where the soils are heavily depleted, "In the end, if we do things right, we'll build a lot of organic into conventional systems," says Don Lotter, the agricultural consultant. Like Bunch, Lotter notes that such an "integrated" approach often out-performs both a strictly organic and chemical-intensive approach in terms of yield, economics, and environmental benefits. Still, Lotter's not sure we'll get there tomorrow, since the world's farming is hardly pointed in the organic direction—which could be the real problem for the world's poor and hungry. "There is such a huge area in sub-Saharan Africa and South America where the Green Revolution has never made an impact and it's unlikely that it will for the next generation of poor farmers," argues Niels Halberg, the Danish scientist who lead the IFPRI study. "It seems that agro-ecological measures for some of these areas have a beneficial impact on yields and food insecurity. So why not seriously try it out?"

---

**BRIAN HALWEIL** is a Senior Researcher at Worldwatch and the author of *Eat Here: Reclaiming Homegrown Pleasures in a Global Supermarket*.

# Where Oil and Water Do Mix

## *Environmental Scarcity and Future Conflict in the Middle East and North Africa*

Jason J. Morrissette and Douglas A. Borer

"Many of the wars of the 20th century were about oil, but wars of the 21st century will be over water."

—Isamil Serageldin
World Bank Vice President

I n the eyes of a future observer, what will characterize the political landscape of the Middle East and North Africa? Will the future mirror the past or, as suggested by the quote above, are significant changes on the horizon? In the past, struggles over territory, ideology, colonialism, nationalism, religion, and oil have defined the region. While it is clear that many of those sources of conflict remain salient today, future war in the Middle East and North Africa also will be increasingly influenced by economic and demographic trends that do not bode well for the region. By 2025, world population is projected to reach eight billion.[1] As a global figure, this number is troubling enough; however, over 90 percent of the projected growth will take place in developing countries in which the vast majority of the population is dependent on local renewable resources. For instance, World Bank estimates place the present annual growth rate in the Middle East and North Africa at 1.9 percent versus a worldwide average of 1.4 percent.[2] In most of these countries, these precious renewable resources are controlled by small segments of the domestic political elite, leaving less and less to the majority of the population. As a result, if present population and economic trends continue, we project that many future conflicts throughout the region will be directly linked to what academic researchers term "environmental scarcity"[3]— the scarcity of renewable resources such as arable land, forests, and fresh water.

The purpose of this article is twofold. In the first section, we conceptualize how environmental scarcity is linked to domestic political unrest and the subsequent crisis of domestic political legitimacy that may ultimately result in conflict. We review the academic literature which suggests that competition over water is the key environmental variable that will play an increasing role in future domestic challenges to governments throughout the region. We then describe how these crises of domestic political legitimacy may result in both intrastate and interstate conflict. Even though the Middle East can generally be characterized as an arid climate, two great river systems, the Nile and the Tigris/Euphrates, serve to anchor the major population centers in the region. Conflict over the water of the Nile may someday come to pass between Egypt, Sudan, and Ethiopia; while Turkey, Syria, and Iraq all are located along the Tigris/Euphrates watershed and compete for its resources. Further conflict over water may embroil Israel, Syria, and the Palestinians.

Despite many existing predictions of war over water, we investigate the intriguing question: How have governments in the Middle East thus far avoided conflict over dwindling water supplies? In the second section of the article, we discuss the concept of "virtual water" and use this concept to illustrate the important linkages between water usage and the global economy, showing how existing tangible water shortages have been ameliorated by a combination of economic factors, which may or may not be sustainable into the future.

## Environmental Scarcity and Conflict: An Overview

Mostafa Dolatyar and Tim Gray identify water resources as "the principal challenge for humanity from the early days of civilization."[4] The 1998 United Nations Development Report estimates that almost a third of the 4.4 billion people currently living in the developing world have no access to clean water. The report goes on to note that the world's population tripled in the 20th century, resulting in a corresponding sixfold increase in the use of water resources. Moreover, infrastructure problems related to water supply abound in much of the developing world; the United Nations estimates that between 30 and 50 percent of the water presently diverted for irrigation purposes is lost through leaking pipes alone. In turn, roughly 20 countries in the developing world presently suffer from water stress (defined as having less than 1,000 cubic meters of available freshwater per capita), and 25 more are expected to join that list by 2050.[5] In response to these trends, the United Nations resolved in 2002 to reduce by half the proportion of people in the developing world who are unable to reach—or afford—safe drinking water.

In turn, numerous scholars in recent years have conceptualized water in security terms as a key strategic resource in many regions of the world. Thomas Naff maintains that water scarcity holds significant potential for conflict in large part because it is fundamentally essential to life. Naff identifies six basic characteristics that distinguish water as a vital and potentially contentious resource. (1) Water is necessary for sustaining life and has no substitute for human or animal use. (2) Both in terms of domestic and international policy, water issues are typically addressed by policymakers in a piecemeal fashion rather than comprehensively. (3) Since countries typically feel compelled by security concerns to control the ground on or under which water flows, by its nature, water is also a terrain security issue. (4) Water issues are frequently perceived as zero-sum, as actors compete for the same limited water resources. (5) As a result of the competition for these limited resources, water presents a constant potential for conflict. (6) International law concerning water resources remains relatively "rudimentary" and "ineffectual."[6] As these factors suggest, water is a particularly volatile strategic issue, especially when it is in severe shortage.

Arguing that environmental concerns have gained prominence in the post-Cold War era, Alwyn R. Rouyer establishes a basic paradigm of contemporary environmental conflict. Rouyer argues that "rapid population growth, particularly in the developing world, is putting severe stress on the earth's physical environment and thus creating a growing scarcity of renewable resources, including water, which in turn is precipitating violent civil and international conflict that will escalate in severity as scarcity increases."[7] Rouyer goes on to assert that this potential conflict over scarce resources will likely be most disruptive in states with rapidly expanding populations in which policymakers lack the political and economic capability to minimize environmental damage.

## "Almost a third of the 4.4 billion people currently living in the developing world have no access to clean water."

Security concerns linked fundamentally to environmental scarcity are far from a contrivance of the post-Cold War era, however. Ulrich Küffner asserts that conflicts over water "have occurred between many countries in all climatic regions, but between countries in arid regions they appear to be unavoidable. Claims over water have led to serious tensions, to threats and counter threats, to hostilities, border clashes, and invasions."[8] Moreover, as Miriam Lowi notes, "Well before the emergence of the nation-state, the arbitrary political division of a unitary river basin . . . led to problems regarding the interests of the states and/or communities located within the basin and the manner in which conflicting interests should be resolved."[9] Lowi fundamentally frames the issue of water scarcity in terms

of a dilemma of collective action and failed cooperation—the archetypal "Tragedy of the Common"—in which communal resources are abused by the greediness of individuals. In many regions of the world, the international agreements and coordinating institutions necessary to lower the likelihood of conflict over water are either inadequate or altogether nonexistent.[10]

Thomas Homer-Dixon argues that the environmental resource scarcity that potentially results in conflict, including water scarcity, fundamentally derives from one of three sources. The first, *supply-induced scarcity*, is caused when a resource is either degraded (for example, when cropland becomes unproductive due to overuse) or depleted (for example, when cropland is converted into suburban housing). Throughout most of the Middle East and North Africa countries, both environmental and resource degradation and depletion are of relevant concern. For instance, many of these countries face significant decreases in the agricultural productivity of their arable soil as a result of ongoing trends of desertification, soil erosion, and pollution. This problem is coupled with the continued loss of croplands to urbanization, as rural dwellers move to cities in search of employment and opportunity. The second source of environmental scarcity, *demand-induced scarcity*, is caused by either an increase in per-capita consumption or by simple population growth. If the supply remains constant, and demand increases by existing users consuming more, or more users each consuming the same amount, eventually scarcity will result as demand overtakes supply. The third type of environmental scarcity is known as *structural scarcity*, a phenomenon that results when resource supplies are unequally distributed. In this case the "haves" in any given society generally control and consume an inordinate amount of the existing supply, which results in the more numerous "have-nots" experiencing the scarcity.[11]

These three sources of scarcity routinely overlap and interact in two common patterns: "resource capture" and ecological marginalization. Resource capture occurs when both demand-induced and supply-induced scarcities interact to produce structural scarcity. In this pattern, powerful groups within society foresee future shortages and act to ensure the protection of their vested interests by using their control of state structures to capture control of a valuable resource. An example of this pattern occurred in Mauritania (one of Algeria's neighbors) in the 1970s and 1980s when the countries bordering the Senegal River built a series of dams to boost agricultural production. As a result of the new dams, the value of land adjacent to the river rapidly increased—an economic development that motivated Mauritanian Moors to abandon their traditional vocation as cattle grazers located in the arid land in the north and, instead, to migrate south onto lands next to the river. However, black Mauritanians already occupied the land on the river's edge. As a result, the Moorish political elite that controlled the Mauritanian government rewrote the legislation on citizenship and land rights to effectively block black Mauritanians from land ownership. By declaring blacks as non-citizens, the Islamic Moors managed to capture the land through nominally legal (structural) means. As a result, high levels of violence later arose between Mauritania and Senegal, where hundreds of thousands of the black Mauritanians had become refugees after being driven from their land.[12]

The second pattern, ecological marginalization, occurs when demand-induced and structural scarcities interact in a way that results in supply-induced scarcity. An example of this pattern comes from the Philippines, a country whose agricultural lands traditionally have been controlled by a small group of dominant landowners who, prior to the election of former President Estrada, have controlled Filipino politics since colonial times. Population growth in the 1960s and 1970s forced many poor peasants to settle in the marginal soils of the upland interior. This more mountainous land could not sustain the lowland slash-and-burn farming practices that they brought with them. As a result, the Philippines suffered serious ecological damage in the form of water pollution, soil erosion, landslides, and changes in the hydrological cycle that led to further hardship for the peasantry as the land's capacity shrank. As a result of their economic marginalization, many upland peasants became increasingly susceptible to the revolutionary rhetoric promoted by the communist-led New People's Army, or they supported the "People Power" movement that ousted US-backed Ferdinand Marcos from power in 1986.[13]

Thus, as shown in the Philippines, social pressures created by environmental scarcity can have a direct influence on the ruling legitimacy of the state, and may cause state power to crumble. Indeed, reductions in agricultural and economic production can produce objective socio-economic hardship; however, deprivation does not necessarily produce grievances against the government that result in serious domestic unrest or rebellion. One can look at the relative stability in famine-stricken North Korea as a poignant example of a polity whose citizens have suffered widespread physical deprivation under policies of the existing regime, but who are unwilling or unable to risk their lives to challenge the state.

This phenomenon is partly explained by conflict theorists who argue that individuals and groups have feelings of "relative deprivation" when they perceive a gap between what they believe they deserve and what in reality they actually have achieved.[14] In other words, can a government meet the expectations of the masses enough to avoid conflict? For example, in North Korea—a regime that tightly controls the information that its people receive—many people understand that they are suffering, but they may not know precisely how much they are suffering relative to others, such as their brethren in the South. The North Korean government indoctrinates its people to expect little other than hardship, which in turn it blames on outside enemies of the state. Thus, the people of North Korea have very low expectations, which their government has been able to meet. More important, then, is the question of whom do the people perceive as being responsible for their plight? If the answer is the people's own government—whether as a result of supply-induced, demand-induced, or structural resource scarcity—then social discord and rebellion are more likely to result in intrastate conflict, as citizens challenge the ruling legitimacy of the state itself. If the answer is someone else's government, then interstate conflict may result.

On numerous occasions, history has shown that governments whose people are suffering can remain in power for long periods of time by pointing to external sources for the people's hardship.[15] As noted above regarding political legitimacy, perception is politically more important than any standard of objective truth.[16] When faced with a crisis of legitimacy derived from environmental resource scarcity, any political regime essentially has a choice of two options in dealing with the situation. The regime may choose temporarily not to respond to looming challenges to its authority because water-induced stress may in fact pass when sufficient heavy rainfall occurs. However, most regimes in the Middle East and North Africa have sought more proactive ways to ensure their survival. Indeed, a people might forgive its government for one drought, but if governmental action is not taken, a subsequent drought-induced crisis of legitimacy could result in significant social upheaval by an unforgiving public. Furthermore, if the government itself is perceived to be the direct source of the scarcity—through structural arrangements, resource capture, or other means—these trends of social unrest are likely to be exacerbated. Thus, in order to survive, most states have developed policies to increase their water supplies and to address issues of environmental scarcity. The problem with doing so throughout most of the Middle East and North Africa, however, is that increasing supply in one state often creates environmental scarcity problems in another. If Turkey builds dams, Iraq and Syria are vulnerable; if Ethiopia or the Sudan builds dams, Egypt feels threatened. Thus far, interstate water problems leading to war have been avoided due to the economic interplay between oil wealth and the importation of "virtual water," which will be discussed at greater length below.

As noted above, resource scarcity issues centered on water are particularly prominent in the Middle East and North Africa. Ewan Anderson notes that resource geopolitics in the Middle East "has long been dominated by one liquid—oil. However, another liquid, water, is now recognized as the fundamental political weapon in the region."[17] Ecologically speaking, water scarcity in the Middle East and North Africa results from four primary causes: fundamentally dry climatic conditions, drought, desiccation (the degradation of land due to the drying up of the soil), and water stress (the low availability of water resulting from a growing population).[18] These resource scarcity problems are exacerbated in the Middle East by such factors as poor water quality and inadequate—and, at times, purposefully discriminatory—resource planning. As a result of these ecological and political trends, Nurit Kliot states, "water, not oil, threatens the renewal of military conflicts and social and economic disruptions" in the Middle East.[19] In the case of the Arab-Israeli conflict, Alwyn Rouyer suggests that "water has become inseparable from land, ideology, and religious prophecy."[20] Martin Sherman echoes these sentiments in the following passage, describing specifically the Arab-Israeli conflict:

> In recent years, particularly since the late 1980s, water has become increasingly dominant as a bone of contention between the two sides. More than one Arab leader, including those considered to be among the most

moderate, such as King Hussein of Jordan and former UN Secretary General, Boutros Boutros-Ghali of Egypt, have warned explicitly that water is the issue most likely to become the cause of a future Israeli-Arab war.[21]

## "Water is a particularly volatile strategic issue, especially when it is in severe shortage."

While Jochen Renger contends that a conflict waged explicitly over water may not lie on the immediate horizon, he notes that "it is likely that water might be used as leverage during a conflict."[22] As a result of such geopolitical trends, managing these water resources in the Middle East and North Africa—and, in turn, managing the conflict over these resources—should be considered a primary concern of both scholars and policymakers.

## Keeping the Peace: The Importance of Virtual Water

The warning signals that war over water may replace war over oil and other traditional sources of conflict are very real in recent history. Yet, for more than 25 years, despite increasing demand, water has not been the primary cause of war in the Middle East and North Africa. The scenarios outlined in the preceding section? have yet to fully address the fundamental questions of why and how governments in the region have thus far avoided major interstate conflict over water. In order to understand the likelihood of war, we must address the foundation of the past peace, testing whether or not this foundation remains strong for the foreseeable future. How have the governments of the region been able to avoid the apparently inevitable consequences of conflict that derive from the interlinked problems of water deficits, population growth, and weak economic performance? In this section of the article, we turn our attention to the important linkages between water usage and the global economy, showing how existing water shortages have been ameliorated by a combination of economic factors.

To understand the politics of water in the Middle East and North Africa, one must first look at the region's most fungible resource: oil. For much of the post-World War II era, the growing need for oil to fuel economic growth has served as the dominant motivating factor in US security policy in the Middle East. Conventional wisdom in the United States holds that US dependency on Middle Eastern oil is a strategic weakness. Indeed, the specter of a regional hegemonic power that controls the oil and that is also hostile to the United States strikes fear into the hearts of policymakers in Washington. Thus, for roughly the past 50 years, the United States has sought to prop up "moderate" (meaning pro-US) regimes while denying hegemony to "radical" (meaning anti-US) regimes.[23] However, we contend that both policymakers and the public at large in the United States generally misunderstand the politics of oil as they relate to water in the Middle East.

| Country | Total Water Resources per Capita (cubic meters) in 2000 | Percent of Population with Access to Adequate, Improved Water Source, 2000 | GDP per Capita, 2000 Estimate |
|---|---|---|---|
| Algeria | 477 | 94% | $5,500 |
| Egypt | 930 | 95% | $3,600 |
| Iran | 2,040 | 95% | $6,300 |
| Iraq | 1,544 | 85% | $2,500 |
| Israel | 180 | 99% | $18,900 |
| Jordan | 148 | 96% | $3,500 |
| Kuwait | 0 | 100% | $15,000 |
| Lebanon | 1,124 | 100% | $5,000 |
| Libya | 148 | 72% | $8,900 |
| Morocco | 1,062 | 82% | $3,500 |
| Oman | 426 | 39% | $7,700 |
| Qatar | — | 100% | $20,300 |
| Saudi Arabia | 119 | 95% | $10,500 |
| Sudan | 5,312 | — | $1,000 |
| Syria | 2,845 | 80% | $3,100 |
| Tunisia | 434 | — | $6,500 |
| Turkey | 3,162 | 83% | $6,800 |
| Yemen | 241 | 69% | $820 |

**Figure 1**  Water Resources and Economics in the Middle East and North Africa.

Sources: World Bank Development Indicators, Country-at-a-Glance Tables, Freshwater Resources, and *CIA World Factbook,* at http://www.worldbank.org and http://www.cia.gov.

In absolute terms, problems arising from US vulnerability to foreign oil are basically true—it would be better to be free of dependency on oil from any foreign source than to be dependent. However, the other side of the equation is often forgotten: oil-producing states are dependent on the United States and other major oil importers for their economic livelihood. More bluntly put, oil-exporting states are dependent on the influx of dollars, euros, and yen to purchase goods, services, and commodities that they lack. Thus, oil-producing countries in the Middle East and North Africa, few of whom have managed to successfully diversify their economies beyond the petroleum sector, exist in an interdependent world economy. The world depends on their oil, and they depend on the world's goods and services—including that most valuable life-sustaining resource, water.

On the surface, this perhaps seems to be a contentious claim. Outgoing oil tankers do not return with freshwater used to grow crops, and Middle East countries do not rely on the importation of bottled water for their daily consumption needs. However, according to hydrologists, each individual needs approximately 100 cubic meters of water each year for personal needs, and an additional 1,000 cubic meters are required to grow the food that person consumes. Thus, every person alive requires approximately 1,100 cubic meters of water every year. In 1970, the water needs of most Middle Eastern and North African countries could be met from sources within the region. During the colonial and early post-colonial eras, regional governments and their engineers had effectively managed supply to deliver new water to meet the requirements of the growing urban populations, industrial requirements, agricultural needs, and other demand-induced factors. What is clear is that in the past 30 years, the status of the region's water resources has significantly worsened as populations have increased (an example of demand-induced scarcity). Since the mid-1970s, most countries have been able to supply daily consumption and industrial needs; however, as indicated in Figure 1, the approximate 1,000 cubic meters of water per capita that is required for self-sufficient agricultural production represents a seemingly impossible challenge for some Middle Eastern and North African economies.

Simply put, many countries of the region cannot presently meet the irrigation requirements needed to feed their own growing populations.[24] Furthermore, for those countries that have sufficient resources to meet this need in aggregate (such as Syria), resource capture and structural distribution problems keep water out of the hands of many citizens. If this situation has been deteriorating for nearly three decades, the question remains: Why has there been no war over water? The answer, according to Tony Allen, lies in an extremely important hidden source of water, which he describes as "virtual water."[25] Virtual water is the water contained in the food that the region imports—from the United States, Australia, Argentina, New Zealand, the countries of the European Union, and other major food-exporting countries. If each person of the world consumes food that requires 1,000 cubic meters of water to grow, plus 100 additional cubic meters for drinking, hygiene, and industrial production, it is still possible that any country that cannot supply the water to produce food may have sufficient water to

meet its needs—if it has the economic capacity to buy, or the political capacity to beg, the remaining virtual water in the form of imported food.

According to Allen, more water flows into the countries of the Middle East and North Africa as virtual water each year than flows down the Nile for Egypt's agriculture. Virtual water obtained in the food available on the global market has enabled the governments of the region's countries to augment their inadequate and declining water resources. For instance, despite its meager freshwater resources of 180 cubic meters per capita, Israel—otherwise self-sufficient in terms of food production—manages its problems of water scarcity in part by importing large supplies of grain each year. As noted in Figure 1, this pattern is replicated by eight other countries in the region that have less than 1,100 cubic meters of water per person. Thus, the global cereal grain commodity markets have proven to be a very accessible and effective system for importing virtual water needs. In the Middle East and North Africa, politicians and resource managers have thus far found this option a better choice than resorting to war over water with their neighbors. As a result, the strategic imperative for maintaining peace has been met through access to virtual water in the form of food imports from the global market.[26]

The global trade in food commodities has been increasingly accessible, even to poor economies, for the past 50 years. During the Cold War, food that could not be purchased was often provided in the form of grants by either the United States or the Soviet Union, and in times of famine, international relief efforts in various parts of the globe have fed the starving. Over time, competition by the generators of the global grain surplus—the United States, Australia, Argentina, and the European Community—brought down the global price of grain. As a result, the past quarter-century, the period during which water conflicts in the Middle East and North Africa have been most insistently predicted, was also a period of global commodity markets awash with surplus grain. This situation allowed the region's states to replace domestic water supply shortages with subsidized virtual water in the form of purchases from the global commodities market. For example, during the 1980s, grain was being traded at about $100 (US) a ton, despite costing about $200 a ton to produce.[27] Thus, US and European taxpayers were largely responsible for funding the cost of virtual water (in the form of significant agricultural subsidies they paid their own farmers) which significantly benefited the countries of the Middle East and North Africa.

For the most part we concur with Allen's evaluation that countries have not gone to war primarily over water, and that they have not done so because they have been able to purchase virtual water on the international market. However, the key question for the future is, Will this situation continue? If the answer is yes, and grain will remain affordable to the countries of the region, then it is relatively safe to conclude that conflict derived from environmental scarcity (in the form of water deficits) will not be a significant problem in the foreseeable future. However, if the answer is no, and grain will not be as affordable as it has been in the past, then future conflict scenarios based on environmental scarcity must be seriously considered.

# Global Economic Restructuring: The World Trade Organization's Impact on Subsidies

Regrettably, a trend toward the answer "no" appears to be gaining some momentum due to ongoing structural changes in the global economy. The year 1995 witnessed a dramatic change in the world grain market, when wheat prices rose rapidly, eventually reaching $250 a ton by the spring of 1996. With the laws of supply and demand kicking in, this increased price resulted in greater production; by 1998, world wheat prices had fallen back to $140 a ton, but had risen again to over $270 by June 2001.[28] These rapid wheat price fluctuations reemphasize the strategic importance and volatility of virtual water. If the global price of food staples remains affordable, many countries in the Middle East and North Africa may struggle to meet the demand-induced scarcity resulting from their growing populations, but they most likely will succeed. However, if basic food staple prices rise significantly in the coming decades and the existing economic growth patterns that have characterized the region's economies over the past 30 years remain constant, an outbreak of war is more likely.

It is clear that recent structural changes in the world economy do not favor the continuation of affordable food prices for the region's countries in the future. As noted above, wheat that costs $200 a ton to produce has often been sold for $100 a ton on world markets. This situation is possible only when the supplier is compensated for the lost $100 per ton in the form of a subsidy. Historically, these subsidies have been paid by the governments of major cereal grain-producing countries, primarily the United States and members of the European Union. Indeed, for the last 100 years, farm subsidies have been a bedrock public policy throughout the food-exporting countries of the first world. However, with the steady embrace of global free-trade economics and the establishment of the World Trade Organization (WTO), agricultural subsidies have come under pressure in most major grain-producing countries. According to a recent US Department of Agriculture (USDA) study, "The elimination of agriculture trade and domestic policy distortions could raise world agriculture prices about 12 percent.[29]

> **"Many countries of the region cannot presently meet the irrigation requirements needed to feed their own growing populations."**

Thus, as the WTO gains systematic credibility over the coming decades, its free-trade policies will further erode the practice of farm price supports, and it is highly unlikely that the aggregate farm subsidies of the past will continue at historic levels in the future. Under the new WTO regime, global food production will be increasingly based on the real cost of production

plus whatever profit is required to keep farmers in business. Therefore, as global food prices rise in the future, and American and European governments are restricted by the new global trading regime from subsidizing their farmers, the price of virtual water in the Middle East and North Africa and throughout the food-importing world will also rise. According to the USDA report mentioned above, both developed and developing countries will gain from WTO liberalization. Developed countries that are major food exporters will gain immediately from the projected $31 billion in increased global food prices, of which they will share $28.5 billion ($13.3 billion to the United States), with $2.6 billion going to food exporters in the developing world. However, the report also claims food-importing countries will gain because global food price increases will spur more efficient production in their own economies, thus enabling them a "potential benefit" of $21 billion.[30] Even if accepted at face value, it is clear that such benefits will occur mostly in those developing countries with an abundance of water resources. Indeed, developing countries that produce fruits, vegetables, and other high-value crops for export to first-world markets may indeed benefit from the reduction of farm subsidies, which today undercut their competitive advantage. But when it comes to basic foodstuffs—wheat, corn, and rice—the cereal grains that sustain life for most people, the developing world cannot compete with the highly efficient mechanized corporate farms of the first world.

In future research, basic intelligence is needed on two fronts. First, we must obtain a clearer understanding of the capacity of global commodity markets to meet future virtual water needs in the form of food. Second, we must identify which Middle East and North Africa governments will most likely have the economic capacity to meet their virtual water needs though food purchases—or, perhaps more important, which ones will not. In short, is there food available in the global market, and can countries afford to buy it? Countries that cannot afford virtual water may choose instead to pursue war as a means of achieving their national interest goals. Clearly the strongest countries, or those least susceptible to intrastate or interstate conflict arising from environmental scarcity, are those that have significant water resources or the economic capacity to purchase virtual water. However, it is also clear that the relative condition of peace that has existed in the Middle East and North Africa has been maintained historically through deeply buried linkages between American and European taxpayers, their massive farm subsidies programs, and world food prices. In the future, it appears that these hidden links may be radically altered if not broken by the World Trade Organization, and, as a result, the likelihood of conflict will increase.

# Conclusion: Why War Will Come

Having moved away from the conventional understanding of water strictly as a zero-sum environmental resource by reconceptualizing it in more fungible economic terms, we nevertheless believe two incompatible social trends will collide to make war in the Middle East and North Africa virtually inevitable in the future. The first trend is economic globalization. As capitalism becomes ever more embraced as the global economic philosophy, and the world increasingly embraces free-trade economics, economic growth is both required and is inevitable. The WTO will facilitate this aggregate global growth, which, on the plus side, will undoubtedly increase the basic standard of living for the average world citizen. However, the global economy will be required to meet the needs of an estimated eight billion citizens in the year 2025. Achieving growth will demand an ever-greater share of the world's existing natural resources, including water. Thus, if present regional economic and demographic trends continue, resource shortfalls will occur, with water being the most highly stressed resource in the Middle East and North Africa.

Globalization is both a cause and a consequence of the rapid spread of information technology. Thus, in the globalized world, the figurative distance between cultures, philosophies of rule, and, perhaps more important, a basic understanding of what is possible in life, becomes much shorter. Personal computers, the internet, cellular phones, fax machines, and satellite television are all working in partnership to rewire the psychological infrastructure of the citizens of the Middle East and North Africa, and the world at large. As a result, by making visible what is possible in the outside world, this cognitive liberation will bring heightened material expectations of a better life, both economically and politically. Consequently, citizens will demand more from their governments. This emerging reality will collide head-on with the second trend—political authoritarianism—that characterizes most Middle East and North Africa governments.

Throughout the region there are few governments that allow for public expression of dissent. Although Turkey, Algeria, Tunisia, and Egypt are democracies in name, these states have exhibited a propensity to revert to authoritarian tactics when deemed necessary to limit political activity among their respective populaces.[31] Likewise, while Israel is institutionally a democracy, ethnic minorities are all but excluded from the democratic process. The remainder of the Middle East and North Africa states can be described only as authoritarian regimes. In retrospect, the most fundamental common denominator of all authoritarian regimes throughout history is their fierce resistance to change. Change is seen as a threat to the regime because most authoritarian regimes base their right to rule in some form of infallibility: the infallibility of the sultan, the king, or the ruling party and its ideology. Any admission that change is needed strikes at the foundation of this inflexible infallibility. Historically, most change has occurred in the Middle East and North Africa during times of intrastate unrest and interstate war. In the coming decades, globalization will bring change that will be resisted by governments of the region. As a result, to the distant observer the future will resemble the past: periods of wholesale peace will be a rare occurrence, intense competition and low-intensity conflict will be the norm, and major wars will occur at sporadic intervals.

The wild card in this equation may be post-2004 Iraq. Operation Iraqi Freedom and the ouster of Saddam Hussein have altered the strategic political landscape. If a sustainable democracy indeed emerges in Iraq, the country may turn away from future conflicts with its neighbors. Potential conflict between Turkey and Iraq over water may now be averted due to the fact that both countries may choose nonviolent solutions to their disputes. If President Bush's vision of a democratic Middle East comes to fruition, war may be averted. After all, there is a rich body of scholarly research regarding the "democratic peace" that suggests liberal democracies are significantly less likely to resort to war to resolve interstate disputes, and post-Saddam Iraq could serve as a key litmus test for the future of democratic reform in the region. However, it is also highly unlikely that regime change will come quickly to the moderate authoritarian states of the region that are also US allies. Decisionmakers in Washington may be able to dictate the political future of Iraq, but even America's mighty arsenal of political, economic, and military power cannot alter the basic demographic and environmental trends in the region.

# Notes

1. Alex Marshall, ed., *The State of World Population 1997* (New York: United Nations Population Fund, 1997), p. 70.

2. "The World Bank: Middle East and North Africa Data Profile," *The World Bank Group Country Data* (2000), http://www.worldbank.org/data/countrydata/countrydata.html.

3. The leading scholar in this area is Thomas Homer-Dixon. For example, see his recent book (coedited with Jessica Blitt), *Ecoviolence: Links Among Environment, Population, and Security* (New York: Rowman & Littlefield, 1998), which focuses on Chiapas, Gaza, South Africa, Pakistan, and Rwanda.

4. Mostafa Dolatyar and Tim S. Gray, *Water Politics in the Middle East: A Context for Conflict or Co-operation?* (New York: St. Martin's Press, 2000), p. 6.

5. *Human Development Report: Consumption for Human Development* (New York: United Nations Development Programme, Oxford Univ. Press, 1998), p. 55; "Water Woes Around the World," MSNBC, 9 September 2002, http://www.msnbc.com/news/802693.asp.

6. Thomas Naff, "Conflict and Water Use in the Middle East," in *Water in the Arab World: Perspectives and Prognoses,* ed. Peter Rogers and Peter Lydon (Cambridge, Mass.: Harvard Univ. Press, 1994), p. 273.

7. Alwyn R. Rouyer, *Turning Water into Politics: The Water Issue in the Palestinian-Israeli Conflict* (New York: St. Martin's Press, 2000), p. 7.

8. Ulrich Küffner, "Contested Waters: Dividing or Sharing?" in *Water in the Middle East: Potential for Conflicts and Prospects for Cooperation,* ed. Waltina Scheumann and Manuel Schiffler (New York: Springer, 1998), p. 71.

9. Miriam R. Lowi, *Water and Power: The Politics of a Scarce Resource in the Jordan River Basin* (Cambridge, Eng.: Cambridge Univ. Press, 1993), p. 1.

10. Ibid., pp. 2ff.

11. Thomas Homer-Dixon and Jessica Blitt, "Introduction: A Theoretical Overview," in *Ecoviolence: Links Among Environment, Population, and Scarcity,* ed. Thomas Homer-Dixon and Jessica Blitt (New York: Rowman & Littlefield, 1998), p. 6.

12. Thomas Homer-Dixon and Valerie Percival, "The Case of Senegal-Mauritania," in *Environmental Scarcity and Violent Conflict: Briefing Book* (Washington: American Association

for the Advancement of Science and the University of Toronto, 1996), pp. 35–38.

13. Douglas Borer witnessed this agricultural problem while visiting rural areas on the Bataan peninsula in late 1985 and early 1986. The members of the New People's Army which he met were uninterested in Marxism, but they were very interested in ridding themselves of the Marcos regime. See Thomas Homer-Dixon and Valerie Percival, "The Case of the Philippines," ibid., p. 49.

14. Ted Gurr, *Why Men Rebel* (Princeton, N.J.: Princeton Univ. Press, 1970).

15. One need only look 90 miles southward from the Florida coast to find proof of this reality in Castro's Cuba.

16. Thus, Saddam Hussein was able to remain in power in Iraq until 2003 due to two essential factors. First, as noted in a recent article by James Quinlivan, Saddam had created "groups with special loyalties to the regime and the creation of parallel military organizations and multiple internal security agencies," that made Iraq essentially a "coup-proof" regime. (See James T. Quinlivan, "Coup-Proofing: Its Practice and Consequences in the Middle East," *International Security,* 24 [Fall 1999], 131–65.) Second, Saddam had convinced a significant portion of his people that the United States (and Britain) were responsible for their suffering. Thus, as long as these perceptions held and Saddam was able to command loyalty of the inner regime, his ouster from power by domestic sources remained unlikely.

17. Ewan W. Anderson, "Water: The Next Strategic Resource," in *The Politics of Scarcity: Water in the Middle East,"* ed. Joyce R. Starr and Daniel C. Stoll (Boulder, Colo.: Westview Press, 1988), p. 1.

18. Hussein A. Amery and Aaron T. Wolf, "Water, Geography, and Peace in the Middle East," in *Water in the Middle East: A Geography of Peace,* ed. Hussein A. Amery and Aaron T. Wolf (Austin: Univ. of Texas Press, 2000).

19. Nubit Kliot, *Water Resources and Conflict in the Middle East* (London and New York: Routledge, 1994), p. v, as quoted in Dolatyar and Gray, p. 9.

20. Rouyer, p. 9.

21. Martin Sherman, *The Politics of Water in the Middle East: An Israeli Perspective on the Hydro-Political Aspects of the Conflict* (New York: St. Martin's Press, 1999), p. xi.

22. Jochen Renger, "The Middle East Peace Process: Obstacles to Cooperation Over Shared Waters," in *Water in the Middle East: Potential for Conflict and Prospects for Cooperation,* ed. Waltina Scheumann and Manuel Schiffler (New York: Springer, 1998), p. 50.

23. Thus, even though the Saudi government is much more Islamized in religious terms than that of the Iraqis or Syrians, as long as the Saudi government is pro-US and serves US interests in supplying cheap oil, it receives the benevolent "moderate" label, while the more secularized Iraqis and Syrians have been labeled with the prerogative labels "radical" or "rogue-states."

24. Tony Allan, "Watersheds and Problemsheds: Explaining the Absence of Armed Conflict over Water in the Middle East," in *MERNIA: Middle East Review of International Affairs Journal,* 2 (March 1998), http://biu.ac.il/SOC/besa/meria/journal/1998/issue1/jv2n1a7.html.

25. Ibid.

26. Ibid.

27. Ibid.

28. Prices from 26 June 2001 quoted at http://www.usafutures.com/commodityprices.htm.

29. "Agricultural Policy Reform in the WTO—The Road Ahead," in *ERS Agricultural Economics Report,* No. 802, ed. Mary E. Burfisher (Washington: US Department of Agriculture, May 2001), p. iii.

30. Ibid., p. 6.

31. For instance, as of 2004, Freedom House (http://www.freedomhouse.org/) classifies Algeria, Egypt, and Tunisia as "not free," and Turkey as only "partly free."

**JASON J. MORRISSETTE** is a doctoral candidate and instructor of record in the School of Public and International Affairs at the University of Georgia. He is currently writing his dissertation on the political economy of water scarcity and conflict.

**DOUGLAS A. BORER** (Ph.D., Boston University, 1993) is an Associate Professor in the Department of Defense Analysis at the Naval Postgraduate School. He recently served as Visiting Professor of Political Science at the US Army War College. Previously he was Director of International Studies at Virginia Tech, and he has taught overseas in Fiji and Australia. Dr. Borer is a former Fulbright Scholar at the University of Kebangsaan Malaysia, and has published widely in the areas of security, strategy, and foreign policy.

From *Parameters,* Winter 2004–2005, pp. 86–101. Published in 2005 by the U.S. Army War College.

# The Irony of Climate

**Archaeologists suspect that a shift in the planet's climate thousands of years ago gave birth to agriculture. Now climate change could spell the end of farming as we know it.**

BRIAN HALWEIL

High in the Peruvian Andes, a new disease has invaded the potato fields in the town of Chacllabamba. Warmer and wetter weather associated with global climate change has allowed late blight—the same fungus that caused the Irish potato famine—to creep 4,000 meters up the mountainside for the first time since humans started growing potatoes here thousands of years ago. In 2003, Chacllabamba farmers saw their crop of native potatoes almost totally destroyed. Breeders are rushing to develop tubers resistant to the "new" disease that retain the taste, texture, and quality preferred by Andean populations.

Meanwhile, old-timers in Holmes County, Kansas, have been struggling to tell which way the wind is blowing, so to speak. On the one hand, the summers and winters are both warmer, which means less snow and less snowmelt in the spring and less water stored in the fields. On the other hand, there's more rain, but it's falling in the early spring, rather than during the summer growing season. So the crops might be parched when they need water most. According to state climatologists, it's too early to say exactly how these changes will play out—if farmers will be able to push their corn and wheat fields onto formerly barren land or if the higher temperatures will help once again to turn the grain fields of Kansas into a dust bowl. Whatever happens, it's going to surprise the current generation of farmers.

Asian farmers, too, are facing their own climate-related problems. In the unirrigated rice paddies and wheat fields of Asia, the annual monsoon can make or break millions of lives. Yet the reliability of the monsoon is increasingly in doubt. For instance, El Niño events (the cyclical warming of surface waters in the eastern Pacific Ocean) often correspond with weaker monsoons, and El Niños will likely increase with global warming. During the El Niño-induced drought in 1997, Indonesian rice farmers pumped water from swamps close to their fields, but food losses were still high: 55 percent for dryland maize and 41 percent for wetland maize, 34 percent for wetland rice, and 19 percent for cassava. The 1997 drought was followed by a particularly wet winter that delayed planting for two months in many areas and triggered heavy locust and rat infestations. According

to Bambang Irawan of the Indonesian Center for Agricultural Socio-Economic Research and Development, in Bogor, this succession of poor harvests forced many families to eat less rice and turn to the less nutritious alternative of dried cassava. Some farmers sold off their jewelry and livestock, worked off the farm, or borrowed money to purchase rice, Irawan says. The prospects are for more of the same: "If we get a substantial global warming, there is no doubt in my mind that there will be serious changes to the monsoon," says David Rhind, a senior climate researcher with NASA's Goddard Institute for Space Studies.

Archaeologists believe that the shift to a warmer, wetter, and more stable climate at the end of the last ice age was key for humanity's successful foray into food production. Yet, from the American breadbasket to the North China Plain to the fields of southern Africa, farmers and climate scientists arc finding that generations-old patterns of rainfall and temperature are shifting. Farming may be the human endeavor most dependent on a stable climate—and the industry that will struggle most to cope with more erratic weather, severe storms, and shifts in growing season lengths. While some optimists are predicting longer growing seasons and more abundant harvests as the climate warms, farmers are mostly reaping surprises.

## Toward the Unknown (Climate) Region

For two decades, Hartwell Allen, a researcher with the University of Florida in Gainesville and the U.S. Department of Agriculture, has been growing rice, soybeans, and peanuts in plastic, greenhouse-like growth chambers that allow him to play God. He can control—"rather precisely"—the temperature, humidity, and levels of atmospheric carbon. "We grow the plants under a daily maximum/minimum cyclic temperature that would mimic the real world cycle," Allen says. His lab has tried regimes of 28 degrees C day/18 degrees C night, 32/22, 36/26, 40/30, and 44/34. "We ran one experiment to 48/38, and got very few surviving plants," he says. Allen found that while a doubling of carbon dioxide and a slightly increased temperature stimulate seeds to germinate and the plants to grow larger and lusher, the higher temperatures are deadly when the plant starts producing pollen. Every stage of the process—pollen transfer, the growth of the tube that links the pollen to the seed, the viability of the pollen itself—is highly sensitive. "It's all or nothing, if pollination isn't successful," Allen notes. At temperatures above 36 degrees C during pollination, peanut yields dropped about six percent per degree of temperature increase. Allen is particularly concerned about the implications for places like India and West Africa, where peanuts are a dietary staple and temperatures during the growing season are already well above 32 degrees C: "In these regions the crops are mostly rain-fed. If global warming also leads to drought in these areas, yields could be even lower."

As plant scientists refine their understanding of climate change and the subtle ways in which plants respond, they are beginning to think that the most serious threats to agriculture will not be the most dramatic: the lethal heatwave or severe drought or endless deluge. Instead, for plants that humans have bred to thrive in specific climatic conditions, it is those subtle shifts in temperatures and rainfall during key periods in the crops' lifecycles that will be most disruptive. Even today, crop losses associated with background climate variability are significantly higher than those caused by disasters such as hurricanes or flooding.

John Sheehy at the International Rice Research Institute in Manila has found that damage to the world's major grain crops begins when temperatures climb above 30 degrees C during flowering. At about 40 degrees C, yields are reduced to zero. "In rice, wheat, and maize, grain yields are likely to decline by 10 percent for every 1 degree C increase over 30 degrees. We are already at or close to this threshold," Sheehy says, noting regular heat damage in Cambodia, India, and his own center in the Philippines, where the average temperature is now 2.5 degrees C higher than 50 years ago. In particular, higher night-time temperatures forced the plants to work harder at respiration and thus sapped their energy, leaving less for producing grain. Sheehy estimates that grain yields in the tropics might fall as much as 30 percent over the next 50 years, during a period when the region's already malnourished population is projected to increase by 44 percent. (Sheehy and his colleagues think a potential solution is breeding rice and other crops to flower early in the morning or at night so that the sensitive temperature process misses the hottest part of the day. But, he says, "we haven't been successful in getting any real funds for the work.") The world's major plants can cope with temperature shifts to some extent, but since the dawn of agriculture farmers have selected plants that thrive in stable conditions.

Climatologists consulting their computer climate models see anything but stability, however. As greenhouse gases trap more of the sun's heat in the Earth's atmosphere, there is also more energy in the climate system, which means more extreme swings—dry to wet, hot to cold. (This is the reason that there can still be severe winters on a warming planet, or that March 2004

was the third-warmest month on record after one of the coldest winters ever.) Among those projected impacts that climatologists have already observed in most regions: higher maximum temperatures and more hot days, higher minimum temperatures and fewer cold days, more variable and extreme rainfall events, and increased summer drying and associated risk of drought in continental interiors. All of these conditions will likely accelerate into the next century.

Cynthia Rosenzweig, a senior research scholar with the Goddard Institute for Space Studies at Columbia University, argues that although the climate models will always be improving, there are certain changes we can already predict with a level of confidence. First, most studies indicate "intensification of the hydrological cycle," which essentially means more droughts and floods, and more variable and extreme rainfall. Second, Rosenzweig says, "basically every study has shown that there will be increased incidence of crop pests." Longer growing seasons mean more generations of pests during the summer, while shorter and warmer winters mean that fewer adults, larvae, and eggs will die off.

Third, most climatologists agree that climate change will hit farmers in the developing world hardest. This is partly a result of geography. Farmers in the tropics already find themselves near the temperature limits for most major crops, so any warming is likely to push their crops over the top. "All increases in temperature, however small, will lead to decreases in production," says Robert Watson, chief scientist at the World Bank and former chairman of the Intergovernmental Panel on Climate Change. "Studies have consistently shown that agricultural regions in the developing world are more vulnerable, even before we consider the ability to cope," because of poverty, more limited irrigation technology, and lack of weather tracking systems. "Look at the coping strategies, and then it's a real double whammy," Rosenzweig says. In sub-Saharan Africa—ground zero of global hunger, where the number of starving people has doubled in the last 20 years—the current situation will undoubtedly be exacerbated by the climate crisis. (And by the 2080s, Watson says, projections indicate that even temperate latitudes will begin to approach the upper limit of the productive temperature range.)

## Coping with Change

"Scientists may indeed need decades to be sure that climate change is taking place," says Patrick Luganda, chairman of the Network of Climate Journalists in the Greater Horn of Africa. "But, on the ground, farmers have no choice but to deal with the daily reality as best they can." Luganda says that several years ago local farming communities in Uganda could determine the onset of rains and their cessation with a fair amount of accuracy. "These days there is no guarantee that the long rains will start, or stop, at the usual time," Luganda says. The Ateso people in north-central Uganda report the disappearance of asisinit, a swamp grass favored for thatch houses because of its beauty and durability. The grass is increasingly rare because farmers have started to plant rice and millet in swampy areas in response to more frequent droughts. (Rice farmers in Indonesia coping with droughts have done the same.) Farmers have also begun to sow a wider diversity of crops and to stagger their plantings to hedge against abrupt climate shifts. Luganda adds that repeated crop failures have pushed many farmers into the urban centers: the final coping mechanism.

The many variables associated with climate change make coping difficult, but hardly futile. In some cases, farmers may need to install sprinklers to help them survive more droughts. In other cases, plant breeders will need to look for crop varieties that can withstand a greater range of temperatures. The good news is that many of the same changes that will help farmers cope with climate change will also make communities more self-sufficient and reduce dependence on the long-distance food chain.

Planting a wider range of crops, for instance, is perhaps farmers' best hedge against more erratic weather. In parts of Africa, planting trees alongside crops—a system called agroforestry that might include shade coffee and cacao, or leguminous trees with corn—might be part of the answer. "There is good reason to believe that these systems will be more resilient than a maize monoculture," says Lou Verchot, the lead scientist on climate change at the International Centre for Research in Agroforestry in Nairobi. The trees send their roots considerably deeper than the crops, allowing them to survive a drought that might damage the grain crop. The tree roots will also pump water into the upper soil layers where crops can tap it. Trees improve the soil as well: their roots create spaces for water flow and their leaves decompose into compost. In other words, a farmer who has trees won't lose everything. Farmers in central Kenya are using a mix of coffee, macadamia nuts, and cereals that results in as many as three marketable crops in a good year. "Of course, in any one year, the monoculture will yield more money," Verchot admits, "but farmers need to work on many years." These diverse crop mixes are all the more relevant since rising temperatures will eliminate much of the traditional coffee- and tea-growing areas in the Caribbean, Latin America, and Africa. In Uganda, where coffee and tea account for nearly 100 percent of agricultural exports, an average temperature rise of 2 degrees C would dramatically reduce the harvest, as all but the highest altitude areas become too hot to grow coffee.

In essence, farms will best resist a wide range of shocks by making themselves more diverse and less dependent on outside inputs. A farmer growing a single variety of wheat is more likely to lose the whole crop when the temperature shifts dramatically than a farmer growing several wheat varieties, or better yet, several varieties of plants besides wheat. The additional crops help form a sort of ecological bulwark against blows from climate change. "It will be important to devise more resilient agricultural production systems that can absorb and survive more variability," argues Fred Kirschenmann, director of the Leopold Center for Sustainable Agriculture at Iowa State University. At his own family farm in North Dakota, Kirschenmann has struggled with two years of abnormal weather that nearly eliminated one crop and devastated another. Diversified farms will cope better with drought, increased pests, and a range of other climate-related jolts. And they will tend to be less reliant on fertilizers and pesticides, and the fossil fuel inputs they require.

Climate change might also be the best argument for preserving local crop varieties around the world, so that plant breeders can draw from as wide a palette as possible when trying to develop plants that can cope with more frequent drought or new pests.

Farms with trees planted strategically between crops will not only better withstand torrential downpours and parching droughts, they will also "lock up" more carbon. Lou Verchot says that the improved fallows used in Africa can lock up 10–20 times the carbon of nearby cereal monocultures, and 30 percent of the carbon in an intact forest. And building up a soil's stock of organic matter—the dark, spongy stuff in soils that stores carbon and gives them their rich smell—not only increases the amount of water the soil can hold (good for weathering droughts), but also helps bind more nutrients (good for crop growth).

Best of all, for farmers at least, systems that store more carbon are often considerably more profitable, and they might become even more so if farmers get paid to store carbon under the Kyoto Protocol. There is a plan, for instance, to pay farmers in Chiapas, Mexico, to shift from farming that involves regular forest clearing to agroforestry. The International Automobile Federation is funding the project as part of its commitment to reducing carbon emissions from sponsored sports car races. Not only that, "increased costs for fossil fuels will accelerate demand for renewable energies," says Mark Muller of the Institute for Agriculture and Trade Policy in Minneapolis, Minnesota, who believes that farmers will find new markets for biomass fuels like switchgrass that can be grown on the farm, as well as additional royalties from installing wind turbines on their farms.

However, "carbon farming is a temporary solution," according to Marty Bender of the Land Institute's Sunshine Farm in Salina, Kansas. He points to a recent paper in *Science* showing that even if America's soils were returned to their pre-plow carbon content—a theoretical maximum for how much carbon they could lock up—this would be equal to only two decades of American carbon emissions. "That is how little time we will be buying," Bender says, "despite the fact that it may take a hundred years of aggressive, national carbon farming and forestry to restore this lost carbon." (Cynthia Rosenzweig also notes that the potential to lock up carbon is limited, and that a warmer planet will reduce the amount of carbon that soils can hold: as land heats up, invigorated soil microbes respire more carbon dioxide.)

"We really should be focusing on energy efficiency and energy conservation to reduce the carbon emissions by our national economy," Bender concludes. That's why Sunshine Farm, which Bender directs, has been farming without fossil fuels, fertilizers, or pesticides in order to reduce its contribution to climate change and to find an inherently local solution to a global problem. As the name implies, Sunshine Farm runs essentially on sunlight. Homegrown sunflower seeds and soybeans become biodiesel that fuels tractors and trucks. The farm raises nearly three-fourths of the feed—oats, grain sorghum, and alfalfa—for its draft horses, beef cattle, and poultry. Manure and legumes in the crop rotation substitute for energy-gobbling nitrogen fertilizers. A 4.5-kilowatt photovoltaic array powers the workshop tools, electric fencing, water pumps, and chick brooding pens. The farm has eliminated an amount of energy equivalent to that used to make and transport 90 percent of its supplies. (Including the energy required to make the farm's machinery lowers the figure to 50 percent, still a huge gain over the standard American farm.)

---

**The atmosphere, which sets the stage for climate dynamics, is a global commons. It has no gatekeeper—everyone can access it—and it has a limited capacity to absorb emissions before climate stability is undermined. Societies have developed a variety of gatekeeping solutions to help manage commons resources. One, privatization, would be difficult to apply to the atmosphere, although carbon-trading schemes might reduce emissions if caps are set low enough. Governments might also act as gatekeepers by means of treaties, such as the Kyoto Protocol, that limit emissions.**

---

But these energy savings are only part of this distinctly local solution to an undeniably global problem, Bender says. "If local food systems could eliminate the need for half of the energy used for food processing and distribution, then that would save 30 percent of the fossil energy used in the U.S. food system," Bender reasons. "Considering that local foods will require some energy use, let's round the net savings down to 25 percent. In comparison, on-farm direct and indirect energy consumption constitutes 20 percent of energy use in the U.S. food system. Hence, local food systems could potentially save more energy than is used on American farms."

In other words, as climate tremors disrupt the vast intercontinental web of food production and rearrange the world's major breadbaskets, depending on food from distant suppliers will be more expensive and more precarious. It will be cheaper and easier to cope with local weather shifts, and with more limited supplies of fossil fuels, than to ship in a commodity from afar.

Agriculture is in third place, far behind energy use and chlorofluorocarbon production, as a contributor to climate warming. For farms to play a significant role, changes in cropping practices must happen on a large scale, across large swaths of India and Brazil and China and the American Midwest. As Bender suggests, farmers will be able to shore up their defenses against climate change, and can make obvious reductions in their own energy use which could save them money.

But the lasting solution to greenhouse gas emissions and climate change will depend mostly on the choices that everyone else makes. According to the London-based NGO Safe Alliance, a basic meal—some meat, grain, fruits, and vegetables—using imported ingredients can easily generate four times the greenhouse gas emissions as the same meal with ingredients from local sources. In terms of our personal contribution to climate change, eating local can be as important as driving a fuel-efficient car, or giving up the car for a bike. As politicians struggle to muster the will power to confront the climate crisis, ensuring that farmers have a less erratic climate in which to raise the world's food shouldn't be too hard a sell.

---

BRIAN HALWEIL is a senior researcher at Worldwatch Institute, and the author of *Eat Here: Reclaiming Homegrown Pleasures in a Global Supermarket*.

# Avoiding Green Marketing Myopia

## Ways to Improve Consumer Appeal for Environmentally Preferable Products

Jacquelyn A. Ottman, Edwin R. Stafford, and Cathy L. Hartman

In 1994, Philips launched the "EarthLight," a super energy-efficient compact fluorescent light (CFL) bulb designed to be an environmentally preferable substitute for the traditional energy-intensive incandescent bulb. The CFL's clumsy shape, however, was incompatible with most conventional lamps, and sales languished. After studying consumer response, Philips reintroduced the product in 2000 under the name "Marathon," to emphasize the bulb's five-year life. New designs offered the look and versatility of conventional incandescent light bulbs and the promise of more than $20 in energy savings over the product's life span compared to incandescent bulbs. The new bulbs were also certified by the U.S. Environmental Protection Agency's (EPA) Energy Star label. Repositioning CFL bulbs' features into advantages that resonated with consumer values—convenience, ease-of-use, and credible cost savings—ultimately sparked an annual sales growth of 12 percent in a mature product market.[1]

Philips' experience provides a valuable lesson on how to avoid the common pitfall of "green marketing myopia." Philips called its original entry "EarthLight" to communicate the CFL bulbs' environmental advantage. While noble, the benefit appealed to only the deepest green niche of consumers. The vast majority of consumers, however, will ask, "If I use 'green' products, what's in it for me?" In practice, green appeals are not likely to attract mainstream consumers unless they also offer a desirable benefit, such as cost-savings or improved product performance.[2] To avoid green marketing myopia, marketers must fulfill consumer needs and interests beyond what is good for the environment.

Although no consumer product has a zero impact on the environment, in business, the terms "green product" and "environmental product" are used commonly to describe those that strive to protect or enhance the natural environment by conserving energy and/or resources and reducing or eliminating use of toxic agents, pollution, and waste.[3] Paul Hawken, Amory Lovins, and L. Hunter Lovins write in their book *Natural Capitalism: Creating the Next Industrial Revolution* that greener, more sustainable products need to dramatically increase the productivity of natural resources, follow biological/cyclical production models, encourage dematerialization, and reinvest in and contribute to the planet's "natural" capital.[4] Escalating energy prices, concerns over foreign oil dependency, and calls for energy conservation are creating business opportunities for energy-efficient products, clean energy, and other environmentally-sensitive innovations and products—collectively known as "cleantech"[5] (see the box on page 85). For example, Pulitzer Prize–winning author and *New York Times* columnist Thomas L. Friedman argues that government policy and industry should engage in a "geo-green" strategy to promote energy efficiency, renewable energy, and other cleantech innovations to help alleviate the nation's dependency on oil from politically conflicted regions of the world.[6] Friedman asserts that such innovations can spark economic opportunity and address the converging global challenges of rising energy prices, terrorism, climate change, and the environmental consequences of the rapid economic development of China and India.

To exploit these economic opportunities to steer global commerce onto a more sustainable path, however, green products must appeal to consumers outside the traditional green niche.[7] Looking at sustainability from a green engineering perspective, Arnulf Grubler recently wrote in *Environment,* "To minimize environmental impacts by significant orders of magnitude requires the blending of good engineering with good economics as well as changing consumer preferences."[8] The marketing discipline has long argued that innovation must consider an intimate understanding of the customer,[9] and a close look at green marketing practices over time reveals that green products must be positioned on a consumer value sought by targeted consumers.

Drawing from past research and an analysis of the marketing appeals and strategies of green products that have either succeeded or failed in the marketplace over the past decade, some important lessons emerge for crafting effective green marketing and product strategies.[10] Based on the evidence, successful green products are able to appeal to mainstream consumers or lucrative market niches and frequently command price premiums by offering "non-green" consumer value (such as convenience and performance).

## Green Marketing Myopia Defined

Green marketing must satisfy two objectives: improved environmental quality and customer satisfaction. Misjudging either or overemphasizing the former at the expense of the latter can be termed "green marketing myopia." In 1960, Harvard business professor Theodore Levitt introduced the concept of "marketing myopia" in a now-famous and influential article in the *Harvard Business Review.*[11] In it, he characterized the common pitfall of companies' tunnel vision, which focused on "managing products" (that is, product features, functions, and efficient production) instead of "meeting customers' needs" (that is, adapting to consumer expectations and anticipation of future desires). Levitt warned that a corporate preoccupation on products rather than consumer needs was doomed to failure because consumers select products and new innovations that offer benefits they desire. Research indicates that many green products have failed because of green marketing myopia—marketers' myopic focus on their products' "greenness" over the broader expectations of consumers or other market players (such as regulators or activists).

## Green marketing must satisfy two objectives: improved environmental quality and customer satisfaction.

For example, partially in response to the 1987 Montreal Protocol, in which signatory countries (including the United States) agreed to phase out ozone-depleting chlorofluorocarbons (CFCs) by 2000, Whirlpool (in 1994) launched the "Energy Wise" refrigerator, the first CFC-free cooler and one that was 30 percent more efficient than the U.S. Department of Energy's highest standard.[12] For its innovation, Whirlpool won the "Golden Carrot," a $30 million award package of consumer rebates from the Super-Efficient Refrigerator Program, sponsored by the Natural Resources Defense Council and funded by 24 electric utilities. Unfortunately, Energy Wise's sales languished because the CFC-free benefit and energy-savings did not offset its $100 to $150 price premium, particularly in markets outside the rebate program, and the refrigerators did not offer additional features or new styles that consumers desired.[13] General Motors (GM) and Ford encountered similar problems when they launched their highly publicized EV-1 and Think Mobility electric vehicles, respectively, in the late 1990s to early 2000s in response to the 1990 zero-emission vehicle (ZEV) regulations adopted in California.[14] Both automakers believed their novel two-seater cars would be market successes (GM offered the EV-1 in a lease program, and Ford offered Think Mobility vehicles as rentals via the Hertz car-rental chain). Consumers, however, found electric vehicles' need for constant recharging with few recharging locations too inconvenient. Critics charged that the automakers made only token efforts to make electric cars a success, but a GM spokesperson recently explained, "We spent more than $1 billion to produce and market the vehicle, [but] fewer than 800 were leased."[15] Most drivers were not willing to drastically change their driving habits and expectations to accommodate electric cars, and the products ultimately were taken off the market.[16]

Aside from offering environmental benefits that do not meet consumer preferences, green marketing myopia can also occur when green products fail to provide credible, substantive environmental benefits. Mobil's Hefty photodegradable plastic trash bag is a case in point. Introduced in 1989, Hefty packages prominently displayed the term "degradable" with the explanation that a special ingredient promoted its decomposition into harmless particles in landfills "activated by exposure to the elements" such as sun, wind, and rain. Because most garbage is buried in landfills that allow limited exposure to the elements, making degradation virtually impossible, the claim enraged environmentalists. Ultimately, seven state attorneys general sued Mobil on charges of deceptive advertising and consumer fraud. Mobil removed the claim from its packaging and vowed to use extreme caution in making environmental claims in the future.[17]

Other fiascos have convinced many companies and consumers to reject green products. Roper ASW's 2002 "Green Gauge Report" finds that the top reasons consumers do not buy green products included beliefs that they require sacrifices—inconvenience, higher costs, lower performance—without significant environmental benefits.[18] Ironically, despite what consumers think, a plethora of green products available in the marketplace are in fact desirable because they deliver convenience, lower operating costs, and/or better performance. Often these are not marketed along with their green benefits, so consumers do not immediately recognize them as green and form misperceptions about their benefits. For instance, the appeal of premium-priced Marathon and other brands of CFL bulbs can be attributed to their energy savings and long life, qualities that make them convenient and economical over time. When consumers are convinced of the desirable "non-green" benefits of environmental products, they are more inclined to adopt them.

Other environmental products have also scored market successes by either serving profitable niche markets or offering mainstream appeal. Consider the Toyota Prius, the gas-electric hybrid vehicle that achieves about 44 miles per gallon of gasoline.[19] In recent years, Toyota's production has hardly kept pace with the growing demand, with buyers enduring long waits and paying thousands above the car's sticker price.[20] Consequently, other carmakers have scrambled to launch their own hybrids.[21] However, despite higher gas prices, analysts assert that it can take 5 to 20 years for lower gas expenses to offset many hybrid cars' higher prices. Thus, economics alone cannot explain their growing popularity.

Analysts offer several reasons for the Prius' market demand. Initially, the buzz over the Prius got a boost at the 2003 Academy Awards when celebrities such as Cameron Diaz, Harrison Ford, Susan Sarandon, and Robin Williams abandoned stretch limousines and oversized sport utility vehicles, arriving in Priuses to symbolize support for reducing America's dependence on foreign oil.[22] Since then, the quirky-looking Prius' badge of "conspicuous conservation" has satisfied many drivers' desires to turn heads and make a statement about their social responsibility, among them Google founders Larry Page and Sergey Brin, columnist Arianna Huffington, comic Bill Maher, and Charles, Prince of Wales.[23] The Prius ultimately was named *Motor Trend's* Car of the Year in 2004. The trendy appeal of the Prius illustrates that some green products can leverage consumer desires for being distinctive. Others say the Prius is just fun to drive—the dazzling digital dashboard that offers continuous feedback on fuel efficiency and other car operations provides an entertaining driving experience. More recently, however, the Prius has garnered fans for more practical reasons. A 2006 Maritz Poll finds that owners purchased hybrids because of the convenience of fewer fill-ups, better performance, and the enjoyment of driving the latest technology.[24] In some states, the Prius and other high-mileage hybrid vehicles, such as Honda's Insight, are granted free parking and solo-occupancy access to high occupancy vehicle (HOV) lanes.[25] In sum, hybrid vehicles offer consumers several desirable benefits that are not necessarily "green" benefits.

Many environmental products have become so common and widely distributed that many consumers may no longer recognize them as green because they buy them for non-green reasons. Green household products, for instance, are widely available at supermarkets and discount retailers, ranging from energy-saving Tide Coldwater laundry detergent to non-toxic Method and Simple Green cleaning products. Use of recycled or biodegradable paper products (such as plates, towels, napkins, coffee filters, computer paper, and other goods) is also widespread. Organic and rainforest-protective "shade grown" coffees are available at Starbucks and other specialty stores and supermarkets. Organic baby food is expected to command 12 percent market share in 2006 as parents strive to protect their children's mental and physical development.[26] Indeed, the organic food market segment has increased 20 percent annually since 1990, five times faster than the conventional food market, spurring the growth of specialty retailers such as Whole Foods Market and Wild Oats. Wal-Mart, too, has joined this extensive distribution of organic products.[27] Indeed, Wal-Mart has recently declared that in North American stores, its non-farm-raised fresh fish will be certified by the Marine Stewardship Council as sustainably harvested.[28]

Super energy-efficient appliances and fixtures are also becoming popular. Chic, front-loading washing machines, for example, accounted for 25 percent of the market in 2004, up from 9 percent in 2001.[29] EPA's Energy Star label, which certifies that products consume up to 30 percent less energy than comparable alternatives, is found on products ranging from major appliances to light fixtures to entire buildings (minimum efficiency standards vary from product to product). The construction industry is becoming increasingly green as government and industry demand office buildings that are "high

# Emerging Age of Cleantech

In a 1960 Harvard Business Review article, Harvard professor Theodore Levitt introduced the classic concept of "marketing myopia" to characterize businesses' narrow vision on product features rather than consumer benefits.[1] The consequence is that businesses focus on making better mousetraps rather than seeking better alternatives for controlling pests. To avoid marketing myopia, businesses must engage in "creative destruction," described by economist Joseph Schumpeter as destroying existing products, production methods, market structures and consumption patterns, and replacing them with ways that better meet ever-changing consumer desires.[2] The dynamic pattern in which innovative upstart companies unseat established corporations and industries by capitalizing on new and improved innovations is illustrated by history. That is, the destruction of Coal Age technologies by Oil Age innovations, which are being destroyed by Information Age advances and the emerging Age of Cleantech—clean, energy- and resource-efficient energy technologies, such as those involving low/zero-emissions, wind, solar, biomass, hydrogen, recycling, and closed-loop processes.[3]

Business management researchers Stuart Hart and Mark Milstein argue that the emerging challenge of global sustainability is catalyzing a new round of creative destruction that offers "unprecedented opportunities" for new environmentally sensitive innovations, markets, and products.[4] Throughout the twentieth century, many technologies and business practices have contributed to the destruction of the very ecological systems on which the economy and life itself depends, including toxic contamination, depletion of fisheries and forests, soil erosion, and biodiversity loss. Recent news reports indicate, however, that many companies and consumers are beginning to respond to programs to help conserve the Earth's natural resources, and green marketing is making a comeback.[5] The need for sustainability has become more acute economically as soaring demand, dwindling supplies, and rising prices for oil, gas, coal, water, and other natural resources are being driven by the industrialization of populous countries, such as China and India. Politically, America's significant reliance on foreign oil has become increasingly recognized as a security threat. Global concerns over climate change have led 141 countries to ratify the Kyoto Protocol, the international treaty requiring the reduction of global warming gases created through the burning of fossil fuels. Although the United States has not signed the treaty, most multinational corporations conducting business in signatory nations are compelled to reduce their greenhouse gas emissions, and many states (such as California) and cities (such as Chicago and Seattle) have or are initiating their own global warming gas emission reduction programs.[6] State and city-level policy incentives and mandates, such as "renewable portfolio standards," requiring utilities to provide increasing amounts of electricity from clean, renewable sources such as wind and solar power, are also driving cleaner technology markets.

While some firms have responded grudgingly to such pressures for more efficient and cleaner business practices, others are seizing the cleantech innovation opportunities for new twenty-first-century green products and technologies for competitive advantage. Toyota, for instance, plans to offer an all-hybrid fleet in the near future to challenge competitors on both performance and fuel economy.[7] Further, Toyota is licensing its technology to its competitors to gain profit from their hybrid sales as well. General Electric's highly publicized "Ecomagination" initiative promises a greener world with a plan to double its investments (to $1.5 billion annually) and revenues (to $20 billion) from fuel-efficient diesel locomotives, wind power, "clean" coal, and other cleaner innovations by 2010.[8] Cleantech is attracting investors looking for the "Next Big Thing," including Goldman Sachs and Kleiner Perkins Caufield & Byers.[9] Wal-Mart, too, is testing a sustainable 206,000-square foot store design in Texas that deploys 26 energy-saving and renewable-materials experiments that could set new standards in future retail store construction.[10] In sum, economic, political, and environmental pressures are coalescing to drive cleaner and greener technological innovation in the twenty-first century, and companies that fail to adapt their products and processes accordingly are destined to suffer from the consequences of marketing myopia and creative destruction.

1. T. Levitt, "Marketing Myopia," *Harvard Business Review* 28, July–August (1960): 24–47.
2. See J. Schumpeter, *The Theory of Economic Development* (Cambridge: Harvard University Press, 1934); and J. Schumpeter, *Capitalism, Socialism and Democracy* (New York: Harper Torchbooks, 1942).
3. "Alternate Power: A Change Is in the Wind," *Business Week,* 4 July 2005, 36–37.
4. S. L. Hart and M. B. Milstein, "Global Sustainability and the Creative Destruction of Industries," *MIT Sloan Management Review* 41, Fall (1999): 23–33.
5. See for example T. Howard, "Being Eco-Friendly Can Pay Economically; 'Green Marketing' Sees Growth in Sales, Ads," *USA Today,* 15 August 2005; and E. R. Stafford, "Energy Efficiency and the New Green Marketing," *Environment,* March 2003, 8–10.
6. J. Ball, "California Sets Emission Goals That Are Stiffer than U.S. Plan," *Wall Street Journal,* 2 June 2005; and J. Marglis, "Paving the Way for U.S. Emissions Trading," *Grist Magazine,* 14 June 2005, www.climatebiz.com/sections/news_print.dfm?NewsID=28255.
7. Bloomberg News, "Toyota Says It Plans Eventually to Offer an All-Hybrid Fleet," 14 September 2005, http://www.nytimes.com/2005/09/14/ automobiles/14toyota.html.
8. J. Erickson, "U.S. Business and Climate Change: Siding with the Marketing?" *Sustainability Radar,* June, www.climatebiz.com/sections/new_ print.cfm?NewsID=28204.
9. *Business Week,* note 3 above.
10. Howard, note 5 above.

performance" (for example, super energy- and resource-efficient and cost-effective) and "healthy" for occupants (for example, well-ventilated; constructed with materials with low or no volatile organic compounds [VOC]). The U.S. Green Building Council's "Leadership in Energy and Environmental Design" (LEED) provides a rigorous rating system and green building checklist that are rapidly becoming the standard for environmentally sensitive construction.[30]

Home buyers are recognizing the practical long-term cost savings and comfort of natural lighting, passive solar heating, and heat-reflective windows, and a 2006 study sponsored by home improvement retailer Lowe's found nine out of ten builders surveyed are incorporating energy-saving features into new homes.[31] Additionally, a proliferation of "green" building materials to serve the growing demand has emerged.[32] Lowe's competitor The Home Depot is testing an 'EcoOptions' product line featuring natural fertilizers and mold-resistant drywall in its Canadian stores that may filter into the U.S. market.[33] In short, energy efficiency and green construction have become mainstream.

The diversity and availability of green products indicate that consumers are not indifferent to the value offered by environmental benefits. Consumers are buying green—but not necessarily for environmental reasons. The market growth of organic foods and energy-efficient appliances is because consumers desire their perceived safety and money savings, respectively.[34] Thus, the apparent paradox between what consumers say and their purchases may be explained, in part, by green marketing myopia—a narrow focus on the greenness of products that blinds companies from considering the broader consumer and societal desires. A fixation on products' environmental merits has resulted frequently in inferior green products (for example, the original EarthLight and GM's EV-1 electric car) and unsatisfying consumer experiences. By contrast, the analysis of past research and marketing strategies finds that successful green products have avoided green marketing myopia by following three important principles: "The Three Cs" of consumer value positioning, calibration of consumer knowledge, and credibility of product claims.

# Consumer Value Positioning

The marketing of successfully established green products showcases non-green consumer value, and there are at least five desirable benefits commonly associated with green products: efficiency and cost effectiveness; health and safety; performance; symbolism and status; and convenience. Additionally, when these five consumer value propositions are not inherent in the green product, successful green marketing programs bundle (that is, add to the product design or market offering) desirable consumer value to broaden the green product's appeal. In practice, the implication is that product designers and marketers need to align environmental products' consumer value (such as money savings) to relevant consumer market segments (for example, cost-conscious consumers).

## Efficiency and Cost Effectiveness

As exemplified by the Marathon CFL bulbs, the common inherent benefit of many green products is their potential energy and resource efficiency. Given sky-rocketing energy prices and tax incentives for fuel-efficient cars and energy-saving home improvements and appliances, long-term savings have convinced cost-conscious consumers to buy green.

Recently, the home appliance industry made great strides in developing energy-efficient products to achieve EPA's Energy Star rating. For example, Energy Star refrigerators use at least 15 percent less energy and dishwashers use at least 25 percent less energy than do traditional models.[35] Consequently, an Energy Star product often commands a price premium. Whirlpool's popular Duet front-loading washer and dryer, for example, cost more than $2,000, about double the price of conventional units; however, the washers can save up to 12,000 gallons of water and $110 on electricity annually compared to standard models (Energy Star does not rate dryers).[36]

Laundry detergents are also touting energy savings. Procter & Gamble's (P&G) newest market entry, Tide Cold-water, is designed to clean clothes effectively in cold water. About 80 to 85 percent of the energy used to wash clothes comes from heating water. Working with utility companies, P&G found that consumers could save an average of $63 per year by using cold rather than warm water.[37] Adopting Tide Coldwater gives added confidence to consumers already washing in cold water. As energy and resource prices continue to soar, opportunities for products offering efficiency and savings are destined for market growth.

## Health and Safety

Concerns over exposure to toxic chemicals, hormones, or drugs in everyday products have made health and safety important choice considerations, especially among vulnerable consumers, such as pregnant women, children, and the elderly.[38] Because most environmental products are grown or designed to minimize or eliminate the use of toxic agents and adulterating processes, market positioning on consumer safety and health can achieve broad appeal among health-conscious consumers. Sales of organic foods, for example, have grown considerably in the wake of public fear over "mad cow" disease, antibiotic-laced meats, mercury in fish, and genetically modified foods.[39] Mainstream appeal of organics is not derived from marketers promoting the advantages of free-range animal ranching and pesticide-free soil. Rather, market positioning of organics as flavorful, healthy alternatives to factory-farm foods has convinced consumers to pay a premium for them.

A study conducted by the Alliance for Environmental Innovation and household products-maker S.C. Johnson found that consumers are most likely to act on green messages that strongly connect to their personal environments.[40] Specifically, findings suggest that the majority of consumers prefer such environmental household product benefits as "safe to use around children," "no toxic ingredients," "no chemical residues," and "no strong fumes" over such benefits as "packaging can be recycled" or "not tested on animals." Seventh Generation, a brand of non-toxic and environmentally-safe household products, derived its name from the Iroquois belief that, "In our every deliberation, we must consider the impact of our decisions on the next seven generations." Accordingly, its products promote the family-oriented value of making the world a safer place for the next seven generations.

Indoor air quality is also a growing concern. Fumes from paints, carpets, furniture, and other décor in poorly ventilated "sick buildings" have been linked to headaches, eye, nose, and throat irritation, dizziness, and fatigue among occupants. Consequently, many manufacturers have launched green products to reduce indoor air pollution. Sherwin Williams, for example, offers "Harmony," a line of interior paints that is low-odor, zero-VOC, and silica-free. And Mohawk sells EverSet Fibers, a carpet that virtually eliminates the need for harsh chemical cleaners because its design allows most stains to be removed with water. Aside from energy efficiency, health and safety have been key motivators driving the green building movement.

## Performance

The conventional wisdom is that green products don't work as well as "non-green" ones. This is a legacy from the first generation of environmentally sensitive products that clearly were inferior. Consumer perception of green cleaning agents introduced in health food stores in the 1960s and 1970s, for example, was that "they cost twice as much to remove half the grime."[41] Today, however, many green products are designed to perform better than conventional ones and can command a price premium. For example, in addition to energy efficiency, front-loading washers clean better and are gentler on clothes compared to conventional top-loading machines because they spin clothes in a motion similar to clothes driers and use centrifugal force to pull dirt and water away from clothes. By contrast, most top-loading washers use agitators to pull clothes through tanks of water, reducing cleaning and increasing wear on clothes. Consequently, the efficiency and high performance benefits of top-loading washers justify their premium prices.

## Market positioning on consumer safety and health can achieve broad appeal among health-conscious consumers.

Homeowners commonly build decks with cedar, redwood, or pressure-treated pine (which historically was treated with toxic agents such as arsenic). Wood requires stain or paint and periodic applications of chemical preservatives for maintenance. Increasingly, however, composite deck material made from recycled milk jugs and wood fiber, such as Weyerhaeuser's ChoiceDek, is marketed as the smarter alternative. Composites are attractive, durable, and low maintenance. They do not contain toxic chemicals and never need staining or chemical preservatives. Accordingly, they command a price premium—as much as two to three times the cost of pressure-treated pine and 15 percent more than cedar or redwood.[42]

Likewise, Milgard Windows' low emissivity SunCoat Low-E windows filter the sun in the summer and reduce heat loss in the winter. While the windows can reduce a building's overall energy use, their more significant benefit comes from helping to create a comfortable indoor radiant temperature climate and protecting carpets and furniture from harmful ultraviolet rays. Consequently, Milgard promotes the improved comfort and performance of its SunCoat Low-E windows over conventional windows. In sum, "high performance" positioning can broaden green product appeal.

### Symbolism and Status

As mentioned earlier, the Prius, Toyota's gas-electric hybrid, has come to epitomize "green chic." According to many automobile analysts, the cool-kid cachet that comes with being an early adopter of the quirky-looking hybrid vehicle trend continues to partly motivate sales.[43] Establishing a green chic appeal, however, isn't easy. According to popular culture experts, green marketing must appear grass-roots driven and humorous without sounding preachy. To appeal to young people, conservation and green consumption need the unsolicited endorsement of high-profile celebrities and connection to cool technology.[44] Prius has capitalized on its evangelical following and high-tech image with some satirical ads, including a television commercial comparing the hybrid with Neil Armstrong's moon landing ("That's one small step on the accelerator, one giant leap for mankind") and product placements in popular Hollywood films and sitcoms (such as *Curb Your Enthusiasm*). More recently, Toyota has striven to position its "hybrid synergy drive" system as a cut above other car makers' hybrid technologies with witty slogans such as, "Commute with Nature," "mpg:)," 

and "There's Nothing Like That New Planet Smell."[45] During the 2006 Super Bowl XL game, Ford launched a similarly humorous commercial featuring Kermit the Frog encountering a hybrid Escape sports utility vehicle in the forest, and in a twist, changing his tune with "I guess it *is* easy being green!"[46]

In business, where office furniture symbolizes the cachet of corporate image and status, the ergonomically designed "Think" chair is marketed as the chair "with a brain and a conscience." Produced by Steelcase, the world's largest office furniture manufacturer, the Think chair embodies the latest in "cradle to cradle" (C2C) design and manufacturing. C2C, which describes products that can be ultimately returned to technical or biological nutrients, encourages industrial designers to create products free of harmful agents and processes that can be recycled easily into new products (such as metals and plastics) or safely returned to the earth (such as plant-based materials).[47] Made without any known carcinogens, the Think chair is 99 percent recyclable; it disassembles with basic hand tools in about five minutes, and parts are stamped with icons showing recycling options.[48] Leveraging its award-winning design and sleek comfort, the Think chair is positioned as symbolizing the smart, socially responsible office. In sum, green products can be positioned as status symbols.

### Convenience

Many energy-efficient products offer inherent convenience benefits that can be showcased for competitive advantage. CFL bulbs, for example, need infrequent replacement and gas-electric hybrid cars require fewer refueling stops—benefits that are highlighted in their marketing communications. Another efficient alternative to incandescent bulbs are light-emitting diodes (LEDs): They are even more efficient and longer-lasting than CFL bulbs; emit a clearer, brighter light; and are virtually unbreakable even in cold and hot weather. LEDs are used in traffic lights due to their high-performance convenience. Recently, a city in Idaho became a pioneer by adopting LEDs for its annual holiday Festival of Lights. "We spent so much time replacing strings of lights and bulbs," noted one city official, "[using LEDs] is going to reduce two-thirds of the work for us."[49]

To encourage hybrid vehicle adoption, some states and cities are granting their drivers the convenience of free parking and solo-occupant access to HOV lanes. A Toyota spokesperson recently told the *Los Angeles Times,* "Many customers are telling us the carpool lane is the main reason for buying now."[50] Toyota highlights the carpool benefit on its Prius Web site, and convenience has become an incentive to drive efficient hybrid cars in traffic-congested states like California and Virginia. Critics have charged, however, that such incentives clog carpool lanes and reinforce a "one car, one person" lifestyle over alternative transportation. In response, the Virginia legislature has more recently enacted curbs on hybrid drivers use of HOV lanes during peak hours, requiring three or more people per vehicle, except for those that have been grandfathered in.[51]

Solar power was once used only for supplying electricity in remote areas (for example, while camping in the wilderness or boating or in homes situated off the power grid). That convenience, however, is being exploited for other applications. In landscaping, for example, self-contained solar-powered outdoor evening lights that recharge automatically during the day eliminate the need for electrical hookups and offer flexibility for reconfiguration. With society's increasing mobility and reliance on electronics, solar power's convenience is also manifest in solar-powered calculators, wrist watches, and other gadgets, eliminating worries over dying batteries. Reware's solar-powered "Juice Bag" backpack is a popular portable re-charger for students, professionals, and outdoor enthusiasts on the go. The Juice Bag's flexible, waterproof solar panel has a 16.6-volt capacity to generate 6.3 watts to recharge PDAs, cell phones, iPods, and other gadgets in about 2 to 4 hours.[52]

### Bundling

Some green products do not offer any of the inherent five consumer-desired benefits noted above. This was the case when energy-efficient and CFC-free refrigerators were introduced in China in the 1990s. While Chinese consumers preferred and were willing to pay about 15 percent more for refrigerators that were "energy-efficient," they did not connect the environmental advantage of "CFC-free" with either energy efficiency or savings. Consequently, the "CFC-free" feature had little impact on purchase decisions.[53] To encourage demand, the CFC-free feature was bundled with attributes desired by Chinese consumers, which included energy efficiency, savings, brand/quality, and outstanding after-sales service.

> **According to popular culture experts, green marketing must appear grass-roots driven and humorous without sounding preachy.**

Given consumer demand for convenience, incorporating time-saving or ease-of-use features into green products can further expand their mainstream acceptance. Ford's hybrid Escape SUV comes with an optional 110-volt AC power outlet suitable for work, tailgating, or camping. Convenience has also enhanced the appeal of Interface's recyclable FLOR carpeting, which is marketed as "practical, goof-proof, and versatile." FLOR comes in modular square tiles with four peel-and-stick dots on the back for easy installation (and pull up for altering, recycling, or washing with water in the sink). Modularity offers versatility to assemble tiles for a custom look. Interface promotes the idea that its carpet tiles can be changed and reconfigured in minutes to dress up a room for any occasion. The tiles come in pizza-style boxes for storage, and ease of use is FLOR's primary consumer appeal.

Finally, Austin (Texas) Energy's "Green Choice" program has led the nation in renewable energy sales for the past three years.[54] In 2006, demand for wind energy outpaced supply so that the utility resorted to selecting new "Green Choice" subscribers by lottery.[55] While most utilities find it challenging to sell green electricity at a premium price on its environmental merit, Austin Energy's success comes from bundling three benefits that appeal to commercial power users: First, Green Choice customers are recognized in broadcast media for their corporate responsibility; second, the green power is marketed as "home grown," appealing to Texan loyalties; and third, the program offers a fixed price that is locked in for 10 years. Because wind power's cost is derived primarily from the construction of wind farms and is not subject to volatile fossil fuel costs, Austin Energy passes its inherent price stability onto its Green Choice customers. Thus, companies participating in Green Choice enjoy the predictability of their future energy costs in an otherwise volatile energy market.

In summary, the analysis suggests that successful green marketing programs have broadened the consumer appeal of green products by convincing consumers of their "non-green" consumer value. The lesson for crafting effective green marketing strategies is that planners need to identify the inherent consumer value of green product attributes (for example, energy efficiency's inherent long-term money savings) or bundle desired consumer value into green products (such as fixed pricing of wind power) and to draw marketing attention to this consumer value.

## Calibration of Consumer Knowledge

Many of the successful green products in the analysis described here employ compelling, educational marketing messages and slogans that connect green product attributes with desired consumer value. That is, the marketing programs successfully calibrated consumer knowledge to recognize the green product's consumer benefits. In many instances, the environmental benefit was positioned as secondary, if mentioned at all. Changes made in EPA's Energy Star logo provide an example, illustrating the program's improved message calibration over the years. One of Energy Star's early marketing messages, "EPA Pollution Preventer," was not only ambiguous but myopically focused on pollution rather than a more mainstream consumer benefit. A later promotional message, "Saving The Earth. Saving Your Money." better associated energy efficiency with consumer value, and one of its more recent slogans, "Money Isn't All You're Saving," touts economic savings as the chief benefit. This newest slogan also encourages consumers to think implicitly about what else they are "saving"—the logo's illustration of the Earth suggests the answer, educating consumers that "saving the Earth" can also meet consumer self-interest.

The connection between environmental benefit and consumer value is evident in Earthbound Farm Organic's slogan, "Delicious produce is our business, but health is our bottom line," which communicates that pesticide-free produce is flavorful and healthy. Likewise, Tide Coldwater's "Deep Clean. Save Green." slogan not only assures consumers of the detergent's cleaning performance, but the term "green" offers a double meaning, connecting Tide's cost saving with its environmental benefit. Citizen's solar-powered Eco-Drive watch's slogan, "Unstoppable Caliber," communicates the product's convenience and performance (that is, the battery will not die) as well as prestige. Table 1 on page 89 shows other successful marketing messages that educate consumers of the inherent consumer value of green.

Some compelling marketing communications educate consumers to recognize green products as "solutions" for their personal needs *and* the environment.[56] When introducing its Renewal brand, Rayovac positioned the reusable alkaline batteries as a solution for heavy battery users and the environment with concurrent ads touting "How to save $150 on a CD player that costs $100" and "How to save 147 batteries from going to landfills." Complementing the money savings and landfill angles, another ad in the campaign featured sports star Michael Jordan proclaiming, "More Power. More Music. And More Game Time." to connect Renewal batteries' performance to convenience.[57] In practice, the analysis conducted here suggests that advertising that draws attention to how the environmental product benefit can deliver desired personal value can broaden consumer acceptance of green products.

## Credibility of Product Claims

Credibility is the foundation of effective green marketing. Green products must meet or exceed consumer expectations by delivering their promised consumer value and providing substantive environmental benefits. Often, consumers don't have the expertise or ability to verify green products' environmental and consumer values, creating misperceptions and skepticism. As exemplified in the case of Mobil's Hefty photodegradable plastic trash bag described earlier, green marketing that touts a product's or a company's environmental credentials can spark the scrutiny of advocacy groups or regulators. For example, although it was approved by the U.S. Food and Drug Administration, sugar substitute Splenda's "Made from sugar, so it tastes like sugar" slogan and claim of being "natural" have been challenged by the Sugar Association and Generation Green, a health advocacy group, as misleading given that its processing results in a product that is "unrecognizable as sugar."[58]

To be persuasive, past research suggests that green claims should be specific and meaningful.[59] Toyota recognizes the ambiguity of the term "green" and discourages its use in its marketing of its gas-electric hybrid cars. One proposed slogan, "Drive green, breathe blue" was dismissed in favor of specific claims about fuel efficiency, such as "Less gas in. Less gasses out."[60] Further, environmental claims must be humble and

## Table 1  Marketing Messages Connecting Green Products with Desired Consumer Value

| Value | Message and business/product |
|---|---|
| Efficiency and cost effectiveness | "The only thing our washer will shrink is your water bill." —ASKO |
| | "Did you know that between 80 and 85 percent of the energy used to wash clothes comes from heating the water? Tide Coldwater—The Coolest Way to Clean." —Tide Coldwater Laundry Detergent "mpg:)" —Toyota Prius |
| Health and safety | "20 years of refusing to farm with toxic pesticides. Stubborn, perhaps. Healthy, most definitely." —Earthbound Farm Organic |
| | "Safer for You and the Environment." —Seventh Generation Household Cleaners |
| Performance | "Environmentally friendly stain removal. It's as simple as $H_2O$." —Mohawk EverSet Fibers Carpet |
| | "Fueled by light so it runs forever. It's unstoppable. Just like the people who wear it." —Citizen Eco-Drive Sport Watch |
| Symbolism | "Think is the chair with a brain and a conscience." —Steelcase's Think Chair |
| | "Make up your mind, not just your face." —The Body Shop |
| Convenience | "Long life for hard-to-reach places." —General Electric's CFL Flood Lights |
| Bundling | "Performance and luxury fueled by innovative technology." —Lexus RX400h Hybrid Sports Utility Vehicle |

Source: Compiled by J.A. Ottman, E.R. Stafford, and C.L. Hartman, 2006.

not over-promise. When Ford Motor Company publicized in *National Geographic* and other magazines its new eco-designed Rouge River Plant that incorporated the world's largest living roof of plants, critics questioned the authenticity of Ford's environmental commitment given the poor fuel economy of the automaker's best-selling SUVs.[61] Even the Prius has garnered some criticism for achieving considerably less mileage (approximately 26 percent less according to *Consumer Reports*) than its government sticker rating claims, although the actual reduced mileage does not appear to be hampering sales.[62] Nonetheless, green product attributes need to be communicated honestly and qualified for believability (in other words, consumer benefits and environmental effectiveness claims need to be compared with comparable alternatives or likely usage scenarios). For example, Toyota includes an "actual mileage may vary" disclaimer in Prius advertising. When Ford's hybrid Escape SUV owners complained that they were not achieving expected mileage ratings, Ford launched the "Fuel-Economy School" campaign to educate drivers about ways to maximize fuel efficiency.[63] Further, EPA is reconsidering how it estimates hybrid mileage ratings to better reflect realistic driving conditions (such as heavy acceleration and air conditioner usage).[64]

## Third Party Endorsements and Eco-Certifications

Expert third parties with respected standards for environmental testing (such as independent laboratories, government agencies, private consultants, or nonprofit advocacy organizations) can provide green product endorsements and/or "seals of approval" to help clarify and bolster the believability of product claims.[65] The "Energy Star" label, discussed earlier, is a common certification that distinguishes certain electronic products as consuming up to 30 percent less energy than comparable alternatives. The U.S. Department of Agriculture's "USDA Organic" certifies the production and handling of organic produce and dairy products.

Green Seal and Scientific Certification Systems emblems certify a broad spectrum of green products. Green Seal sets specific criteria for various categories of products, ranging from paints to cleaning agents to hotel properties, and for a fee, companies can have their products evaluated and monitored annually for certification. Green Seal-certified products include Zero-VOC Olympic Premium interior paint and Johnson Wax professional cleaners. Green Seal has also certified the Hyatt Regency in Washington, DC, for the hotel's comprehensive energy and water conservation, recycling programs, and environmental practices. By contrast, Scientific Certification Systems (SCS) certifies specific product claims or provides a detailed "eco-profile" for a product's environmental impact for display on product labels for a broad array of products, from agricultural products to fisheries to construction. For example, Armstrong hard surface flooring holds SCS certification, and SCS works with retailers like The Home Depot to monitor its vendors' environmental claims.[66]

Although eco-certifications differentiate products and aid in consumer decisionmaking, they are not without controversy. The science behind eco-seals can appear subjective and/or complex, and critics may take issue with certification criteria.[67] For example, GreenOrder, a New York-based environmental consulting firm, has devised a scorecard to evaluate cleantech products marketed in General Electric's "Ecomagination" initiative, which range from fuel-efficient aircraft engines to wind turbines to water treatment technologies. Only those passing GreenOrder's criteria are marketed as Ecomagination products, but critics have questioned GE's inclusion of "cleaner coal" (that is, coal gasification for cleaner burning and sequestration of carbon dioxide emissions) as an "Ecomagination" product.[68]

## Although eco-certifications differentiate products and aid in consumer decisionmaking, they are not without controversy.

Consequently, when seeking endorsements and eco-certifications, marketers should consider the environmental tradeoffs and complexity of their products and the third parties behind endorsements and/or certifications: Is the third party respected? Are its certification methodologies accepted by leading environmentalists, industry experts, government regulators, and other key stakeholders? Marketers should educate their customers about the meaning behind an endorsement or an eco-seal's criteria. GE recognizes that its cleaner coal technology is controversial but hopes that robust marketing and educational outreach will convince society about cleaner coal's environmental benefits.[69] On its Web site, GE references U.S. Energy Information Administration's statistics that coal accounts for about 24 percent of the world's total energy consumption, arguing that coal will continue to be a dominant source of energy due to its abundance and the increasing electrification of populous nations such as China and India.[70] In response to GE's commitment to clean coal, Jonathan Lash, president of the World Resources Institute, said, "Five years ago, I had to struggle to suppress my gag response to terms like 'clean coal,' but I've since faced the sobering reality that every two weeks China opens a new coal-fired plant. India is moving at almost the same pace. There is huge environmental value in developing ways to mitigate these plants' emissions."[71]

### Word-of-Mouth Evangelism and the Internet

Increasingly, consumers have grown skeptical of commercial messages, and they're turning to the collective wisdom and experience of their friends and peers about products.[72] Word-of-mouth or "buzz" is perceived to be very credible, especially as consumers consider and try to comprehend complex product innovations. The Internet, through e-mail and its vast, accessible repository of information, Web sites, search engines, blogs, product ratings sites, podcasts, and other digital platforms, has opened significant opportunities for tapping consumers' social and communication networks to diffuse credible "word-of-mouse" (buzz facilitated by the Internet) about green products. This is exemplified by one of the most spectacular product introductions on the Web: Tide Coldwater.

In 2005, Proctor & Gamble partnered with the non-profit organization, the Alliance to Save Energy (ASE), in a "viral marketing" campaign to spread news about the money-saving benefits of laundering clothes in cold water with specially formulated Tide Coldwater.[73] ASE provided credibility for the detergent by auditing and backing P&G's claims that consumers could save an average of $63 a year if they switched from warm to cold water washes. ASE sent e-mail promotions encouraging consumers to visit Tide.com's interactive Web site and take the "Coldwater Challenge" by registering to receive a free sample. Visitors could calculate how much money they would save by using the detergent, learn other energy-saving laundry tips, and refer e-mail addresses of their friends to take the challenge as well. Tide.com offered an engaging map of the United States where, over time, visitors could track and watch their personal networks grow across the country when their friends logged onto the site to request a free sample.

Given the immediacy of e-mail and the Internet, word-of-mouse is fast becoming an important vehicle for spreading credible news about new products. According to the Pew Internet & American Life Project, 44 percent of online U.S. adults (about 50 million Americans)

are "content creators," meaning that they contribute to the Internet via blogs, product recommendations, and reviews.[74] To facilitate buzz, however, marketers need to create credible messages, stories, and Web sites about their products that are so compelling, interesting, and/or entertaining that consumers will seek the information out and forward it to their friends and family.[75] The fact that P&G was able to achieve this for a low-involvement product is quite remarkable.

International online marketing consultant Hitwise reported that ASE's e-mail campaign increased traffic at the Tide Coldwater Web site by 900 percent in the first week, and then tripled that level in week two.[76] Within a few months, more than one million Americans accepted the "Coldwater Challenge," and word-of-mouse cascaded through ten degrees of separation across all 50 states and more than 33,000 zip codes.[77] In October 2005, Hitwise reported that Tide.com ranked as the twelfth most popular site by market share of visits in the "Lifestyle—House and Garden" category.[78] No other laundry detergent brand's Web site has gained a significant Web presence in terms of the number of visits.

P&G's savvy implementation of "The Three Cs"—consumer value positioning on money savings, calibration of consumer knowledge about cold wash effectiveness via an engaging Web site, and credible product messages dispatched by a respected non-profit group and consumers' Internet networks—set the stage for Tide Coldwater's successful launch.

## The Future of Green Marketing

Clearly, there are many lessons to be learned to avoid green marketing myopia (see the box)—the short version of all this is that effective green marketing requires applying good marketing principles to make green products desirable for consumers. The question that remains, however, is, what is green marketing's future? Historically, green marketing has been a misunderstood concept. Business scholars have viewed it as a "fringe" topic, given that environmentalism's acceptance of limits and conservation does not mesh well with marketing's traditional axioms of "give customers what they want" and "sell as much as you can." In practice, green marketing myopia has led to ineffective products and consumer reluctance. Sustainability, however, is destined to dominate twenty-first century commerce. Rising energy prices, growing pollution and resource consumption in Asia, and political pressures to address climate change are driving innovation toward healthier, more-efficient, high-performance products. In short, all marketing will incorporate elements of green marketing.

As the authors of *Natural Capitalism* argue, a more sustainable business model requires "product dematerialization"—that is, commerce will shift from the "sale of goods" to the "sale of services" (for example, providing illumination rather than selling light bulbs).[79] This model is illustrated, if unintentionally, by arguably the twenty-first century's hottest product—Apple's iPod. The iPod gives consumers the convenience to download, store, and play tens of thousands of songs without the environmental impact of manufacturing and distributing CDs, plastic jewel cases, and packaging.

Innovations that transform material goods into efficient streams of services could proliferate if consumers see them as desirable. To encourage energy and water efficiency, Electrolux piloted a "pay-per-wash" service in Sweden in 1999 where consumers were given new efficient washing machines for a small home installation fee and then were charged 10 Swedish kronor (about $1) per use. The machines were connected via the Internet to a central database to monitor use, and Electrolux maintained ownership and servicing of the washers. When the machines had served their duty, Electrolux took them back for remanufacturing. Pay-per-wash failed, however, because

# Summary of Guideposts for the "Three C'S"

Evidence indicates that successful green products have avoided green marketing myopia by following three important principles: consumer value positioning, calibration of consumer knowledge, and the credibility of product claims.

## Consumer Value Positioning

- Design environmental products to perform as well as (or better than) alternatives.
- Promote and deliver the consumer-desired value of environmental products and target relevant consumer market segments (such as market health benefits among health-conscious consumers).
- Broaden mainstream appeal by bundling (or adding) consumer-desired value into environmental products (such as fixed pricing for subscribers of renewable energy).

## Calibration of Consumer Knowledge

- Educate consumers with marketing messages that connect environmental product attributes with desired consumer value (for example, "pesticide-free produce is healthier"; "energy-efficiency saves money"; or "solar power is convenient").

- Frame environmental product attributes as "solutions" for consumer needs (for example, "rechargeable batteries offer longer performance").
- Create engaging and educational Internet sites about environmental products' desired consumer value (for example, Tide Coldwater's interactive Web site allows visitors to calculate their likely annual money savings based on their laundry habits, utility source (gas or electricity), and zip code location).

## Credibility of Product Claims

- Employ environmental product and consumer benefit claims that are specific, meaningful, unpretentious, and qualified (that is, compared with comparable alternatives or likely usage scenarios).
- Procure product endorsements or eco-certifications from trustworthy third parties, and educate consumers about the meaning behind those endorsements and eco-certifications.
- Encourage consumer evangelism via consumers' social and Internet communication networks with compelling, interesting, and/or entertaining information about environmental products (for example, Tide's "Coldwater Challenge" Web site included a map of the United States so visitors could track and watch their personal influence spread when their friends requested a free sample).

---

consumers were not convinced of its benefits over traditional ownership of washing machines.[80] Had Electrolux better marketed pay-per-wash's convenience (for example, virtually no upfront costs for obtaining a top-of-the-line washer, free servicing, and easy trade-ins for upgrades) or bundled pay-per-wash with more desirable features, consumers might have accepted the green service. To avoid green marketing myopia, the future success of product dematerialization and more sustainable services will depend on credibly communicating and delivering consumer-desired value in the marketplace. Only then will product dematerialization steer business onto a more sustainable path.

# Notes

1. G. Fowler, "'Green Sales Pitch Isn't Moving Many Products," *Wall Street Journal,* 6 March 2002.

2. See, for example, K. Alston and J. P. Roberts, "Partners in New Product Development: SC Johnson and the Alliance for Environmental Innovation," *Corporate Environmental Strategy* 6, no. 2: 111–28.

3. See, for example, J. Ottman, *Green Marketing: Opportunity for Innovation* (Lincolnwood [Chicago]: NTC Business Books, 1997).

4. P. Hawken, A. Lovins, and L. H. Lovins, *Natural Capitalism: Creating the Next Industrial Revolution* (Boston: Little, Brown, and Company, 1999).

5. See, for example, *Business Week,* "Alternate Power: A Change in the Wind," 4 July 2005, 36–37.

6. See T. L. Friedman, "Geo-Greening by Example," *New York Times,* 27 March 2005; and T. L. Friedman, "The New 'Sputnik' Challenges: They All Run on Oil," *New York Times,* 20 January 2006.

7. There is some debate as to how to define a "green consumer." Roper ASW's most recent research segments American consumers by their propensity to purchase environmentally sensitive products into five categories, ranging from "True Blue Greens," who are most inclined to seek out and buy green on a regular basis (representing 9 percent of the population), to "Basic Browns," who are the least involved group and believe environmental indifference is mainstream (representing 33 percent of the population); see Roper ASW "Green Gauge Report 2002: Americans Perspective on Environmental Issues—Yes . . . But," November 2002, http://www.windustry.com/conferences/november2002/nov2002_proceedings/plenary/greenguage2002.pdf(accessed 7 February 2006). Alternatively, however, some marketers view green consumers as falling into three broad segments concerned with preserving the planet, health consequences of environmental problems, and animal welfare; see Ottman, note 3 above, pages 19–44. Because environmental concerns are varied, ranging from resource/energy conservation to wildlife protection to air quality, marketing research suggests that responses to green advertising appeals vary by consumer segments. For example, in one study, young college-educated students were found to be drawn to health-oriented green appeals, whereas working adults were more responsive toward health, waste, and energy appeals; see M. R. Stafford, T. F. Stafford, and J. Chowdhury, "Predispositions

Toward Green Issues: The Potential Efficacy of Advertising Appeals," *Journal of Current Issues and Research in Advertising* 18, no. 2 (1996): 67–79. One of the lessons from the study presented here is that green products must be positioned on the consumer value sought by targeted consumers.

8. A. Grubler, "Doing More with Less: Improving the Environment through Green Engineering," *Environment 48,* no. 2 (March 2006): 22–37.

9. See, for example, L.A. Crosby and S. L. Johnson, "Customer-Centric Innovation," *Marketing Management* 15, no. 2 (2006): 12–13.

10. The methodology for this article involved reviewing case descriptions of green products discussed in the academic and business literature to identify factors contributing to consumer acceptance or resistance. Product failure was defined as situations in which the green product experienced very limited sales and ultimately was either removed from the marketplace (such as General Motor's EV1 electric car and Electrolux's "pay-per-wash" service) or re-positioned in the marketplace (such as Philips' "EarthLight"). Product success was defined as situations in which the green product attained consumer acceptance and was widely available at the time of the analysis. Particular attention centered on the market strategies and external market forces of green products experiencing significant growth (such as gas-electric hybrid cars and organic foods), and the study examined their market context, pricing, targeted consumers, product design, and marketing appeals and messages.

11. See T. Levitt, "Marketing Myopia," *Harvard Business Review* 28, July–August (1960): 24–47.

12. A. D. Lee and R. Conger, "Market Transformation: Does it Work? The Super Energy Efficient Refrigerator Program," *ACEEE Proceedings,* 1996, 3.69–3.80.

13. Ibid.

14. The California Air Resources Board (CARB) adopted the Low-Emission Vehicle (LEV) regulations in 1990. The original LEV regulations required the introduction of zero-emission vehicles (ZEVs) in 1998 as 2 percent of all vehicles produced for sale in California, and increased the percentage of ZEVs from 2 percent to 10 percent in 2003. By 1998, significant flexibility was introduced through partial ZEV credits for very-low-emission vehicles. For a review, see S. Shaheen, "California's Zero-Emission Vehicle Mandate," *Institute of Transporation Studies,* Paper UCD-ITS-RP-04-14, 2 September 2004.

15. C. Palmeri, "Unplugged," *Business Week,* 20 March 2006, 12.

16. "Think Tanks," *Automotive News,* 6 March 2006, 42; J. Ottman, "Lessons from the Green Graveyard," *Green@Work,* April 2003, 62–63.

17. J. Lawrence, "The Green Revolution: Case Study," *Advertising Age,* 29 January 1991, 12.

18. See Roper ASW, note 7 above.

19. "Fuel Economy: Why You're Not Getting the MPG You Expect," *Consumer Reports,* October 2005, 20–23.

20. J. O'Dell, "Prices Soar for Hybrids with Rights to Fast Lane," *Los Angeles Times,* 27 August 2005.

21. M. Landler and K. Bradsher, "VW to Build Hybrid Minivan with Chinese," *New York Times,* 9 September 2005.

22. K. Carter, "'Hybrid' Cars Were Oscars' Politically Correct Ride," *USA Today,* 31 March 2003.

23. See, for example, H. W. Jenkins, "Dear Valued Hybrid Customer . . . ," *Wall Street Journal,* 30 November 2005; E. R. Stafford, "Conspicuous Conservation," *Green@Work,* Winter 2004, 30–32. A recent Civil Society Institute poll found that 66 percent of survey participants agreed that driving fuel efficient vehicles was "patriotic"; see Reuters, "Americans See Fuel Efficient Cars as 'Patriotic,' " 18 March 2005, http://www.planetark.com/avantgo/dailynewsstory.cfm?newsid=29988.

24. "Rising Consumer Interest in Hybrid Technology Confirmed by Maritz Research," PRNewswire, 5 January 2006.

25. O'Dell, note 20 above.

26. J. Fetto, "The Baby Business," *American Demographics,* May 2003, 40.

27. See D. McGinn, "The Green Machine," *Newsweek,* 21 March 2005, E8–E12; and J. Weber, "A Super-Natural Investing Opportunity," *Business 2.0,* March 2005, 34.

28. A. Murray, "Can Wal-Mart Sustain a Softer Edge?" *Wall Street Journal,* 8 February 2006.

29. C. Tan, "New Incentives for Being Green," *Wall Street Journal,* 4 August 2005.

30. For an overview of the Leadership in Energy and Environmental Design Green Building Rating System, see http://www.usgbc.org. The 69-point LEED rating system addresses energy and water use, indoor air quality, materials, siting, and innovation and design. Buildings can earn basic certification or a silver, gold, or platinum designation depending on the number of credits awarded by external reviewers. Critics charge, however, that the costly and confusing administration of the LEED system is inhibiting adoption of the program and impeding the program's environmental objectives; see A. Schendler and R. Udall, "LEED is Broken; Let's Fix It," *Grist Magazine,* 16 October 2005, http://www.grist.com/comments/soapbox/2005/10/26/leed/index1.html.

31. GreenBiz.com, "Survey: Home Builders Name Energy Efficiency as Biggest Industry Trend," 26 January 2006, http://www.greenerbuildings.com/news_details.cfm?NewsID=30221.

32. D. Smith, "Conservation: Building Grows Greener in Bay Area," *San Francisco Chronicle,* 1 June 2005.

33. E. Beck, "Earth-Friendly Materials Go Mainstream," *New York Times,* 5 January 2006, 8.

34. J. M. Ginsberg and P. N. Bloom, "Choosing the Right Green Marketing Strategy," *MIT Sloan Management Journal,* Fall 2004: 79–84.

35. Tan, note 29 above.

36. Tan, note 29, above.

37. C. C. Berk, "P&G Will Promote 'Green' Detergent," *Wall Street Journal,* 19 January 2005.

38. K. McLaughlin, "Has Your Chicken Been Drugged?" *Wall Street Journal,* 2 August 2005; and E. Weise, "Are Our Products Our Enemy?" *USA Today,* 13 August 2005.

39. McLaughlin, ibid.

40. Alston and Roberts, note 2 above.

41. R. Leiber, "The Dirt on Green Housecleaners," *Wall Street Journal,* 29 December 2005.

42. M. Alexander, "Home Improved," *Readers Digest,* April 2004, 77–80.

43. For example, see D. Leonhardt, "Buy a Hybrid, and Save a Guzzler," *New York Times,* 8 February 2006.

44. See, for example, D. Cave, "It's Not Sexy Being Green (Yet)," *New York Times,* 2 October 2005.

45. G. Chon, "Toyota Goes After Copycat Hybrids; Buyers are Asked to Believe Branded HSD Technology is Worth the Extra Cost," *Wall Street Journal,* 22 September 2005.

46. B. G. Hoffman, "Ford: Now It's Easy Being Green," *Detroit News,* 31 January 2006.

47. See W. McDonough and M. Braungart, *Cradle to Cradle: Remaking the Way We Make Things* (New York: North Point Press, 2002).

48. R. Smith, "Beyond Recycling: Manufacturers Embrace 'C2C' Design," *Wall Street Journal*, 3 March 2005.

49. K. Hafen, "Preston Festival Goes LED," *Logan Herald Journal,* 21 September 2005.

50. O'Dell, note 20 above.

51. A. Covarrubias, "In Carpool Lanes, Hybrids Find Cold Shoulders," *Los Angeles Times,* 10 April 2006.

52. M. Clayton, "Hot Stuff for a Cool Earth," *Christian Science Monitor,* 21 April 2005.

53. See Ogilvy & Mather Topline Report, *China Energy-Efficient CFC-Free Refrigerator Study* (Beijing: Ogilvy & Mather, August 1997); E. R. Stafford, C. L. Hartman, and Y. Liang, "Forces Driving Environmental Innovation Diffusion in China: The Case of Green-freeze," *Business Horizons* 9, no. 2 (2003): 122–35.

54. J. Baker, Jr., K. Denby, and J. E. Jerrett, "Market-based Government Activities in Texas," *Texas Business Review,* August 2005, 1–5.

55. T. Harris, "Austinites Apply to Save With Wind Power," *KVUE News,* 13 February 2006.

56. J. Ottman, note 3 above.

57. J. Ottman, note 3 above.

58. Generation Green, "Splenda Letter to Federal Trade Commission," 13 January 2005, http://www.generationgreen.org/2005_01-FTC-letter.htm (accessed 7 February 2006).

59. J. Davis, "Strategies for Environmental Advertising," *Journal of Consumer Marketing* 10, no. 2 (1993): 23–25.

60. S. Farah, "The Thin Green Line," CMO Magazine, 1 December 2005, http://www.cmomagazine.com/read/120105/green_line.html (accessed 9 February 2006).

61. Ibid.

62. See Jenkins, note 23 above.

63. See Farah, note 60 above.

64. M. Maynard, "E.P.A. Revision is Likely to Cut Mileage Ratings," *New York Times,* 11 January 2006.

65. For a more comprehensive overview of eco-certifications and labeling, see L. H. Gulbrandsen, "Mark of Sustainability? Challenges for Fishery and Forestry Eco-labeling," *Environment* 47, no. 5 (2005): 8–23.

66. For a comprehensive overview of other eco-certifications, see Consumers Union's Web site at http://www.eco-labels.org/home.cfm.

67. Gulbrandsen, note 65 above, pages 17–19.

68. Farah, note 60 above.

69. Farah, note 60 above.

70. See GE Global Research, *Clean Coal,* http://ge.com/research/grc_2_1_3.html (accessed 16 April 2006).

71. A. Griscom Little, "It Was Just My Ecomagination," *Grist Magazine,* 10 May 2005, http://grist.org/news/muck/2005/05/10/little-ge/index.html.

72. E. Rosen, *The Anatomy of Buzz: How to Create Word-of-Mouth Marketing* (New York: Doubleday, 2000).

73. Viral marketing is a form of "word-of-mouse" buzz marketing defined as "the process of encouraging honest communication among consumer networks, and it focuses on email as the channel." See J. E. Phelps, R. Lewis, L. Mobilio, D. Perry, and N. Raman, "Viral Marketing or Electronic Word-of-Mouth Advertising: Examining Consumer Responses and Motivation to Pass Along Email," *Journal of Advertising Research* 44, no. 4 (2004): 333–48.

74. G. Ramsey, "Ten Reasons Why Word-of-Mouth Marketing Works," Online Media Daily, 23 September 2005, http://publications.mediapost.com/index.cfm?fuseaction=Articles.san&s=34339&Nid=15643&p=114739 (accessed 16 February 2006).

75. See Rosen, note 72 above.

76. Ramsey, note 74 above.

77. Tide press release, "ColdWater Challenge Reaches One Million," http://www.tide.com/tidecoldwater/challenge.html (accessed 13 September 2005).

78. L. Prescott, "Case Study: Tide Boosts Traffic 9-fold," iMedia Connection, 30 November 2005, http://www.imdiaconnection.com/content/7406.asp.

79. Hawken, Lovins, and Lovins, note 4 above; see also A. B. Lovins, L. H. Lovins, and P. Hawken, "A Road Map for Natural Capitalism," *Harvard Business Review,* May–June 1999, 145–58.

80. J. Makower, "Green Marketing: Lessons from the Leaders," Two Steps Forward, September 2005, http://makower.typepad.com/joel_makower/2005/09/green_marketing.html.

JACQUELYN A. OTTMAN is president of J. Ottman Consulting, Inc. in New York and author of *Green Marketing: Opportunity for Innovation,* 2nd edition (NTC Business Books, 1997). She can be reached at jaottman@greenmarketing.com. EDWIN R. STAFFORD is an associate professor of marketing at Utah State University, Logan. He researches the strategic marketing and policy implications of clean technology (also known as "cleantech") and is the co-principal investigator for a $1 million research grant from the U.S. Department of Energy on the diffusion of wind power in Utah. He may be reached at ed.stafford@usu.edu. CATHY L. HARTMAN is a professor of marketing at Utah State University, Logan. Her research centers on how interpersonal influence and social systems affect the diffusion of ideas and clean products and technology. She is principal investigator on a $1 million U.S. Department of Energy grant for developing wind power in the state of Utah. She can be contacted at cathy.hartman@usu.edu.

From *Environment,* Vol. 48, no. 5, June 2006, pp. 23–36. Reprinted by permission of the Helen Dwight Reid Educational Foundation. Published by Heldref Publications, 1319 Eighteenth St., NW, Washington, DC 20036-1802. Copyright © 2006. www.heldref.org

# UNIT 3

# Energy: Present and Future Problems

## Unit Selections

## Key Points to Consider

- Why are some energy experts rethinking the long moratorium on the development of new nuclear power plants to create electricity? Have the lessons of Three Mile Island and Chernobyl been forgotten or ignored, or are there new technologies that make nuclear energy more attractive?

- How can conservation measures actually increase the economic benefits of energy production? What impact would improving energy efficiency have on fossil fuel consumption and on potential climate change?

- What are some of the major benefits of such alternate energy sources as solar power and wind power? Do these energy alternatives really have a chance at competing with fossil fuels for a share of the global energy market?

- Why have renewable energy sources like wind or solar power not been competitive with fossil fuels in the generation of electricity? Is this situation changing and, if so, for what reasons?

## Student Web Site
www.mhcls.com/online

## Internet References
Further information regarding these Web sites may be found in this book's preface or online.

**Alliance for Global Sustainability (AGS)**
*http://globalsustainability.org*

**Alternative Energy Institute, Inc.**
*http://www.altenergy.org*

**Energy and the Environment: Resources for a Networked World**
*http://zebu.uoregon.edu/energy.html*

**Institute for Global Communication/EcoNet**
*http://www.igc.org*

**Nuclear Power Introduction**
*http://library.thinkquest.org/17658/pdfs/nucintro.pdf*

**U.S. Department of Energy**
*http://www.energy.gov*

There has been a tendency, particularly in the developed nations of the world, to view the present high standards of living as exclusively the benefit of a high-technology society. In the "techno-optimism" of the post–World War II years, prominent scientists described the technical-industrial civilization of the future as being limited only by a lack of enough trained engineers and scientists to build and maintain it. This euphoria reached its climax in July 1969 when American astronauts walked upon the surface of the Moon, an accomplishment brought about solely by American technology—or so it was supposed. It cannot be denied that technology has been important in raising standards of living and permitting Moon landings, but how much of the growth in living standards and how many outstanding and dramatic feats of space exploration have been the results of technology alone? The answer is few—for in many of humankind's recent successes, the contributions of technology to growth have been no more important than the availability of incredibly cheap energy resources, particularly petroleum, natural gas, and coal.

As the world's supply of recoverable (inexpensive) fossil fuels dwindles and becomes—as evidenced by recent international events—more important as a factor in international conflict, it becomes increasingly clear that the energy dilemma is the most serious economic and environmental threat facing the Western world and its high standard of living. With the exception of the specter of global climate change, the scarcity and cost of conventional (fossil fuel) energy is probably the most serious threat to economic growth and stability in the rest of the world as well. The economic dimensions of the energy problem are rooted in the instabilities of monetary systems produced by and dependent on inexpensive energy. The environmental dimensions of the problem are even more complex, ranging from the inability of farmers in the developing world to purchase necessary fertilizer produced from petroleum, which has suddenly become very costly, to the enhanced greenhouse effect created by fossil fuel consumption and emissions. The only answers to the problems of dwindling and geographically vulnerable, inexpensive energy supplies are conservation and alternative and sustainable energy technology. All of these require a massive readjustment of thinking, away from the exuberant notion that technology can solve any problem. The difficulty with conservation, of course, is a philosophical one that grows out of the still-prevailing optimism about high technology. Conservation is not as exciting as putting a man on the Moon. Its tactical applications—caulking windows and insulating attics—are dog-paddle technologies to people accustomed to the crawl stroke. Does a solution to this problem entail the technological fixes of which many are so enamored? Probably not, as it appears that the accelerating energy demands of the world's developing nations will most likely be met first by increased reliance on the traditional (and still relatively cheap) fossil fuels. Although there is a need to reduce this reliance, there are few ready alternatives available to the poorer developing countries. It would appear that conservation is the only option. But conservation requires social and economic change in order to solve the environmental problems related to fossil fuel use, as do alternate strategies for acquiring energy.

Alternate approaches—technological, social, and economic—to the dominant forms of the extraction and use of energy form the basis for nearly all the articles in this section. In "More Profit with Less Carbon," energy expert Amory B. Lovins describes the links between energy extraction and economics. Lovins suggests that experts on both sides of the debate about the relationship between fossil fuel combustion and climate change have it wrong in claiming that protecting Earth's environment will require a trade-off between the environment and the economy. It need not be necessary, according to Lovins, that burning less fossil fuel to slow global warming will put intolerable pressures on people's pocketbooks to pay the increased cost of more expensive energy. Rather, Lovins claims, since burning fossil fuels is inefficient, increasing the efficiency of consumption of energy in factories, buildings, vehicles, and consumer products will dramatically reduce the demand for and consumption of coal and oil. Not only would decreased demand and consumption have a positive impact on the environment but it would also save immense amounts of money for both commercial and domestic energy consumers. In other words, conservation saves more than just energy—it saves money as well.

The next two selections in the section deal with some of the most promising renewable or alternative energy strategies. In "Wind Power: Obstacles and Opportunities," Martin J. Pasqualetti of Arizona State University suggests that one of the world's first energy sources to be harnessed for human use is also one of the most promising alternative energy sources. While the most traditional use of wind power has been to pump water, the same kinds of environments that produce efficient windmill water pumps also produce efficiency wind turbines that "pump energy" rather than water. The use of wind-driven turbines to produce electricity, while still a developing technology—and one that often encounters strident local opposition when it interferes with land use or aesthetics of scenery—holds great promise for the future. Particularly in the Great Plains of the United States, the dedication of areas of land to electrical energy production could slow a decades-long trend toward farm abandonment and consolidation and restore the economic vitality of small, former agricultural communities. In "Sunrise for Renewable Energy?" the environmental editors of The Economist argue that while many renewable energy sources such as wind and solar power may not seem to be competitive economically with fossil fuels, the gap is closing dramatically. Part of what is closing the gap is the fact that renewable energy has regulatory, commercial, and technological trends on its side and this drives increased investment in renewable energy sources. The increasing cost and shortage of oil also tends to make renewable energy sources more competitive. It is revealing that several of the world's largest oil companies, including British Petroleum and Shell, now have large and well-funded renewable energy divisions. It is even more predictive of the promise of renewable energy that a number of off-shore oil platforms in the North Sea are being converted to platforms for wind turbines.

# More Profit with Less Carbon

**Focusing on energy efficiency will do more than protect Earth's climate—it will make businesses and consumers richer**

AMORY B. LOVINS

A basic misunderstanding skews the entire climate debate. Experts on both sides claim that protecting Earth's climate will force a trade-off between the environment and the economy. According to these experts, burning less fossil fuel to slow or prevent global warming will increase the cost of meeting society's needs for energy services, which include everything from speedy transportation to hot showers. Environmentalists say the cost would be modestly higher but worth it; skeptics, including top U.S. government officials, warn that the extra expense would be prohibitive. Yet both sides are wrong. If properly done, climate protection would actually *reduce costs,* not raise them. Using energy more efficiently offers an economic bonanza—not because of the benefits of stopping global warming but because saving fossil fuel is a lot cheaper than buying it.

The world abounds with proven ways to use energy more productively, and smart businesses are leaping to exploit them. Over the past decade, chemical manufacturer DuPont has boosted production nearly 30 percent but cut energy use 7 percent and greenhouse gas emissions 72 percent (measured in terms of their carbon dioxide equivalent), saving more than $2 billion so far. Five other major firms—IBM, British Telecom, Alcan, NorskeCanada and Bayer—have collectively saved at least another $2 billion since the early 1990s by reducing their carbon emissions more than 60 percent. In 2001 oil giant BP met its 2010 goal of reducing carbon dioxide emissions 10 percent below the company's 1990 level, thereby cutting its energy bills $650 million over 10 years. And just this past May, General Electric vowed to raise its energy efficiency 30 percent by 2012 to enhance the company's shareholder value. These sharp-penciled firms, and dozens like them, know that energy efficiency improves the bottom line and yields even more valuable side benefits: higher quality and reliability in energy-efficient factories, 6 to 16 percent higher labor productivity in efficient offices, and 40 percent higher sales in stores skillfully designed to be illuminated primarily by daylight.

The U.S. now uses 47 percent less energy per dollar of economic output than it did 30 years ago, lowering costs by $1 billion a day. These savings act like a huge universal tax cut that also reduces the federal deficit. Far from dampening global development, lower energy bills accelerate it. And there is plenty more value to capture at every stage of energy production, distribution and consumption. Converting coal at the power plant into incandescent light in your house is only 3 percent efficient. Most of the waste heat discarded at U.S. power stations—which amounts to 20 percent more energy than Japan uses for everything—could be lucratively recycled. About 5 percent of household electricity in the U.S. is lost to energizing computers, televisions and other appliances that are turned off. (The electricity wasted by poorly designed standby circuitry is equivalent to the output of more than a dozen 1,000-megawatt power stations running full-tilt.) In all, preventable energy waste costs Americans hundreds of billions of dollars and the global economy more than $1 trillion a year, destabilizing the climate while producing no value.

If energy efficiency has so much potential, why isn't everyone pursuing it? One obstacle is that many people have confused efficiency (doing more with less) with curtailment, discomfort or privation (doing less, worse or without). Another obstacle is that energy users do not recognize how much they can benefit from improving efficiency, because saved energy comes in millions of invisibly small pieces, not in obvious big chunks. Most people lack the time and attention to learn about modern efficiency techniques, which evolve so quickly that even experts cannot keep up. Moreover, taxpayer-funded subsidies have made energy seem cheap. Although the U.S. government has declared that bolstering efficiency is a priority, this commitment is mostly rhetorical. And scores of ingrained rules and habits block efficiency efforts or actually reward waste. Yet relatively simple changes can turn all these obstacles into business opportunities.

Enhancing efficiency is the most vital step toward creating a climate-safe energy system, but switching to fuels that emit less carbon will also play an important role. The world economy is already decarbonizing: over the past two centuries, carbon-rich fuels such as coal have given way to fuels with less carbon (oil and natural gas) or with none (renewable sources such as solar and wind power). Today less than one third of the fossil-fuel atoms burned are carbon; the rest are climate safe hydrogen. This decarbonization trend is reinforced by greater efficiencies in converting, distributing and using energy; for example, combining the production of heat and electricity can extract twice as much useful work from each ton of carbon emitted into the atmosphere. Together these advances could dramatically reduce total carbon emissions by 2050 even as the global economy expands. This article focuses on the biggest prize: wringing more work from each unit of energy delivered to businesses and consumers. Increasing end-use efficiency can yield huge savings in fuel, pollution and capital costs because large amounts of energy are lost at every stage of the journey from production sites to delivered services. So even small reductions in the power used at the downstream end of the chain can enormously lower the required input at the upstream end.

# The Efficiency Revolution

Many energy-efficient products, once costly and exotic, are now inexpensive and commonplace. Electronic speed controls, for example, are mass-produced so cheaply that some suppliers give them away as a free bonus with each motor. Compact fluorescent lamps cost more than $20 two decades ago but only $2 to $5 today; they use 75 to 80 percent less electricity than incandescent bulbs and last 10 to 13 times longer. Window coatings that transmit light but reflect heat cost one fourth of what they did five years ago. Indeed, for many kinds of equipment in competitive markets—motors, industrial pumps, televisions, refrigerators—some highly energy-efficient models cost no

---

## Crossroads for Energy

### The Problem

- The energy sector of the global economy is woefully inefficient. Power plants and buildings waste huge amounts of heat, cars and trucks dissipate most of their fuel energy, and consumer appliances waste much of their power (and often siphon electricity even when they are turned off).
- If nothing is done, the use of oil and coal will continue to climb, draining hundreds of billions of dollars a year from the economy as well as worsening the climate, pollution and oil-security problems.

### The Plan

- Improving end-use efficiency is the fastest and most lucrative way to save energy. Many energy-efficient products cost no more than inefficient ones. Homes and factories that use less power can be cheaper to build than conventional structures. Reducing the weight of vehicles can double their fuel economy without compromising safety or raising sticker prices.
- With the help of efficiency improvements and competitive renewable energy sources, the U.S. can phase out oil use by 2050. Profit-seeking businesses can lead the way.

---

more than inefficient ones. Yet far more important than all these better and cheaper technologies is a hidden revolution in the design that combines and applies them.

For instance, how much thermal insulation is appropriate for a house in a cold climate? Most engineers would stop adding insulation when the expense of putting in more material rises above the savings over time from lower heating bills. But this comparison omits the capital cost of the heating system—the furnace, pipes, pumps, fans and so on—which may not be necessary at all if the insulation is good enough. Consider my own house, built in 1984 in Snowmass, Colo., where winter temperatures can dip to –44 degrees Celsius and frost can occur any day of the year. The house has no conventional heating system; instead its roof is insulated with 20 to 30 centimeters of polyurethane foam, and its 40-centimeter-thick masonry walls sandwich another 10 centimeters of the material. The double-pane windows combine two or three transparent heat reflecting films with insulating krypton gas, so that they block heat as well as eight to 14 panes of glass. These features, along with heat recovery from the ventilated air, cut the house's heat losses to only about 1 percent more than the heat gained from sunlight, appliances and people inside the structure. I can offset this tiny loss by playing with my dog (who generates about 50 watts of heat, adjustable to 100 watts if you throw a ball to her) or by burning obsolete energy studies in a small woodstove on the coldest nights.

Eliminating the need for a heating system reduced construction costs by $1,100 (in 1983 dollars). I then reinvested this money, plus another $4,800, into equipment that saved half the water, 99 percent of the water-heating energy and 90 percent of the household electricity. The 4,000-square-foot structure—which also houses the original headquarters of Rocky Mountain Institute (RMI), the nonprofit group I co-founded in 1982—consumes barely more electricity than a single 100-watt lightbulb. (This amount excludes the power used by the institute's office equipment.) Solar cells generate five to six times that much electricity, which I sell back to the utility. Together all the efficiency investments repaid their cost in 10 months with 1983 technologies; today's are better and cheaper.

In the 1990s Pacific Gas & Electric undertook an experiment called ACT² that applied smart design in seven new and old buildings to demonstrate that large efficiency improvements can be cheaper than small ones. For example, the company built a new suburban tract house in Davis, Calif., that could stay cool in the summer without air-conditioning. PG&E estimated that such a design, if widely adopted, would cost about $1,800 less to build and $1,600 less to maintain over its lifetime than a conventional home of the same size. Similarly, in 1996 Thai architect Soontorn Boonyatikarn built a house near steamy Bangkok that required only one-seventh the air-conditioning capacity usually installed in a structure of that size; the savings in equipment costs paid for the insulating roof, walls and windows that keep the house cool. In all these cases, the design approach was the same: optimize the whole building for multiple benefits rather than use isolated components for single benefits.

**Using energy more efficiently offers an economic bonanza—not because of the benefits of stopping global warming but because saving fossil fuel is a lot cheaper than buying it.**

Such whole-system engineering can also be applied to office buildings and factories. The designers of a carpet factory built in Shanghai in 1997 cut the pumping power required for a heat-circulating loop by 92 percent through two simple changes. The first change was to install fat pipes rather than thin ones, which greatly reduced friction and hence allowed the system to use smaller pumps and motors. The second innovation was to lay out the pipes before positioning the equipment they connect. As a result, the fluid moved through short, straight pipes instead of tracing circuitous paths, further reducing friction and capital costs.

This isn't rocket science; it's just good Victorian engineering rediscovered. And it is widely applicable. A practice team at RMI has recently developed new-construction designs offering energy savings of 89 percent for a data center, about 75 percent for a chemical plant, 70 to 90 percent for a supermarket and about 50 percent for a luxury yacht, all with capital costs lower than those of conventional designs. The team has also proposed retrofits for existing oil refineries, mines and microchip factories that would reduce energy use by 40 to 60 percent, repaying their cost in just a few years.

# Vehicles of Opportunity

Transportation consumes 70 percent of U.S. oil and generates a third of the nation's carbon emissions. It is widely considered the most intractable part of the climate problem, especially as hundreds of millions of people in China and India buy automobiles. Yet transportation offers enormous efficiency opportunities. *Winning the Oil Endgame,* a 2004 analysis written by my team at RMI and co-sponsored by the Pentagon, found that artfully combining lightweight materials with innovations in propulsion and aerodynamics could cut oil use by cars, trucks and planes by two thirds without compromising comfort, safety, performance or affordability.

Despite 119 years of refinement, the modern car remains astonishingly inefficient. Only 13 percent of its fuel energy even reaches the wheels—the other 87 percent is either dissipated as heat and noise in the engine and drivetrain or lost to idling and accessories such as air conditioners. Of the energy delivered to the wheels, more than half heats the tires, road and air. Just 6 percent of the fuel energy actually accelerates the car (and all this energy converts to brake heating when you stop). And, because 95 percent of the accelerated mass is the car itself, less than 1 percent of the fuel ends up moving the driver.

Yet the solution is obvious from the physics: greatly reduce the car's weight, which causes three fourths of the energy losses at the wheels. And every unit of energy saved at the wheels by lowering weight (or cutting drag) will save an additional seven units of energy now lost en route to the wheels. Concerns about cost and safety have long discouraged attempts to make lighter cars, but modern light-but-strong materials—new metal alloys and advanced polymer composites—can slash a car's mass without sacrificing crashworthiness. For example, carbon-fiber composites can absorb six to 12 times as much crash energy per kilogram as steel does, more than offsetting the composite car's weight disadvantage if it hits a steel vehicle that is twice as heavy. With such novel materials, cars can be big, comfortable and protective without being heavy, inefficient and hostile, saving both oil *and* lives. As Henry Ford said, you don't need weight for strength; if you did, your bicycle helmet would be made of steel, not carbon fiber.

Advanced manufacturing techniques developed in the past two years could make carbon-composite car bodies competitive with steel ones. A lighter body would allow automakers to use smaller (and less expensive) engines. And because the assembly of carbon-composite cars does not require body or paint shops, the factories would be smaller and cost 40 percent less to build than conventional auto plants. These savings would offset the higher cost of the carbon-composite materials. In all, the introduction of ultralight bodies could nearly double the fuel efficiency of today's hybrid-electric vehicles—which are already twice as efficient as conventional cars—without raising their sticker prices. If composites prove unready, new ultralight steels offer a reliable backstop. The competitive marketplace will sort out the winning materials, but, either way, super efficient ultralight vehicles will start pulling away from the automotive pack within the next decade.

What is more, ultralight cars could greatly accelerate the transition to hydrogen fuel-cell cars that use no oil at all [see "On the road to Fuel-Cell Cars," by Steven Ashley; SCIENTIFIC AMERICAN, March]. A midsize SUV whose halved weight and drag cut its needed power to the wheels by two thirds would have a fuel economy equivalent to 114 miles per gallon and thus require only a 35-kilowatt fuel cell—one third the usual

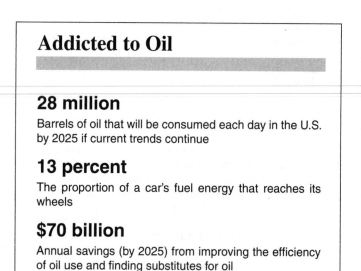

## Addicted to Oil

**28 million**
Barrels of oil that will be consumed each day in the U.S. by 2025 if current trends continue

**13 percent**
The proportion of a car's fuel energy that reaches its wheels

**$70 billion**
Annual savings (by 2025) from improving the efficiency of oil use and finding substitutes for oil

size and hence much easier to manufacture affordably. And because the vehicle would need to carry only one third as much hydrogen, it would not require any new storage technologies; compact, safe, off-the-shelf carbon-fiber tanks could hold enough hydrogen to propel the SUV for 530 kilometers. Thus, the first automaker to go ultralight will win the race to fuel cells, giving the whole industry a strong incentive to become as boldly innovative in materials and manufacturing as a few companies now are in propulsion.

RMI's analysis shows that full adoption of efficient vehicles, buildings and industries could shrink projected U.S. oil use in 2025—28 million barrels a day—by more than half, lowering consumption to pre-1970 levels. In a realistic scenario, only about half of these savings could actually be captured by 2025 because many older, less efficient cars and trucks would remain on the road (vehicle stocks turn over slowly). Before 2050, though, U.S. oil consumption could be phased out altogether by doubling the efficiency of oil use and substituting alternative fuel supplies. Businesses can profit greatly by making the transition, because saving each barrel of oil through efficiency improvements costs only $12, less than one fifth of what petroleum sells for today. And two kinds of alternative fuel supplies could compete robustly with oil even if it sold for less than half the current price. The first is ethanol made from woody, weedy plants such as switchgrass and poplar. Corn is currently the main U.S. source of ethanol, which is blended with gasoline, but the woody plants yield twice as much ethanol per ton as corn does and with lower capital investment and far less energy input.

The second alternative is replacing oil with lower-carbon natural gas, which would become cheaper and more abundant as efficiency gains reduce the demand for electricity at peak periods. At those times, gas-fired turbines generate power so wastefully that saving 1 percent of electricity would cut U.S. natural gas consumption by 2 percent and its price by 3 or 4 percent. Gas saved in this way and in other uses could then replace oil either directly or, even more profitably and efficiently, by converting it to hydrogen.

The benefits of phasing out oil would go far beyond the estimated $70 billion saved every year. The transition would lower U.S. carbon emissions by 26 percent and eliminate all the social and political costs of getting and burning petroleum—military conflict, price volatility, fiscal and diplomatic distortions, pollution and so on. If the country becomes oil-free, then petroleum will no longer be worth fighting over. The Pentagon would also reap immediate rewards from raising energy efficiency because it badly needs to reduce the costs and risks of supplying fuel to its troops. Just as the U.S. Department of Defense's research efforts transformed civilian industry by creating the Internet and the Global Positioning System, it should now spearhead the development of advanced ultralight materials.

The switch to an oil-free economy would happen even faster than RMI projected if policymakers stopped encouraging the perverse development patterns that make people drive so much. If federal, state and local governments did not mandate and subsidize suburban sprawl, more of us could live in neighborhoods where almost everything we want is within a five-minute walk. Besides saving fuel, this New Urbanist design builds stronger communities, earns more money for developers and is much less disruptive than other methods of limiting vehicle traffic (such as the draconian fuel and car taxes that Singapore uses to avoid Bangkok-like traffic jams).

# Renewable Energy

Efficiency improvements that can save most of our electricity also cost less than what the utilities now pay for coal, which generates half of U.S. power and 38 percent of its fossil-fuel carbon emissions. Furthermore, in recent years alternatives to coal-fired power plants—including renewable sources such as wind and solar power, as well as decentralized cogeneration plants that produce electricity and heat together in buildings and factories—have begun to hit their stride. Worldwide the collective generating capacity of these sources is already greater than that of nuclear power and growing six times as fast. This trend is all the more impressive because decentralized generators face many obstacles to fair competition and usually get much lower subsidies than centralized coal-fired or nuclear plants.

Wind power is perhaps the greatest success story. Mass production and improved engineering have made modern wind turbines big (generating two to five megawatts each), extremely reliable and environmentally quite benign. Denmark already gets a fifth of its electricity from wind, Germany a tenth. Germany and Spain are each adding more than 2,000 megawatts of wind power each year, and Europe aims to get 22 percent of its electricity and 12 percent of its total energy from renewables by 2010. In contrast, global nuclear generating capacity is expected to remain flat, then decline.

The most common criticism of wind power—that it produces electricity too intermittently—has not turned out to be a serious drawback. In parts of Europe that get all their power from wind on some days, utilities have overcome the problem by diversifying the locations of their wind turbines, incorporating wind forecasts into their generating plans and integrating wind power with hydroelectricity and other energy sources. Wind and solar power work particularly well together, partly because the conditions that are bad for wind (calm, sunny weather) are good for solar, and vice versa. In fact, when properly combined, wind and solar facilities are more reliable than conventional power stations—they come in smaller modules (wind turbines, solar cells) that are less likely to fail all at once, their costs do not swing wildly with the prices of fossil fuels, and terrorists are much more likely to attack a nuclear reactor or an oil terminal than a wind farm or a solar array.

Most important, renewable power now has advantageous economics. In 2003 U.S. wind energy sold for as little as 2.9 cents a kilowatt-hour. The federal government subsidizes wind power with a production tax credit, but even without that subsidy, the price—about 4.6 cents per kilowatt-hour—is still cheaper than subsidized power from new coal or nuclear plants. (Wind power's subsidy is a temporary one that Congress has repeatedly allowed to expire; in contrast, the subsidies for the fossil-fuel and nuclear industries are larger and permanent.) Wind power is also abundant: wind farms occupying just a few percent of the available land in the Dakotas could cost-effectively meet all of America's electricity needs. Although solar cells currently cost more per kilowatt-hour than wind turbines do, they can still be profitable if integrated into buildings, saving the cost of roofing materials. Atop big, flat-roofed commercial buildings, solar cells can compete without subsidies if combined with efficient use that allows the building's owner to resell the surplus power when it is most plentiful and valuable—on sunny afternoons. Solar is also usually the cheapest way to get electricity to the two billion people, mostly in the developing world, who have no access to power lines. But even in rich countries, a house as efficient as mine can get all its electricity from just a few square meters of solar cells, and installing the array costs less than connecting to nearby utility lines.

## Cheaper to Fix

Inexpensive efficiency improvements and competitive renewable sources can reverse the terrible arithmetic of climate change, which accelerates exponentially as we burn fossil fuels ever faster. Efficiency can outpace economic growth if we pay attention: between 1977 and 1985, for example, U.S. gross domestic product (GDP) grew 27 percent, whereas oil use fell 17 percent. (Over the same period, oil imports dropped 50 percent, and Persian Gulf imports plummeted 87 percent.) The growth of renewables has routinely outpaced GDP; worldwide, solar and wind power are doubling every two and three years, respectively. If both efficiency and renewables grow faster than the economy, then carbon emissions will fall and global warming will slow—buying more time to develop even better technologies for displacing the remaining fossil-fuel use, or to

master and deploy ways to capture combustion carbon before it enters the air [see "Can We Bury Global Warming?" by Robert H. Socolow; SCIENTIFIC AMERICAN, July].

In contrast, nuclear power is a slower and much more expensive solution. Delivering a kilowatt-hour from a new nuclear plant costs at least three times as much as saving one through efficiency measures. Thus, every dollar spent on efficiency would displace at least three times as much coal as spending on nuclear power, and the efficiency improvements could go into effect much more quickly because it takes so long to build reactors. Diverting public and private investment from market winners to losers does not just distort markets and misallocate financial capital—it worsens the climate problem by buying a less effective solution.

The good news about global warming is that it is cheaper to fix than to ignore. Because saving energy is profitable, efficient use is gaining traction in the marketplace. U.S. Environmental Protection Agency economist Skip Laitner calculates that from 1996 to mid-2005 prudent choices by businesses and consumers, combined with the shift to a more information and service-based economy, cut average U.S. energy use per dollar of GDP by 2.1 percent a year—nearly three times as fast as the rate for the preceding 10 years. This change met 78 percent of the rise in demand for energy services over the past decade (the remainder was met by increasing energy supply), and the U.S. achieved this progress without the help of any technological breakthroughs or new national policies. The climate problem was created by millions of bad decisions over decades, but climate stability can be restored by millions of sensible choices—buying a more efficient lamp or car, adding insulation or caulk to your home, repealing subsidies for waste and rewarding desired outcomes (for example, by paying architects and engineers for savings, not expenditures).

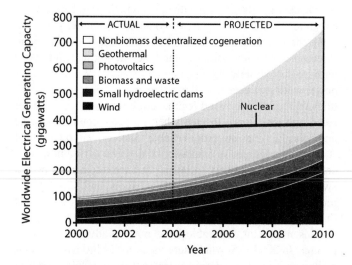

**Figure 1 Electricity Alternatives.** Decentralized sources of electricity—cogeneration (the combined production of electricity and heat, typically from natural gas) and renewables (such as solar and wind power)—surpassed nuclear power in global generating capacity in 2002. The annual output of these low- and no-carbon sources will exceed that of nuclear power this year.

Source: Rocky Mountain Institute

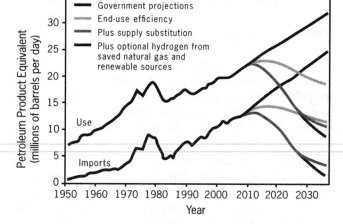

**Figure 2 An Oil-Free America.** U.S. oil consumption and imports can be profitably slashed by doubling the efficiency of vehicles, buildings and industries. The U.S. can achieve further reductions by replacing oil with competitive substitutes such as advanced biofuels and saved natural gas and with hydrogen fuel.

Source: RMI (graph); Projections to 2025 from U.S. Energy Information Administration; estimates after 2025 from RMI

The proper role of government is to steer, not row, but for years officials have been steering our energy ship in the wrong direction. The current U.S. energy policy harms the economy and the climate by rejecting free-market principles and playing favorites with technologies. The best course is to allow every method of producing or saving energy to compete fairly, at honest prices, regardless of which kind of investment it is, what technology it uses, how big it is or who owns it. For example, few jurisdictions currently let decentralized power sources such as rooftop solar arrays "plug and play" on the electric grid, as modern technical standards safely permit. Although 31 U.S. states allow net metering—the utility buys your power at the same price it charges you—most artificially restrict or distort this competition. But the biggest single obstacle to electric and gas efficiency is that most countries, and all U.S. states except California and Oregon, reward distribution utilities for selling more energy and penalize them for cutting their customers' bills. Luckily, this problem is easy to fix: state regulators should align incentives by decoupling profits from energy sales, then letting utilities keep some of the savings from trimming energy bills.

Superefficient vehicles have been slow to emerge from Detroit, where neither balance sheets nor leadership has supported visionary innovation. Also, the U.S. lightly taxes gasoline but heavily subsidizes its production, making it cheaper than bottled water. Increasing fuel taxes may not be the best solution, though; in Europe, stiff taxes—which raise many countries' gasoline prices to $4 or $5 a gallon—cut driving more than they make new cars efficient, because fuel costs are diluted by car owners' other expenses and are then steeply discounted (most car buyers count only the first few years' worth of fuel savings). Federal standards adopted in the 1970s helped to lift the fuel economy of new cars and light trucks from 16 miles per gallon in 1978 to 22 miles per gallon in 1987, but the average has slipped to 21 mpg since then. The government projects that the auto industry will spend the next 20 years getting its vehicles to be just 0.5 mile per gallon more efficient than they were in 1987. Furthermore, automakers loathe the standards as restrictions on choice and have become adept at gaming the system by selling more vehicles classified as light trucks, which are allowed to have lower fuel economy than cars. (The least efficient light trucks even get special subsidies.)

The most powerful policy response is "feebates"—charging fees on inefficient new cars and returning that revenue as rebates to buyers of efficient models. If done separately for each size class of vehicle, so there is no bias against bigger models, feebates would expand customer choice instead of restricting it. Feebates would also encourage innovation, save customers money and boost automakers' profits. Such policies, which can be implemented at the state level, could speed the adoption of advanced-technology cars, trucks and planes without mandates, taxes, subsidies or new national laws.

In Europe and Japan, the main obstacle to saving energy is the mistaken belief that their economies are already as efficient as they can get. These countries are up to twice as efficient as the U.S., but they still have a long way to go. The greatest opportunities, though, are in developing countries, which are on average three times less efficient than the U.S. Dreadfully wasteful motors, lighting ballasts and other devices are freely traded and widely bought in these nations. Their power sector currently devours one quarter of their development funds, diverting money from other vital projects. Industrial countries are partly responsible for this situation because many have exported inefficient vehicles and equipment to the developing world. Exporting inefficiency is both immoral and uneconomic; instead the richer nations should help developing countries build an energy-efficient infrastructure that would free up capital to address their other pressing needs. For example, manufacturing efficient lamps and windows takes 1,000 times less capital than building power plants and grids to do the same tasks, and the investment is recovered 10 times faster.

China and India have already discovered that their burgeoning economies cannot long compete if energy waste continues to squander their money, talent and public health. China is setting ambitious but achievable goals for shifting from coal-fired power to decentralized renewable energy and natural gas. (The Chinese have large supplies of gas and are expected to tap vast reserves in eastern Siberia.) Moreover, in 2004 China announced an energy strategy built around "leapfrog technologies" and rapid improvements in the efficiency of new buildings, factories and consumer products. China is also taking steps to control the explosive growth of its oil use; by 2008 it will be illegal to sell many inefficient U.S. cars there. If American automakers do not innovate quickly enough, in another decade you may well be driving a superefficient Chinese-made car. A million U.S. jobs hang in the balance.

Today's increasingly competitive global economy is stimulating an exciting new pattern of energy investment. If governments can remove institutional barriers and harness the dynamism of free enterprise, the markets will naturally favor choices that generate wealth, protect the climate and build real security by replacing fossil fuels with cheaper alternatives. This technology-driven convergence of business, environmental and security interests—creating abundance by design—holds out the promise of a fairer, richer and safer world.

**Amory B. Lovins** is co-founder and chief executive of Rocky Mountain Institute, an entrepreneurial nonprofit organization based in Snowmass, Colo., and chairman of Fiberforge, an engineering firm in Glenwood Springs, Colo. A physicist, Lovins has consulted for industry and governments worldwide for more than 30 years, chiefly on energy and its links with the environment, development and security. He has published 29 books and hundreds of papers on these subjects and has received a MacArthur Fellowship and many other awards for his work.

# Wind Power

## Obstacles and Opportunities

MARTIN J. PASQUALETTI

To know the wind is to respect nature. You ride with the wind when it fills your sails, but pay its power no heed and risk inconvenience, expense, even death. Drive through calm air in Los Angeles one moment only to encounter 30 minutes later Santa Anas whipping wildfires across mountaintops and pushing tractor-trailers into ditches. Lounge on the beach on Kauai one day, but find yourself huddling for protection the next day as a hurricane levels entire forests.[1] Sit on the porch during a quiet and muggy Oklahoma night when suddenly a mass of debris, once a house, swirls past, before dropping nearby as a pile of kindling and shattered dreams. More than any other force of nature, we have little defense against the wind. The wind keeps us on our toes.

If we cannot control the wind, perhaps we can put it to our use. It is a challenge with which we have had some success. Historically, we have used the wind to help us with work that would otherwise fall heavily upon our own backs. The wind helped humans explore the world; they had no other energy source. It continues to help us prepare foods and pump water. In some places, the wind is such a part of daily life that in its absence, silence blankets the landscape and puts us out of sorts. When it picks up again, flags flutter, well water rises, and grains are again ground to flour.

The wind machines humans developed were among the earliest icons of civilization. We can see them in early sketches from the Orient, scrolls from Persia, paintings from the Low Countries, photographs from the Dust Bowl, and even movies from Hollywood. Putting the wind to work was our first conscious use of solar power.

Perhaps the most widespread use of wind machines, at least in the United States, has been to pump water. Dotting the Great Plains by the hundreds of thousands, farm windmills—along with grain silos—were once as characteristic of the landscape as coal spoils were of Appalachia. Spinning whenever air moved, they brought to the surface the water that allowed ranches and settlements to flourish in an area otherwise too dry for either to exist for long. Most of these ingenious whirling devices eventually gave way to powerful compact motors that ran on fossil fuels, and as a result, wind energy landscapes largely disappeared. Before they all were removed, some folks preserved a few of them, drawn to their quaint beauty and the nostalgia they evoked as symbols of a Great Plains lifestyle. By then, however, most people considered the era of wind machines dead.

As it has happened, the epitaphs were premature. Today, wind machines are back. It has not been a quiet resurrection but rather one with substantial notoriety and publicity, plus a controversial mix of support and resistance. The new devices look and act little like their ancestors: Instead of the creaking, wooden machines of the past, those of the new species are made of metal and fiberglass—and are bigger, quieter, sleeker, and more powerful than ever. Instead of pumping water, the moving blades spin generators housed with an assemblage of gears in the nacelle, which is located behind the hub where all the blades meet. Instead of a stream of water, modern wind machines are pumping a stream of electrons, a product proving to be a valuable asset to farmers who are trying to address present day economic realities of living off the land.

The new appearance and mechanics of wind machines reflects their different role. Instead of producing mechanical power for the purposes of pumping and grinding, the new machines convert mechanical energy into electricity. Instead of being erected here and there in splendidly independent isolation, many are being clustered in symmetrically interdependent neighborhoods, designed to work together as parts of a larger organism. Nor are they just generating electricity: Unexpectedly, modern wind machines are prompting us to consider how best to weigh the energy we need against the environmental quality we want. All the while they are continuing their transformation from public indifference to public curiosity, from an overlooked energy supplier to alternative energy's "holy grail," one possible way to get most of what we want and little of which we do not.

## An Old Resource with a New Mission

Compared to the variety of uses that stretch back millennia, converting wind energy to electricity is a recent application. Although a few people were trying to accomplish this at the same time Thomas Edison opened his coal-fired Pearl Street generating plant in the latter years of the nineteenth century, it would be another 80 years before such proof-of-concept machines would

**Table 1**  Historical Wind Turbines

| Turbine, Country | Date in Service | Diameter (meters) | Swept Area (meters) | Power (kilowatts) | Specific Power (kilowatts per square meter) | Number of Blades | Tower Height (meters) |
|---|---|---|---|---|---|---|---|
| Poul la Cour, Denmark | 1891 | 23 | 408 | 18 | 0.04 | 4 | — |
| Smith-Putnam, United States | 1941 | 53 | 2,231 | 1,250 | 0.56 | 2 | 34 |
| F. L. Smidth, Denmark | 1941 | 17 | 237 | 50 | 0.21 | 3 | 24 |
| F. L. Smidth, Denmark | 1942 | 24 | 456 | 70 | 0.15 | 3 | 24 |
| Gedser, Denmark | 1957 | 24 | 452 | 200 | 0.44 | 3 | 25 |
| Hütter, Germany | 1958 | 34 | 908 | 100 | 0.11 | 2 | 22 |

SOURCE: P. Gipe, *Wind Energy Comes of Age* (New York: John Wiley & Sons, 1995), 78.

evolve into the commercial generators that started sprouting in the California landscape in 1981 (see Table 1). Indeed, the beginning of the modern era of wind power bore few similarities to earlier water pumping. The vision of modern wind power was much grander in scale, one that has evolved to row upon row of machines spreading over hundreds of acres, contributing enough electricity to power an entire city but—and this is the big difference—without undesirable side effects that accompanied the use of conventional resources.

Obviously, the machinery of the late twentieth century differs both in form and function from the equipment that nineteenth century ranchers and farmers developed to help them wrest a living from the dry lands that predominate west of the 100th meridian. For them, it was enough that machines were turning when the air was on the move. In the new era, such simple fulfillment is not enough; wind power today is viewed less from the living room and more from the boardroom. Wind power is big business, and the managers of that business must be sophisticated not just in the ways of making money, but in several disciplines that lead to success. Even something as seemingly innocent as turbine placement can no longer be considered just from the perspectives of convenience, necessity, or whim. Instead, the new wind barons must understand meteorology, metallurgy, physics, aerodynamics, capacity factors, land ownership, planning, zoning, and the influence of public perception.

Wind power's popularity is widespread and growing, a result of its increasing profitability and the perceived environmental benefits it engenders. It also results from the simple fact that, unlike fossil and nuclear fuels, wind is a widely available, familiar element of the environment. The first step is seemingly the easiest: finding it.

## Wind and the Family Farm

The initial step in developing any resource, be it gold or wind, is locating it. While it is often an uncomplicated step, not every

place is attractive. Just like gold, wind is not evenly distributed in the richness of its product. Subtropical deserts, such as the Sonoran Desert surrounding Phoenix, are created by persistent high pressure and are often unsuitable for wind development. For obvious reasons, forested areas are unattractive for wind turbines, as are equatorial areas with their characteristically light and variable winds. The rest of the world is more promising, although detailed data collection must precede full-fledged capital investment.

Finding windy places is a relatively easy step: Unlike fossil or nuclear fuels, it does not require drilling rigs, seismic gear, or Geiger counters. Wind power, in most places, is simply "out there." This means that the most obvious early task of wind prospectors is to determine where winds are strong enough. Once identified, such areas reveal several common characteristics, including exposed terrain, colliding air masses, and, in particular circumstances, topographic funneling, as through mountain passes. In the United States, several areas meet such criteria, including sites in California, southeastern Washington, central Wyoming, east-central New Mexico, and most notably the Great Plains (see Figure 1). In Europe, strong winds are significant along the west coast of Ireland, Great Britain, the eastern North Sea, the southern Baltic Sea, the Pyrénées Mountains, and the Rhone Valley (see Figure 2). Many of these areas are "nuggets" that we are plucking first, and they have been stimulating further prospecting and the development of grand plans for the future. By the end of 2003, the total installed capacity in the European Union was 28,440 megawatts (MW).[2]

Like the gold rush of the 1850s, the modern wind rush started in California. California still leads the way, with 2,042 MW installed by January 2004, principally in four locations: Altamont Pass, on the edge of the Central Valley east of San Francisco; Tehachapi Pass at the southern end of the Sierra Nevada; San Gorgonio Pass, near Palm Springs; and in the rolling hills between San Francisco and Sacramento. Although first, the primacy of California is certainly temporary: With a

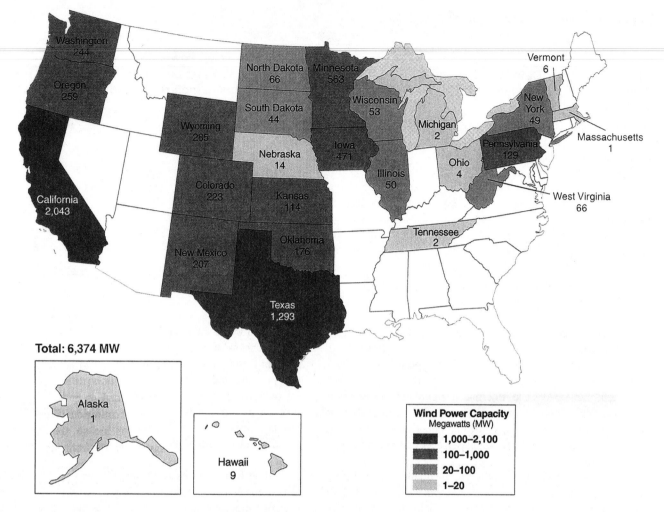

**Figure 1**    2003 Year-End Installed Wind Power Capacity (in megawatts).

Source: American Wind Energy Association, *Wind Energy Projects Throughout the United States of America,* http://www.awea.org/projects (accessed 30 June 2004).

potential for 6,770 MW, it ranks only seventeenth among the 50 states in potential (see Table 2).[3]

Among other states attracting interest, Texas has been the pacesetter with more than $1 billion in new wind investment and 1,293 MW installed, mostly in the western part of the state near such communities as Big Springs and McCamey. Coincidentally, many of these developments are positioned among the now derelict oil equipment that helped bring great wealth to this part of the state and has underpinned many of its towns and cities. At the end of January 2004, California, Texas, and 24 additional states held within their borders an installed capacity of 6,374 MW.[4]

Another 2,000 MW has been proposed for development in the near future, with some of the largest projects planned for the states of Washington and Massachusetts. However, the greatest potential remains where wind machines once so dominated the landscape—midway between these extremes in the Great Plains.

The wind of the Great Plains is as obvious as is its treeless expanse. When Francisco Vásquez de Coronado and his men crossed this region searching for the Seven Cities of Cíbola

in 1540, they found no gold but two other resources instead. The most obvious and most useful to Coronado were the great herds of bison—totaling perhaps 50 million head—that were scattered across a million square miles of grassland. Not only did they provide food, but their droppings helped guide the expedition across otherwise indistinct landscapes. The other resource was the wind, but three centuries would pass before it was appreciated.

By the late 1800s, the general pattern had reversed; bison were being decimated for sport, and wind power was lifting water for irrigation. Today, only this second resource remains, yet it is being used for a different mission: to generate electricity and make money.[5] It is a realistic ambition: The winds of the Great Plains are so abundant that the energy potential from just three states (North Dakota, Texas, and Kansas), were they fully developed, would match the electrical needs of the entire country. These and several other Great Plains states hold the largest expanse of class 4 (400–500 watts per square meter) lands in the country (see the box).[6]

Although weather on the Great Plains is often viewed as being inhospitable—farming families there endure swirling

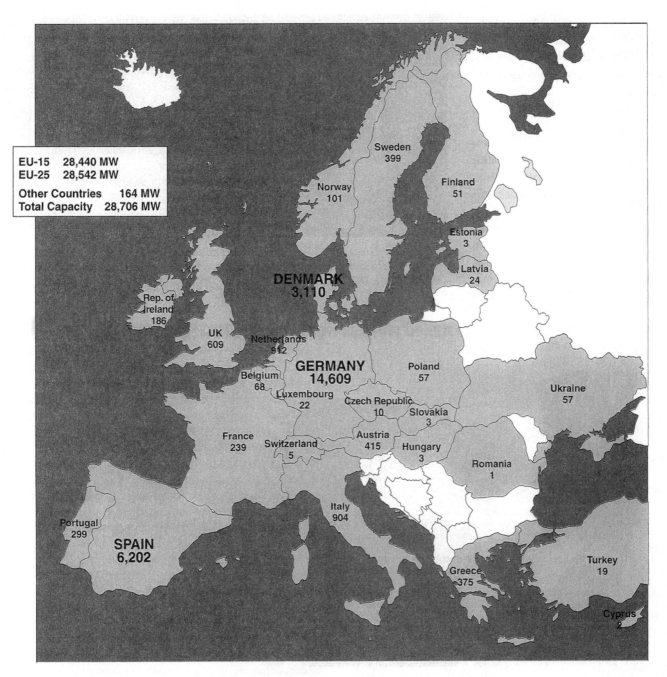

EU-15    28,440 MW
EU-25    28,542 MW

Other Countries    164 MW
Total Capacity    28,706 MW

Sweden
399

Norway
101

Finland
51

Estonia
3

Latvia
24

DENMARK
3,110

Rep. of
Ireland
186

UK
609

Netherlands
912

GERMANY
14,609

Poland
57

Ukraine
57

Belgium
68

Luxembourg
22

Czech Republic
10

Slovakia
3

France
239

Switzerland
5

Austria
415

Hungary
3

Romania
1

Italy
904

Portugal
299

SPAIN
6,202

Greece
375

Turkey
19

Cyprus
2

**Figure 2**    2003 Year-End European Installed Wind Power Capacity (in megawatts).

Source: Adapted from a map compiled by the European Wind Energy Association, http://www.ewea.org, 3 February 2004.

snow in winter and blowing dust in summer—attitudes toward the frequent tempests are lately bending in a new direction. Always alert for new sources of income to ease their financial volatility, locals are turning to wind developers with equanimity and even enthusiasm. They are finding that the same winds that strip soil from the fields and bury houses in snow can fuel rural economic development.[7]

Construction of a typical 100 MW wind farm produces more than 50,000 days (approximately 419,020 manhours) of employment. In Prowers County, Colorado, the recently completed wind development is each year providing $764,000 in new revenues,

$917,000 in school general funds, $203,000 in school bond funds, $189,000 to the Prowers Medical Center, and $189,000 in additional revenue to the county tax base.[8] Meanwhile, a 250 MW project in Iowa is providing $2 million in property taxes and $5.5 million in operation and maintenance income. The leases on offer to farmers in this area commonly provide yearly royalties of more than $2,000 per turbine: For a single Iowa project, local farmers are receiving $640,000 annually. Other projects return $4,000–$5,000 per turbine. In some cases, a one-megawatt turbine could generate revenues for the owner of $150,000 per year once the debt for purchase is repaid.[9] Though appreciable at the

**Table 2**  Top 20 States for Wind Energy Potential (measured by annual energy potential in billions of kilowatt hours)

| | | |
|---|---|---|
| 1 | North Dakota | 1,200 |
| 2 | Texas | 1,190 |
| 3 | Kansas | 1,070 |
| 4 | South Dakota | 1,030 |
| 5 | Montana | 1,020 |
| 6 | Nebraska | 868 |
| 7 | Wyoming | 747 |
| 8 | Oklahoma | 725 |
| 9 | Minnesota | 657 |
| 10 | Iowa | 551 |
| 11 | Colorado | 481 |
| 12 | New Mexico | 435 |
| 13 | Idaho | 73 |
| 14 | Michigan | 65 |
| 15 | New York | 62 |
| 16 | Illinois | 61 |
| 17 | California | 59 |
| 18 | Wisconsin | 58 |
| 19 | Maine | 56 |
| 20 | Missouri | 52 |

NOTE: As of July 2004, reevaluation of the wind potential has been done for 28 states by the U.S. Department of Energy's Windpowering America program; reevaluation of additional states is still in process. Once complete, the numbers for each state might change, as might the relative rankings. In many cases, the potential will increase. The Great Plains will continue to dominate the rest of the country in terms of potential.

SOURCE: Pacific Northwest Laboratory, *An Assessment of the Available Wind Land Area and Wind Energy Potential in the Contiguous United States* (Richland, WA: Pacific Northwest National Laboratory, 1991).

individual level, such royalties are only a small part of the variable costs for wind developers (see Figure 3). To them, such largesse seems good business. To farmers, such revenues can mean the difference between bankruptcy and prosperity.

**In spite of a weighty list of environmental attributes, wind power carries some unexpectedly heavy baggage.**

In spite of promising trends, it would be an overstatement to claim that wind power offers an energy panacea or a reversal of the national trend toward increasingly concentrated generation of electricity: It is, at least so far, a relatively small enterprise. Taken together, all the wind developments in the country

contribute less than 1 percent of our current needs.[10] However, there is a real attraction to wind power's promise for the future: Its estimated generating potential in the United States alone is 10,777 billion kilowatts per hour (kWh) annually, or three times the electricity generated in the entire country today.[11] In recognition of such potential for pollution-free electricity, the U.S. government is sponsoring a program called Wind Powering America (WPA) to tap more deeply this vast natural resource. WPA's new goal is to increase to 30 the number of states with more than 100 megawatts of wind-generating capacity by 2010.[12] The program also aims to increase rural economic development and, to some degree, local energy independence.

## The Environmental Irony

Wind power attracts many adherents from the environmental community. These organizations focus on its solar roots, emphasizing that it requires no mining, drilling, or pumping, no pipelines, port facilities, or supply trains. It produces no air pollution or radioactive waste, and it neither dirties water nor requires water for cooling. Wind power is relatively benign, simple, modular, affordable, and domestic. It is, in short, an environmental golden goose.

However, in spite of such a weighty list of attributes, wind power carries some unexpectedly heavy baggage. In England, anti-wind epithets have been particularly colorful: Developers have suffered their machines being called everything—including "lavatory brushes in the air" for their busy top ends. This leads us to an irony of wind power: While we usually consider wind power environmentally friendly, most of the objections to its expansion have had environmental origins.

We can follow the thread of such reactions most clearly to Palm Springs, California, in the mid-1980s. Soon after installing thousands of turbines in windy San Gorgonio Pass just north of the city limits, developers were battered with complaints that the machines interfered with television reception, produced annoying and inconsistent noise, posed risks to wildlife and aircraft, and represented incompatible land-use practices.[13] The most troubling, bitter, and outraged complaint, however, was that the wind machines destroyed the aesthetic appeal of the landscape, thereby threatening the very attribute that most attracts tourists to the area's fancy resorts.[14]

Developers and bureaucrats of the day, all of whom had expected a warmer welcome, were startled by such reactions. It was apparent to everyone with hopes for contributions from wind power that any future success would have to rest on greater environmental compatibility and a more complete respect for public attitudes and opinions. The initial experience provoked industry musing as to what actions might better attract support. Soon, manufacturers began making improvements to design and engineering. Locally, the concerns over wind impacts led to stricter planning rules and more uniform standards. These adjustments softened the problems, but they could not eliminate them. Turbines remained unavoidably visible and the center of a classic example of incompatible land use. The very characteristic that had long kept residential development in the

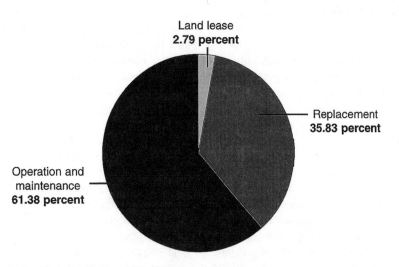

**Figure 3**  Variable Costs of Wind Energy Projects.

Source: E. DeMeo and B. Parsons, "Some Common Misconceptions about Wind Power," presented at the All States Wind Summit, Austin, TX, 22 May 2003. See U.S. Department of Energy, *State Wind Energy Handbook,* http://www.eere.energy.gov/windpoweringamerica/pdfs/wpa/34600_wind_handbook.pdf (accessed 30 June 2004), 90.

## Wind Classes

Developers need to know the average wind speed at a particular site to design and build the most appropriate turbines. It would be no more prudent to size the turbines for the slowest speed than it would be to size them from the fastest, but infrequently occurring speed. One can get a good idea of this relationship by using the Weibull distribution, a plot of frequency against speed. This distribution helps identify various classes of wind (see the table). The higher the number, the stronger the average speed. A good wind speed is 7 meters per second (mps); 20 mps may be excessive and cause damage to equipment. Turbine manufacturers have a "rated wind speed" for all models and sizes of turbines they sell. Typical rated wind speed requirements are in the range of 8–13 mps, but many machines will produce some power with much slower speeds.[1] Currently, developers are concentrating on class 4 and above as the most promising areas.

1. For an excellent and detailed description of these and other principles, see P. Gipe, *Wind Energy Comes of Age* (New York: John Wiley & Sons, 1995).

## Wind Power Classification

| Wind power class | Resource potential | Wind power density at 50 meters (in watts per square meter) | Wind speed at 50 meters (in meters per second) | Wind speed at 50 meters (in miles per hour) |
|---|---|---|---|---|
| 2 | Marginal | 200–300 | 5.6–6.4 | 12.5–14.3 |
| 3 | Fair | 300–400 | 6.4–7.0 | 14.3–15.7 |
| 4 | Good | 400–500 | 7.0–7.5 | 15.7–16.8 |
| 5 | Excellent | 500–600 | 7.5–8.0 | 16.8–17.9 |
| 6 | Outstanding | 600–800 | 8.0–8.8 | 17.9–19.7 |
| 7 | Superb | 800–1600 | 8.8–11.1 | 19.7–24.8 |

San Gorgonio Pass minimal—the strong wind—was the same characteristic that prompted developers to fill it with machines. There was very little compromise potential.

So strong was the backlash against wind development that the City of Palm Springs sued the U.S. Bureau of Land Management and the County of Riverside, claiming that developers had not followed proper environmental procedures. Although the suit was eventually abandoned, it was not before the local jurisdictions, including Palm Springs and Riverside County, enacted a long list of required adjustments, stipulating (for example) height limitations, the use of nonglinting paint, reporting mechanisms for endangered species, and the establishment of decommissioning bonds.

A few years later, in an unpredictable turnaround, attitudes changed. This happened once Palm Springs, led by its mayor Sonny Bono, began eyeing wind machines as generators of tax revenue, as well as electricity. With a financial windfall in mind, the city annexed several square miles of land in the middle of the windiest part of the pass, thereby enlarging the city limits and sweeping additional tax revenues into the municipal treasury. Also, counter to the early intuition and opposition of city officials, the wind turbines have become something of a tourist attraction. Organized tours are available, images of wind farms adorn many local postcards, and brochures advertise the Palm Springs wind industry. Even Hollywood producers have incorporated the striking wind energy landscapes in movies and advertisements. These changes reflect the progression in public attitudes toward greater acceptance, although a closer look still finds disgruntled residents who have the original objections. As they point out, we can paint them, size them, sculpt them, and engineer them to a fine edge, but we cannot make them disappear.

# The Aesthetic Core

Reactions to wind power tend to be both quick and subjective. While one group fights intrusion, another is organizing visits for enthusiastic tourists. Where one person loathes turbines, that person's neighbors find them fascinating. Whichever reaction prevails in any given location, wind turbines cannot be ignored, for they do not fit naturally upon the land. They are, to apply Massachusetts Institute of Technology historian Leo Marx's famous phrase, "machines in the garden."[15]

**The spread of wind power encounters the most strident opposition where it interferes with local land use.**

Wind power's development contained a surprise: Among its corps of supporters, no one anticipated the need to defend wind projects. Why did no one foresee objections? We can only speculate, but it seems that the advantages were considered by adherents to be so obvious, especially when compared to nuclear power, that developing a defensive strategy for this new technology seemed superfluous. Supporters failed to recognize how opposite is the signature between the two: Nuclear power is compact and quiet, whereas wind power is expansive and obvious. Reflecting on this difference, resistance to nuclear power accumulated slowly only after a long educational process that culminated with accidents at Three Mile Island and Chernobyl, while resistance to wind power was immediate and instinctive.

Although this difference suggests the heft of visual aesthetics in shaping public opinion, it masks two other ingredients of

equal importance. One is the immobility of the resource: Wind moves, windy sites do not. In this way, wind differs from coal and most other fuels, because its nature does not allow it to be extracted and transported for use at a distant site. For wind power to be successful, turbines must be installed where sufficient wind resources exist or not at all. Thus, just like two other resources—geothermal energy and hydropower—the site-specific nature of wind developments intrinsically invites conflicts with existing or planned land uses. This is true even in deserts, the common dumping ground of society.[16]

The second ingredient in helping form public attitudes toward wind power is the landscape itself. Simply put, some landscapes are more valued than others. Place turbines in sensitive areas—perhaps along the coast or in a national park—and prepare for an uproar. Place them out of view or in low-value areas—sanitary landfills, for example—and opposition diminishes.

These characteristics produce wind power's most intractable challenges. First, owing to resource immobility and the subjectivity of its aesthetic impact, total mitigation is impossible. Second, because environmental competition changes from place to place and from one time to another, generic solutions are few and elusive. Third, because nothing can make turbines invisible, little we do will make them more acceptable to those perceiving land-use interference. There is no escaping the essence of wind turbines: They will always be spinning, pulsing, exoskeletal contraptions that naturally attract the eye.

The foregoing notwithstanding, the future of wind power remains both promising and substantial, if we can identify and follow the appropriate path. Two general strategies suggest them-

## Larger and Larger Turbines

Wind turbines are getting larger and larger. What is driving this trend? To answer this question, we need to know that movement obtains its impetus from the sun; as solar energy strikes the surface of the Earth, it creates differences in pressure. The wind, in turn, moves "downhill" along the pressure gradients that are formed, from higher to lower pressure. Speed increases as the horizontal distance between different pressures is shortened. Wind also typically accelerates when it is constricted, as when it moves through a mountain pass. The faster the wind moves, the more energy it carries, but it is not in a linear function. Rather, it increases with the cube of the wind speed, usually written $x^3$. This means that a wind speed of 8 meters per second (mps), yields 314 ($8^3$) watts for every square meter exposed to the wind, while at 16 mps, we get 2,509 ($16^3$) watts per square meter, again eight times as much. This relationship puts a premium on sites having the strongest winds. This relationship also explains why the area "swept" by the turbine blades is so important and why the wind industry has been striving fervently to increase the scale of the equipment it installs. A one-megawatt turbine at a typical European site would produce enough electricity annually to meet the needs of 700 typical European households.

selves: work to bend public opinion in favor of wind power, or install the turbines out of view. The first approach is under way but slow. The second approach can be quicker and would seem to hold promise, but it is being met with mixed results, especially when projects are proposed for offshore locations—the newest tactic to avoid public criticism and maximize profits.

## Moving Offshore

The spread of wind power encounters the most strident opposition where it interferes with local land use. Tourism, recreation, entertainment, resorts, and a host of other outdoor activities create most of the challenge because their function is to help people escape reality. For relief from this dilemma, developers are looking for sites offshore, and they have been finding them, especially in the shallow, wind-swept waters of the Irish Sea and North Sea. Denmark, which is characteristically leading the way with this strategy, has already installed and activated several fields of this type. Other projects are in place or planned off Ireland, the United Kingdom, the Netherlands, and several other countries.

In addition to prohibiting wind projects in populated areas, positioning them offshore offers several operational advantages. For example, winds passing over water tend to be stronger than those passing over land; offshore placement removes no land from existing or planned uses; any noise produced at sea is muffled by

that of the surf; road use is largely a moot issue; and negotiation with multiple landowners is unnecessary. Nonetheless, offshore placement requires some tradeoffs. For example, offshore equipment is more costly to construct and maintain, and it inherently increases the potential for conflict with any recreational use of the seashore. It also tends to encourage the installation of larger turbines (see the box).

Strictly from a public perspective, offshore placement has the presumed advantage of mitigating complaints about aesthetic intrusion. It has not, however, turned out to be the expected universal remedy. Indeed, moving offshore is increasing rather than diminishing the enmity of wind power in some quarters, especially in the northeastern United States.

Tempted by the strong offshore winds of coastal Massachusetts and responding to the hostility to wind developments witnessed in California, several entrepreneurs advocated placing wind turbines on the shallow offshore banks. The proposal itself may go down as the most foolhardy miscalculation in renewable energy history. The problem, as usual, is incompatible use of space. Called Cape Wind, the project is proposed for Nantucket Sound, a site between the popular vacation spots on Cape Cod and the exclusive holiday retreats of Martha's Vineyard and Nantucket Island. Like development near Palm Springs, Cape Wind is colliding with the wishes of a prosperous and politically astute residential corps bent on protecting existing scenic and recreational qualities that it has come to cherish.

---

## The Cape Wind Project: A Wind Power Lightning Rod

Cape Wind is a proposed $500–$750 million wind development project for Horseshoe Shoal in Nantucket Sound. If approved, the turbines will come within 5 miles of land, spread over an area of 24 square miles, and consist of 130, 417-foot wind turbines connected to a central service platform that includes a helicopter pad and crew quarters. Each turbine blade will be 164 feet long with a total diameter of 328 feet. Each turbine will have a base diameter of 16 feet and an above-water profile taller than the Statue of Liberty.

The proposal has become a lightning rod for the wind industry. The Alliance to Protect Nantucket Sound (the Alliance),[1] which strongly opposes the project, has been accumulating arguments against it. They point out that

- each turbine will have about 150 gallons of hydraulic oil, and the service platform will have at least 30,000 gallons of dielectric oil, and diesel fuel;
- the project will be within the flight path of thousands of small planes; and
- the turbines will pose a navigation hazard to the commercial ferry lines in the area.

These and other objections, however, take a secondary position to the Alliance's primary objection, that of aesthetic intrusion. The Alliance claims that the turbines will be visible for farther than 20 miles, that they will be lighted at night, and that they will flicker with changing sun angle. The Alliance has developed many computer visualizations of how

they would appear. (Some proponents might point out that the Visualizations illustrate how inconsequential the turbines would appear from the beach.)

The pro-development side has not been idle. Cape Wind has its own Web site,[2] which identifies the many benefits of the project, including that it offsets the need for 113 million gallons of oil yearly and creates approximately 600–1,000 new jobs. They also refer to many studies attesting to the benefits of such projects as Cape Wind. One of the most recent references the positive impacts on sea creatures around the wind turbines off the southern Swedish coast.[3] Other Web sites provide many testimonials to the good sense of offshore wind power.[4] The controversy over Cape Wind's offshore proposal is just the beginning of many other anticipated projects along the East Coast, such as off Long Island.[5]

1. http://www.saveoursound.org.
2. http://www.capewind.org.
3. L. Nordstrom, "Windmills off Swedish Coast are Providing Unexpected Benefit for Marine Life, Scientists Say," *Environmental News Network*, 11 February 2004, http://www.enn.com/news/200402-11/s_13011.asp.
4. windfarm@cleanpowernow.org; http://www.safewind. info.
5. See the Safe Wind Coalition's Web site, http://www. safewind.info/wind_farms_where.htm; and the Long Island Offshore Wind Initiative's site at http://www. lioffshorewindenergy.org/.

# Riding a Roller Coaster

Wind energy has experienced a wild ride over the past 20 years, one where initial enthusiasm soured quickly with the perception that generous incentives and lax oversight were allowing virtually any wind farm development, no matter how carelessly designed or operated, to be financially tenable. An undertow that quickly started to pull against the early currents of promise was a perception that wind developments were being installed without sufficient public notification, due consideration, or individual benefit. By the late 1980s, it was clear that improved turbines and business situations were going to be necessary if wind power was to develop a significant position in the alternative energy mix in the United States or abroad.

Some of the earliest advances first came into view in Europe, where even casual inspection spotted substantial differences from early installations in California. For example, instead of large clusters of turbines spread haphazardly upon the land, deployment of European turbines was more sensitively organized into smaller groupings that were carefully integrated into the landscape. This was partly a result of a higher sensitivity to existing conditions and partly a measured response to the experiences in California that had dulled the promise of wind power and threatened its future.[17]

Reflecting improvements and continued support, wind power's trajectory is once again upward. Today it is the fastest-growing renewable energy resource in the world.[18] Wind power is especially popular outside the United States: In countries like Germany, it is welcomed, encouraged, and promoted as one way to reduce greenhouse gas emissions. In Denmark, the value of wind power to the economy now exceeds that of its economic mainstay, ham. Spain's development of wind power is currently growing at a faster pace than it is in any other country.

The roller coaster ride is not over, however: Even amid news of improvements and quickened growth, wind power continues to have its critics. The more determined of these opponents work to keep wind machines from their property and out of their view. They hire public relations experts, make abundant use of the Internet to promote their view and attract adherents, and invite the support of prominent citizens to their cause. The group Save Our Sound is perhaps the most visible example of such techniques.[19] Such determined resistance was never envisioned when the champions of wind power came calling more than two decades ago. Today, despite progress in assuaging public apprehensions, a measure of uncertainty still hangs over wind's future.

# From Incentives to Independence

What is to be made of the many incentives that wind power enjoys? Tax incentives, utility portfolio standards, feed-in laws,[20] and many other aids currently help make it an economically viable alternative energy provider. Some would say that the requirement for these incentives demonstrates that wind power is not a legitimate competitor for our energy dollars. Others might argue that the mere existence of these aids suggests how narrow the economic gap is between a present need for subsidies and independent viability. While its increasingly competitive status results partly from a rising cost of conventional energy, it also reflects the declining costs of all alternatives, including wind. The message is this: Even without incentives, wind power has been moving toward economic independence, and it seems destined to reach parity with conventional sources soon.

It is often the smallest margin of help that wins the day for an emerging technology. One way to demonstrate this is to examine the impact of higher conventional energy cost. In one study of 12 Midwestern states, where electricity sold at 4.5 cents per kWh, the regional potential for cost-effective wind power was about 7 percent of current total generation in the United States.[21] If the market would support a price of 5.0 cents per kWh, however, the potential would grow to 177 percent of current generation. If one additional penny is added to the price, the potential blossoms to 14 times current levels.[22]

Until conventional energy makes this inevitable jump, wind operators need another way to bridge the gap. This brings us to the U.S. Production Tax Credit (PTC). This credit originally provided for an inflation-adjusted 1.5 cents per kilowatt-hour for electricity generated with wind turbines. With PTC now at about 1.9 cents, wind projects are economically favored. In its absence, however, development of new projects virtually ceases. This occurred, for instance, when the credit expired at the end of 2001, before it was reinstated some months later. PTC lapsed again on 31 December 2003, and discussions in Congress are once again under way as to whether to extend it for a five-year period.[23]

Without such a credit, the U.S. wind industry will suffer. According to Craig Cox, executive director of the Interwest Energy Alliance, "The lapse of the PTC has created uncertainty in the wind energy marketplace, and interest in new developments has slowed."[24] Renewal of the credit is part of the $31 billion energy bill that stalled in Congress at the end of 2003, again putting the wind energy industry back on its "roller coaster" in the United States. The world's major wind turbine manufacturer, Vestas Group, delayed its decision to build a wind turbine plant in Oregon because of the uncertainty of the credit. Ultimately, such uncertainty spreads to all phases of wind energy development, not just deployment of turbines. "We've been looking to establish a manufacturing facility in the U.S. but have not done that only because of the boom and bust cycle of the wind energy industry in the U.S.," Scott Kringen of Vestas told Reuters.[25] Other spokespersons have made similar observations: "Today, a wide range of U.S. companies are interested in the wind industry, but many are staying on the sidelines because of the on-again, off-again nature of the market produced by frequent expirations of the PTC," said Randall Swisher, executive director of the Washington, DC-based American Wind Energy Association.[26] Most countries offer more stable, longer-term policy support for wind than does the United States, and they use mechanisms that are inherently more pluralistic and egalitarian. This helps explain why wind power is on such a fast track in countries such as Germany, Denmark, and Spain.

# Wind Power and Bird Mortality

There is a persistent public impression that birds and windmills don't mix very well. Particularly for the smaller turbines, the spinning blades are hard to see during the day and are invisible at night. Many of those who campaign against wind power expansion cite this concern as part of their argument.

Concerns about turbine-related bird mortality stem largely from the experience at Altamont Pass, California, where approximately 7,000 wind turbines are located on rolling grassland 50 miles east of San Francisco Bay.[1] Between 1989 and 1991, 182 dead birds were found in study plots associated with wind turbines, including approximately 39 golden eagles killed per year by the turbines.[2] Golden eagles, red-tailed hawks, and American kestrels had higher mortality than more common American ravens and turkey vultures.[3] Deaths of eagles and potential danger to endangered California condors are the biggest issues at Altamont Pass. Bird mortality at comparably sized wind facilities has been reported as being similar or lower than those at Altamont Pass.[4]

While such fatalities are regrettable, there is serious question as to whether they are sufficient to slow or halt the use of wind power. One environmental group, The Defenders of Wildlife, recommends that bird mortality should be "kept in perspective."[5] For comparison glass windows kill 100–900 million birds per year; house cats, 100 million; cars and trucks, 50–100 million; transmission line collisions, up to 175 million; agriculture, 67 million; and hunting, more than 100 million.[6] Clean Power Now, an advocacy group encouraging wind development in Nantucket Sound, answers the question "Do wind turbines kill birds?" by stating "Very few and not always."[7] Altamont Pass, where much of the concern for avian safety originated, appears to be more of the exception than the rule. Data show the actual numbers killed in the pass do not exceed one bird per turbine per year, and for raptors, reported kill rates are 0.05 per turbine per year.[8] Nevertheless, the wind power industry has made several adjustments. For example, perch guards are being installed and a program to replace the old machines with modern turbines on high monopoles is ongoing (One modern turbine replaces seven older machines).[9] More study on this matter would be welcome.

1. W. G. Hunt, R. E. Jackman, T. L. Hunt, D. E. Driscoll, and L. Culp, *A Population Study of Golden Eagles in the Altamont Pass Wind Resource Area: Population Trend Analysis 1997,* report prepared for the National Renewable Energy Laboratory (NREL), Subcontract XAT-6-16459-01 (Santa Cruz, CA: Predatory Bird Research Group, University of California, 1998).
2. S. Orloff and A. Flannery, *Wind Turbine Effects on Avian Activity, Habitat Use, and Mortality in Altamont Pass and Solano County WRAs,* report prepared by BioSystems Analysis, Inc. for the California Energy Commission, 1992.
3. C. G. Thelander and L. Rugge, *Avian Risk Behavior and Fatalities at the Altamont Pass Wind Resource Area,* report prepared for the National Renewable Energy Laboratory: SR-500-27545, (Santa Cruz, CA: Predatory Bird Research Group, University of California, 2000).
4. M. D. McCrary et al., *Summary of Southern California Edison's Bird Monitoring Studies in the San Gorgonio Pass,* unpublished data; and R. L. Anderson, J. Tom, N. Neumann, J. A. Cleckler, and J. A. Brownell, *Avian Monitoring and Risk Assessment at Tehachapi Pass Wind Resource Area, California* (Sacramento, CA: California Energy Commission, 1996).
5. Defenders of Wildlife, *Renewable Energy: Wind Energy Resources, Principles and Recommendations,* http://www. defenders.org/habitat/renew/wind.html.
6. Curry & Kerlinger, *What Kills Birds?* http://www. currykerlinger.com/birds.htm.
7. Clean Power Now, *Do Wind Turbines Kill Birds?* http://www.cleanpowernow.org/birdkills.php.
8. P. Kerlinger, *An Assessment of the Impacts of Green Mountain Power Corporation's Wind Power Facility on Breeding and Migrating Birds in Searsburg, Vermont, July 1996–July 1998,* report prepared for the Vermont Department of Public Service, NREL/SR-500-28591 (Golden, CO: NREL, March 2002), page 64.
9. R. C. Curry and P. Kerlinger, *Avian Mitigation Plan: Kenetech Model Wind Turbines, Altamont Pass WRA, California,* report presented at the National Avian Wind Power Planning Meeting III, San Diego, CA, May 1998, page 26.

Also playing an important role in helping wind power gain a competitive advantage are Renewable Energy Credits, or "green tags."[27] These tags result from laws currently in force in 13 states that require electricity providers to include a prescribed amount of renewable electricity in the electric power-supply portfolio they offer to their customers. Electricity providers meet this requirement through several possible approaches. They can generate the necessary amount of renewable electricity themselves, purchase it from someone else, or buy credits from other providers who have excess. The green tags rely almost entirely on private market forces. Taken together with production tax credits and various industry improvements, they are helping wind power continue its trend toward independent profitability. Such status, coupled with reduced public resistance, will move wind power from the realm of alternative to the position of mainstream energy resource. This might be possible if we can move from NIMBY to PIMBY.

## From NIMBY to PIMBY

When plans encounter resistance, developers usually make amended suggestions to attract greater support. This would seem to be a sensible tactic for wind developers, but it was not a part of planning for wind power in the mid-1980s. Instead,

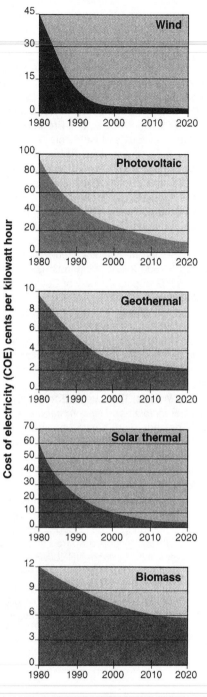

**Figure 4** Renewable Energy Cost Trends.

Note: These graphs are reflections of historical cost trends, not precise annual historical data.
Source: National Renewable Energy Laboratory, 2002.

wind developers (and various government officials) received notices of lawsuits for their trouble. NIMBY, even then a battlescarred acronym (for "Not In My Back Yard"), emerged in the headlines.

The ingredients mixed in the cauldron of subsequent wind power development made for a rich and complex brew. On the positive and promising side, developers learned to appreciate the power of public opinion and to work to inform it more completely. This also applied to regulators and policymakers. All parties began ascribing primacy to cooperation over imposition. By the mid-1990s, wind companies successfully improved efficiency and design, and jurisdictional authorities made zoning codes more appropriate if not more restrictive. The use of focus groups and public hearings became common elements in wind development planning procedures. As a result, controversy and press attention subsided, and projects continued to come on line with little fanfare or public notice in Iowa, Kansas, Minnesota, and Texas.

Then came Cape Wind, and much of the old debate began anew. Developers who had forgotten or never fully appreciated the power of public opinion started retreating. Despite many improvements and increasing experience, Cape Wind planners had devoted insufficient attention to considering the combination of factors that make one place unique from another. They reasoned that if offshore installations were meeting with success in Europe, why should they not find acceptance in "green" Massachusetts? However, they failed to realize the poor comparability between the mindset of people in the United States, who live in a spacious and largely post-industrial country, and their European contemporaries who have been living with industrial landscapes, greater population density, and much less personal space for centuries. In making their calculations, they neglected to note that the coastal areas of Massachusetts, heavily utilized for recreation, is not comparable with the lightly settled coastal areas of Europe.

In many ways, the Cape Wind episode is an East Coast version of the California experience 20 years earlier. Admittedly, the setting is different—desert versus ocean—but the underlying problem is the same: wind turbines—immovable and numerous—interfering with the aesthetics of a valued recreational resource.

The experience of Cape Wind suggests the need for a fresh approach to wind power development. The key element of this new approach is simplicity itself: Avoid sites having a high potential for conflict. Making this assessment would involve two steps. The first step would be to assign sites "compatibility rankings," starting with the most compatible sites.

- Rank #1 properties would be those where it is not only suitable but overtly requested for wind development, such as farms in Iowa or Kansas.
- Rank #2 properties would likely be acceptable, such as in southeastern Washington.
- Rank #3 properties might be acceptable in certain circumstances, such as near Palm Springs.
- Rank #4 properties would be completely off-limits, for example, on the top of Mt. Rushmore.

a naive impression prevailed that wind power would attract unquestioned support. However, the public resisted the blatant placement of wind turbines on the landscape. It was a particularly unexpected experience because California was known as a state where many of the most ardent environmentalists held forth. Instead of receiving congratulatory handshakes,

Ranks would be determined according to points assigned to site-specific characteristics, including lines of site, type and color tone of terrain, ownership, bird flyways, endangered species, competitive economic value, transmission lines/corridors, protected status (such as national parks), economic development, energy security, and so forth. This should be part of the process of environmental impact assessment, and it should be initiated at any location with a strong, class 4 or above wind resource. Without such rankings, the current ad hoc and contentious approach will continue. It would be akin to a general plan for a city: Variances could be granted, but there would be a broad guidance document in place.

Part two of this plan would be to concentrate our attention on Rank #1 sites. In the United States, this means the Great Plains. There are two simple reasons to emphasize this region. First, the United States' greatest wind resource is there. Second, the small-scale farmers in the area generally welcome the turbines. The message is this: When contentious sites breed contempt, avoid them, at least for now, even if the resource base is attractive and the load centers are nearby. Admit that the wind power alternative is uniquely visible and interferes with scenic vistas and cease trying to force-feed developments down the throats of a resistant public. This is not good for the future of wind power.

On the other hand, there are places where wind power development is welcome. Small farms of the Great Plains have been losing ground for decades to consolidation and the vagaries of weather; they need an economic boost to stay viable. The owners of these farms have put out the welcome mat for wind developers in places such as along Buffalo Ridge, on the border between northwest Iowa and Minnesota, and even farther west in places like Lamar, Colorado. As Chris Rundell, a local rancher, phrased it: "The wind farm has installed a new spirit of community in Lamar… it's intangible but very real." They are embracing a new acronym, PIMBY—Please In My Back Yard.

Seeing the wind development in the Great Plains in recent years is a continuation of history, if in a slightly different form: Where a century ago hundreds of thousands of farm windmills made the local agricultural life possible, wind power is again proving its worth to those who would live there. It is bringing needed cash into the local economy and slowing a multiyear trend of farm abandonment and consolidation. The same lands that early wind machines helped develop, new wind machines are helping preserve.[28]

# Notes

1. See http://www.state.hi.us/dbedt/ert/wwg/windy.html#molokai.

2. European Wind Energy Association, "Wind Power Expands 23% in Europe but Still Is Only a 3-Member State Story," press release, 3 February 2004, www.ewea.org/documents/0203_EU2003 figures&x005F; final6.pdf.

3. Pacific Northwest Laboratory, *An Assessment of the Available Windy Land Area and Wind Energy Potential in the Contiguous United States* (Washington, DC: U.S. Department of Energy (DOE), 1991).

4. American Wind Energy Association, "Wind Energy: An Untapped Resource," fact sheet, 13 January 2004, http://www.awea.org/pubs/factsheets/WindEnergyAnUntappedResource.pdf.

5. P. Gipe, "More Than First Thought? Wind Report Stirs Minor Tempest," *Renewable Energy World* 6, no. 5 (2003), available at http://www.jxj.com/magsandj/rew/2003_05/wind_report.html. See also C. L. Archer and M. Z. Jacobson, "The Spatial and Temporal Distributions of U.S. Winds and Wind Power at 80 M Derived from Measurements," *Journal of Geophysical Research* 108, no. D9 (2003): 4289.

6. North Dakota has the capacity to produce 138,000 MW; Texas, 136,000; Kansas, 122,000; South Dakota, 117,000 MW; and Montana, 116,000 MW.

7. American Wind Energy Association; see also the National Wind Technology Center Web site (http://www.nrel.gov/wind/).

8. Craig Cox, senior associate, Interwest Energy Alliance, personal communication with author, 27 April 2004.

9. Paul Gipe, executive director, Ontario Sustainable Energy Association, personal communication with author, 20 April 2004.

10. American Wind Energy Association, note 4 above.

11. American Wind Energy Association, note 4 above.

12. Lawrence Flowers, technical director, National Renewable Energy Laboratory, personal communication with author, 22 June 2004.

13. M. J. Pasqualetti and E. Butler, "Public Reaction to Wind Development in California," *International Journal of Ambient Energy* 8, no. 3 (1987): 83–90.

14. M. J. Pasqualetti, "Accommodating Wind Power in a Hostile Landscape," in M. J. Pasqualetti, P. Gipe, and R. Righter, eds., *Wind Power in View: Energy Landscapes in a Crowded World,* (San Diego, CA: Academic Press, 2002), 153–71; M. J. Pasqualetti, "Morality, Space, and the Power of Wind-Energy Landscapes," *The Geographical Review* 90, no. 3 (2001): 381–94; and M. J. Pasqualetti, "Wind Energy Landscapes: Society and Technology in the California Desert," *Society and Natural Resources* 14, no. 8 (2001): 689–99.

15. L. Marx, *The Machine in the Garden* (Oxford, UK: Oxford University Press, 1964).

16. C. C. Reith and B. M. Thomson, eds., *Deserts As Dumps? The Disposal of Hazardous Materials in Arid Ecosystems* (Albuquerque: University of New Mexico Press, 1992).

17. Pasqualetti, "Wind Energy Landscapes: Society and Technology in the California Desert," note 14 above.

18. Total worldwide wind energy installations were 1,000 MW in 1985, 18,000 MW in 2000, and nearly 40,000 MW in 2003, growing at about 35 percent per annum. See Solarbuzz, *Fast Solar Energy Facts: Solar Energy Global,* http://www.solarbuzz.com/FastFactsIndustry.htm.

19. See http://www.saveoursound.org/windspin.html.

20. Today's support started with Public Utility Regulatory Policies Act (PURPA), the 1978 law that promoted alternative energy sources and energy efficiency by requiring utilities to buy power from independent companies that could produce power for less than what it would have cost for the utility to generate the power, the so-called "avoided cost." In the past 20 years, electricity feed-in laws have been popular in Denmark, Germany, Italy, France, Portugal, and Spain. Private generators, or producers, charge a feed-in tariff for the price-per-unit of electricity the suppliers or utility buy. The rate of the tariff is determined by the federal government. In other words, the government sets the price for electricity in the country. Because the producer is guaranteed a price for the electricity, if he or she meets certain criteria, feed-

in laws help attract new generation capacity. During the past decade, Germany became the world leader in wind development. Much of this success is due to *Stromeinspeisungsgesetz,* literally meaning the "Law on Feeding Electricity from Renewable Sources into the Public Network." The original Electricity Feed Law set the price for renewable electricity sources at 90 percent the retail residential price. In 2001, the German feed law was modified to a simple, fixed price for each renewable technology. See http://www.geni.org/globalenergy/policy/renewableenergy/electricityfeed-inlaws/germany/index.shtml.

21. Union of Concerned Scientists, "How Wind Power Works," briefing, http://www.ucsusa.org/CoalvsWind/brief.wind.html.

22. Ibid.

23. The U.S. Production Tax Cut, by its emphasis on the actual generation and transmission of electricity, not just the construction of equipment, provides additional incentives for greater technical efficiency.

24. See http://www.interwestenergy.org.

25. Reuters went on to report that the "Wind industry backers say the gaps have created a roller coaster in U.S. wind production growth because companies become fearful of investing in the alternative energy source. They say the tax-break gaps hamper wind power growth in the United States which grew last year at a rate of only 10 percent compared to global growth of 28 percent." See "Wind Power Tax Credit Expires in December,"

*Reuters World Environment News,* 27 November 2003, http://www.planetark.com/dailynewsstory.cfm/newsid/22956/newsDate/27-Nov2003/story.htm.

26. *First Quarter Report: Wind Industry Trade Group Sees Little To No Growth In 2004, Following Near-Record Expansion In 2003,* http://www.awea.org/news/news0405121qt.html

27. For more information about green tags for wind power, see http://www.sustainablemarketing.com/wind.php?google.

28. For several economic summaries, see S. Clemmer, *The Economic Development Benefits of Wind Power,* presentation at Harvesting Clean Energy Conference, Boise, Idaho, 10 February 2003, available at http://www.eere.energy.gov/windpoweringamerica/pdfs/wpa/34600_wind_handbook.pdf.

**MARTIN J. PASQUALETTI** is a professor of geography at Arizona State University in Tempe. His research interests include renewable energy and the landscape impacts of energy development and use. He is a coeditor of and contributor to *Wind Power in View: Energy Landscapes in a Crowded World* (Academic Press, 2002). He also contributed articles on wind power to the *Encyclopedia of Energy* (Academic Press, 2004). His research has appeared in *The Geographical Review and Society and Natural Resources.* He thanks Paul Gipe and Robert Righter for reading the manuscript for this article and offering many helpful suggestions. Pasqualetti may be reached at (480) 965-7533 or via e-mail at pasqualetti@asu.edu.

From *Environment,* Vol. 46, No. 7, September 2004, pp. 23–38. Reprinted by permission of the Helen Dwight Reid Educational Foundation. Published by Heldref Publications, 1319 Eighteenth St., NW, Washington, DC 20036-1802. Copyright © 2004. www.heldref.org

# Personalized Energy

## *The Next Paradigm*

**In the future, energy will be more in the control of neighborhoods and homeowners. But for that to happen, new technologies need to be developed that bring efficiency and reliability up and costs down, says a technology futurist.**

STEPHEN M. MILLETT

Consumers want more control over their energy, and they want it cheaper, cleaner, more convenient, and more reliable. They want energy, like other commodities, to be more personalized. Technological improvements will focus on meeting those demands, but they won't happen quickly. Current forecasts for energy supply and demand are limited by a mind-set stuck in the past. But new technologies and new consumer imperatives will spawn new ideas about energy that could get us off the grid and bring power generation into neighborhoods and even into homes.

The primary question we need to ask about the future of energy is whether the old supply-and-demand paradigm of fossil fuels still applies. In that case, the key to solving our energy woes lies in finding ways to increase production of traditional hydrocarbon fuels (such as oil, natural gas, and coal) and promote consumer conservation. But if the old paradigm is out, there may be a whole new paradigm emerging, where new technologies, for instance, could change the whole energy picture.

Right now, we hear too many discussions about drilling more oil, conserving energy, and other actions based on old-paradigm thinking. Indeed, statistics show a big gap between projected energy demand and supplies in the United States: Oil and natural gas consumption are going up and available quantities are going down, so we're going to have a big projected shortfall.

The biggest jump in American energy consumption in the twentieth century was the use of petroleum, and that's almost exclusively transportation. Transportation relies on petroleum to meet 95% of its energy needs, according to the U.S. Bureau of Transportation Statistics. The story on coal is a little bit different. Developed economies such as the United States used to use coal in homes for heating, but that's done almost nowhere anymore. Americans are using more and more coal, but it's to generate electricity in large power plants. Coal is now rarely used at the individual level.

The 2001 report of the president's National Energy Policy Development Group stated, "Renewable and alternative fuels offer hope for America's energy future but they supply only a small fraction of present energy needs. The day they fulfill the bulk of our [energy] needs is still years away. Until that day comes we must continue meeting the nation's energy requirements by the means available to us."

This assertion assumes no changes to the existing energy paradigm: no new technological breakthroughs, no shifts in people's values or consumers' demands, no surprising events—natural or manmade—to alter the energy picture. But this paradigm-blinder limits our thinking—and our forecasts. Paradigms and social systems are rarely permanent, and new technology often drives a transition to other paradigms.

My thesis is that we have just begun the shift away from what I call the "carbon-combustion paradigm" to a new "electro-hydrogen paradigm." The shift is going to be very dramatic in the next 20 years, but the full integration is going to take easily a hundred years. We're going to see a lot of exciting technology innovations in the laboratory and in prototype systems in the next 10 to 20 years, but to go from our current paradigm and all its infrastructure to a new paradigm and all of its infrastructures is going to take a very long time.

## Energy and the Consumer

Changes in consumer behavior are driving many trends. In the U.S. market, baby boomers seek convenience, while the elderly put heavy emphasis on the reliability and affordability of power. The question for policy makers and the energy industry is how reliable the electric grid will be in the future. Both baby boomers and Generation X'ers value customization—the personalization of products, especially computers and cell phones. Consumers also want more mobility and longevity in their products. And of course we all want inexpensive energy.

Along with consumer behavior, there are marketplace trends working toward this paradigm shift, including the effects of more-stringent environmental-quality regulations. The bad news is that no energy system will ever be 100% environmentally friendly; the good news is that the next paradigm will be a lot friendlier than the last one was.

(Millions of barrels per day)

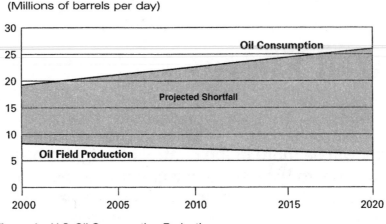

**Figure 1**   U.S. Oil Consumption Projections.

Sources: National Energy Policy Report: Battelle.

U.S. energy policy calls for greater energy self-sufficiency. An additional need, a new element since September 11, 2001, is security of the energy infrastructure. Before then, people didn't even worry about infrastructure security, but now it's a major issue. The electric grid system is absolutely vulnerable to weather and potential security compromises.

Another marketplace issue is the need for energy-cost stability and continued necessity for economic growth. If you take a strictly conservationist approach to this issue and say this cost stability and continued economic growth can be realized by voluntary simplicity, then you are limited by the old energy paradigm. This low-growth scenario has a low probability of occurring. We need more energy for continued economic growth, and that is still what most people in the world want.

Also impacting marketplace trends are emerging technologies such as those behind a gradual shift now beginning from central-station generation to decentralized, or distributed, generation of power from local sources. The paradigm shift to distributed generation, or distributed resources, parallels the paradigm shift from fossil fuels to hydrogen. For example, one of the biggest needs we have is gasification of coal, though surprisingly little work is yet being done in this area. The current gasification technology has been in existence for at least 20 years, it isn't really very good, and it's very expensive. It's just not competitive.

## The Real Hydrogen Future

It's easy to speculate about the hydrogen economy's potential and to go off into science-fiction scenarios. But because Battelle deals with the real-world challenges of governments and corporations, we spend a lot of time separating science from science fiction. So here's a little reality check: All forms of energy are going to have some negative environmental consequences. We need to recognize that fact and then try to make the energy system a lot better in the future than it is today. Another reality check is that no fuels will be free: There will always be costs for both the fuels and their infrastructure of

production and distribution. The challenge is to find the new ways that improve the value (benefits/costs) relationship.

The challenge in this transitional period to the hydrogen future is to extract hydrogen from hydrocarbon sources in an affordable way. There are many potential avenues being explored today. The approach at Battelle is to develop a universal reformer for the fuel cell, where we would take methane, methanol, and even gasoline and convert it into hydrogen at the point of burning it. The ability to extract sufficiently pure hydrogen from methane, methanol, or gasoline means that we could continue to use the existing infrastructure (such as all of those gas stations) to distribute safe liquid fuels without the expense and hazards of storing hydrogen today. Avoiding new infrastructure costs in the short run would greatly help the transition to the electrohydrogen paradigm of fuel cells for both transportation and stationary power generation applications.

Economics really favor the current, fading energy paradigm; economics do not yet favor the next one. We're going to have to see a lot of economic and regulatory changes, as well as technological changes. The challenge is cost. We can make fuel cells, we can produce hydrogen, but we can't do it at competitive prices relative to the existing hydrocarbon system, the carbon-combustion system. Electric utilities use a benchmark of $1,500 per kilowatt capacity; anything costing more than that is simply not competitive enough. Researchers at United Technologies, for example, are getting the cost of the fuel cell down below $3,000 per kilowatt—it has been as high as $15,000—but the price is still too high for general commercialization.

And what about solar cells? There's no question that there's a market today for solar cells, but it's largely a vanity technology for people who put it on their houses. If you take a pocket calculator that has solar panels, generating power measured in milliwatts, and normalize that to a kilowatt, the cost might be as high as $17,000. So, clearly, there's a long way to go before solar technology can replace the carbon-combustion system.

Alternative fuels like wind power, which is now growing, have been attractive because of a number of government incentives and subsidies at both the federal and the state levels.

---

## About Battelle

The Battelle Memorial Institute was established in 1929 in Columbus, Ohio, and now manages or co-manages four of the 16 U.S. national labs. We are in the business of technology development, management, and commercialization. We do mostly government work, but we also have industrial clients. We're independent, meaning we have stakeholders but no stockholders. We are technically not for profit but, as we're reminded daily by our CEO, we are not for loss. We do more than $1.7 billion in business a year, and that's a lot of R&D. Battelle scientists have contributed to a wide range of breakthroughs, such as copy machines, optical digital recording, and bar codes; Battelle's R&D yields between 50 and 100 patented inventions a year.

All corporations and organizations face the challenge of keeping up with and anticipating change. At Battelle, we do futuring. I like to use the word *futuring* because the participle adds the action of making or doing something. We do trend analysis, expert focus groups, and expert judgment, and we have our own process of scenario analysis based on cross-impact analysis. We also have our own scenarios software, which we've been using for 20 years for our work with corporations.

Battelle studies "consumer value zones," where marketplace trends, new customer demands, and emerging technologies all converge. The study of energy's consumer value zone leads us to conclude, for example, that the future of energy is personal; that is, energy production will increasingly move from large, centralized power plants to distributed power.

Among its outreach projects, Battelle does an annual technology forecast and maintains a separate Web site for our scenarios and trends: www.dr-futuring.com.

For more information, see Battelle's Web site, www.battelle.org.

—*Stephen M. Millett*

---

## Technologies to Watch

The technologies to watch include:

- Innovations in materials for batteries and fuel cells, especially PEM (polymer electrolyte membrane) and solid-oxide fuel cells.
- Breakthroughs in reducing diesel emissions.
- Innovations for reconfiguring backup and emergency power generation into distributed generation systems.
- Biofuel development.
- New approaches to the gasification of coal.
- Global warming and carbon-dioxide management.

For fuel cells and batteries, the biggest challenge is in materials development. Exciting new developments in battery technology include the sodium sulfur battery, which the American Electric Power Company is working on in Columbus, Ohio.

For PEM and solid-oxide fuel cells, the name of the game is the materials and getting their costs down. For instance, the current membrane material used in PEM fuel cells now costs as much as $800 per square meter. We need to get the cost down to $8 to make this transition to the next paradigm. So we need technology breakthroughs that bring costs down. In addition, the current use of platinum as the catalyst is obviously very expensive and needs to be changed.

Reducing diesel emissions is another significant area of research at Battelle and other institutions. Biofuel blending with diesel and other fuels is a very exciting growth area.

Bringing diesel emissions down will promote the transition of current backup generation to distributed resources coordinated with the power grid. Most utilities now dismiss customer-driven backup generation as simply being irrelevant to the grid, but if you can make all of those diesel generators environmentally compliant, and if you can coordinate them with the grid, then you've got a prototype distributed-generation system already in place.

Biofuel development, not just bioblending, is another breakthrough area. The DNA revolution in agriculture is very exciting because we could design plants—not just corn, but also chickweed or garbage grass, for instance—that could be engineered for high-starch content to be more easily converted into methanol.

Gasification of coal is a huge area of research and development. Affordable and efficient coal gasification would enable us to break down the constituent parts of coal and get the hydrogen atoms out of it. In an ideal process, we would be able to separate sulfur and other undesirable constituents out of the coal and extract pure hydrogen. We could also separate out the carbon content that produces carbon-dioxide emission from stacks. An innovative coal gasification technology would be a tremendous boon for the American economy—and the economies of Germany, Russia, China, and India, to mention just a few others. Hydrogen from coal would be a major step in the transition to fuel cells.

The new energy paradigm is also about the environment, and we at Battelle are concerned about global climate change and carbon-dioxide management. To that end, Battelle is actively pursuing approaches like carbon sequestration. We are currently working with the U.S. Department of Energy to evaluate how to capture carbon dioxide and store it underground so that it cannot escape into the atmosphere.

# Toward a Distributed Power System

We're not suddenly going to do away with our coal-burning plants, but there are emerging opportunities to use large fuel cells and batteries in conjunction with central generation. This could produce emergency and peak power at the generation site as well as provide supplementary power at distributed sites. By "distributed," I don't mean we're going from the big power plant to the home all in one jump. Energy will be distributed first at the level of neighborhoods and districts, and then we'll work it on down to homes generating their own power. It'll go step by step, but the trend favors personalized energy.

We're going to see some exciting technologies developed in the next 10 years, but it's going to be a slow process toward full-blown commercialization. If we're on a low technology-development trajectory, it will take more time. If we get a couple of breakthroughs in technology or some regulatory changes, then we can be on a faster track, but no sooner than 2008 or even 2010 at the earliest unless a desperate need for power drives the trends faster. Slow progress favors the "tracker" and "adapter" companies and organizations, while fast progress favors the early innovators. Many companies are now agonizing over whether to be the progress leaders or the followers (or fast followers).

Who's going to lead this energy paradigm shift? Who really is going to provide the thought leadership and the breakthroughs? The Japanese are clearly ahead of the Americans in fuel cells. Honda and Toyota are ahead of the Big Three auto manufacturers in Detroit on energy breakthroughs for transportation. Honda in particular is the world leader in thinking through distributed power generation.

As for regulatory leadership, the question is who is going to provide the standards. There's a dearth of leadership for the new energy paradigm in the United States. Neither the federal government nor the states are showing signs of leadership, and there are very few progressive electric and gas utilities out there in the United States. Wherever the leadership comes from for the new energy paradigm, that's who will likely succeed and capture the largest market share.

But for now, those who should lead seem to be saying, "Change is good. You go first."

STEPHEN M. MILLETT is a thought leader at Battelle and co-author of *A Manager's Guide to Technology Forecasting and Strategy Analysis Methods.* His address is Battelle, 505 King Avenue, Columbus, Ohio 43201. E-mail milletts@battelle.org; Web site www.battelle.org.

Originally published in the July/August 2004 issue of *The Futurist,* pp. 44–48. Copyright © 2004 by World Future Society, 7910 Woodmont Avenue, Suite 450, Bethesda, MD 20814. Telephone: 301/656-8274; Fax: 301/951-0394; http://www.wfs.org. Used with permission from the World Future Society.

# Hydrogen: Waiting for the Revolution

**Everybody agrees it's the future fuel of choice. Why hasn't the future arrived?**

BILL KEENAN

Here's how you'll live in the Hydrogen Age: Your car, powered by hydrogen fuel cells and electric motors, quietly drives along smog-free highways. At night, when you return your vehicle to the garage, you hook up its fuel cell to a worldwide distributed-energy network; the central power grid automatically purchases your battery's leftover energy, offsetting your overall energy costs.

In the garage, you also have a suitcase-sized electrolyzer, or other conversion device, plugged into the electrical system to pump a fresh batch of hydrogen into your car. (The fuel cell uses hydrogen to produce electricity, which powers the motor.) If you need a refill as you're driving along one of the nation's highways, you pull up to a clean, quiet hydrogen fueling station to top off in less time than it takes today to fill a car with gasoline.

The electricity in your home will also come from hydrogen, either via small local fuel-cell power plants or residential fuel cells in your basement. "Moreover," says Jeremy Rifkin, president of the Foundation on Economic Trends and author of *The Hydrogen Economy*, "sensors attached to every appliance or machine powered by electricity—refrigerators, air-conditioners, washing machines, security alarms—will provide up-to-the-minute information on energy prices, as well as on temperature, light, and other environmental conditions, so that factories, offices, homes, neighborhoods, and whole communities can continuously and automatically adjust their energy consumption to one another's needs and to the energy load flowing through the system."

The U.S. Department of Energy is only slightly less enthusiastic, maintaining in a report that in the hydrogen economy, "America will enjoy a secure, clean, and prosperous energy sector that will continue for generations to come. It will be produced cleanly, with near-zero net carbon emissions, and it will be transported and used safely. [Hydrogen] will be the fuel of choice for American businesses and consumers."

The new energy regime will have economic and political ramifications as well. Oil companies and utility companies will merge and morph into "energy companies" with a focus on generating renewable energy and local power distribution, including purchasing power from residential customers. Distributed energy production will also result in a worldwide "democratization of energy," bringing low-cost power to underdeveloped areas.

## Oil and Hydrogen Don't Mix

Driving the interest in a hydrogen-based energy system: threats to the economy, the environment, and national security. Oil production, by current estimates, will likely peak sometime between 2020 and 2040. At this point, the world's economies will have consumed half of the known oil reserves, with two-thirds of the remaining oil in the volatile Middle East. As a result, prices will rise dramatically, and global consumers will experience increasingly frequent shortages.

Global warming is another significant threat that a shift to hydrogen might ameliorate. The release of carbon dioxide into the atmosphere from the burning of fossil fuels such as coal, oil, and natural gas makes up about 85 percent of greenhouse-gas emissions in the United States. This increase has resulted in an unprecedented rate of global warming, according to most scientific experts. The thinning of the polar ice caps, the retreat of glaciers around the world, the spread of tropical diseases to more temperate climates, and the rising of global sea levels are all evidence of global warming. Says Rifkin: "Weaning the world away from a fossil-fuel energy regime will limit carbon-dioxide emissions to only twice their pre-industrial levels and mitigate the effects of global warming on the Earth's already beleaguered biosphere."

---

### The Mechanics of Hydrogen

While still untested on a large scale, the promise of a hydrogen economy is based on a number of undeniable realities. Hydrogen can be burned or converted into electricity in a way that creates virtually no pollution. It is also Earth's most abundant element, available everywhere in the world. While hydrogen is scarce naturally in pure form, it can be generated easily by reforming gasoline, methanol, natural gas, and other readily available resources. It can also be created by electrolysis, a process by which electricity is run through water to separate the oxygen and hydrogen molecules.

The fuel cell, which combines oxygen in the air with hydrogen to create electricity and water, is the vital link in the hydrogen vision. It closes the energy loop and allows electricity to be stored and transported via hydrogen and then reconverted back into electricity.

In an ideal future, renewable-energy sources such as wind, solar, or water power will be used to create hydrogen through electrolysis. The hydrogen can be converted again to electricity locally by means of a fuel cell to power a car, provide energy for a home, power a laptop, or operate any number of other products.

—B.K

Add to these threats the burden of growing world populations, an increasingly unstable political situation in the Middle East, and the likelihood of longer and more frequent blackouts and brownouts resulting from an aging and vulnerable power grid in the United States, and the promise of a safe, pollution-free, and distributed power system based on hydrogen becomes increasingly attractive.

# Pathways and Roadblocks

Does all this sound too good to be true? It is: The hydrogen economy faces serious obstacles. More than 90 percent of the hydrogen produced today comes from reformulated natural gas generated through a process that creates a significant amount of carbon dioxide. Energy for this process, or for electrolysis, a more expensive way of generating hydrogen would also come from power plants fueled by oil or natural gas. So in the near term, a shift to hydrogen will not greatly reduce the world's dependence on fossil fuels and, in fact, may well hasten the greenhouse effect and global warming by increasing carbon-dioxide emissions.

Consequently, a lot of discussion about the hydrogen economy revolves around the various "pathways," or means of producing hydrogen. Atakan Ozbek, director of energy research at ABI Research, a technology-research think tank, points out that while hydrogen can come from virtually any fuel, energy from oil and gas is currently cheaper and more efficient than energy from renewable resources such as wind, sun, or water. Then, too, in the event of an oil crisis and resultant electricity shortage, coal will likely be pressed into service, regardless of the environmental cost. Nuclear power plants can also provide electricity to create hydrogen, but nuclear energy's high cost—plus the still-hot controversy over waste disposal—make such a pathway less than certain.

"What we're trying to find out right now," Ozbek says, "is how to get hydrogen to the fuel cell in a way that is economically feasible and makes sense engineering-wise."

Environmental considerations are paramount: If coal is reintroduced in a large way into our "energy portfolio"—whether to produce hydrogen or as part of our existing energy plan to replace oil—carbon-dioxide emissions will rise significantly.

The Department of Energy roadmap anticipates this, and the DOE is funding research into the "sequestration" of carbon-dioxide gases created by coal processing and natural-gas reformation. This would involve capturing these gases at some point in the energy process and permanently storing them underground or in the ocean.

To many, this is unrealistic. Jon Ebacher, vice president of power-systems technology for GE Energy, won't say that sequestration is impossible, but his comments fall short of an endorsement. Even a fairly efficient coal plant, Ebacher says, produces millions of tons of carbon dioxide each year. "So if you're going to sequester carbon dioxide from all of the plants that use hydrocarbon fuels," he says, "that's a pretty massive undertaking."

Only a hydrogen economy based 100 percent on renewable power would result in zero emissions—the vision that has captured so many imaginations. And that vision remains decades away. In the meantime, Ebacher says, "natural gas can see us through a transition period until we get solar and other renewable-energy efficiencies up to a much higher level." That transition period, he suggests, might last twenty-five to fifty years.

Another potential roadblock: transport and storage of hydrogen. Less dense than other fuels, the gas must be compressed or liquefied to be stored or moved efficiently, adding to costs and inconvenience. While the existing natural-gas infrastructure would seem to offer a convenient pathway to hydrogen delivery, this can't be done without a major retrofit. Indeed, Ebacher says, almost all of the country's existing natural-gas pipeline would have to be modified to handle hydrogen.

Finally, fuel-cell researchers must make significant advances. The power produced by a fuel cell is significantly more expensive per unit than that produced by an internal-combustion engine. Fuel-cell vehicle development is also beset by problems and costs related to type of fuel, storage, and performance. A number of prototype and "concept car" fuel-cell vehicles have been produced and displayed at auto shows and fuel-cell conferences around the world—but at a development cost of about $250,000 or more per vehicle. GM estimates that it spent between $1 million and $2 million to develop its Hy-wire fuel-cell concept car. A consumer version would cost far less, obviously, but likely would still take sticker shock to a whole new dimension.

# Putting a Brake on Hydrogen Cars

Linking the hydrogen age to cars could be a critical policy mistake, according to Joseph Romm, former acting assistant secretary for the DOE's Office of Energy Efficiency and Renewable Energy and author of *The Hype About Hydrogen*. Despite car-company promises to have fuel-cell vehicles in dealer showrooms by 2010, if not sooner, Romm argues that the cost of fuel cells, problems with onboard storage of hydrogen in vehicles, and the issues related to creating a hydrogen delivery infrastructure are likely to push the market for hydrogen fuel-cell vehicles well into the future.

The focus on hydrogen as an immediate goal in the transportation sector amounts to confusing a means (hydrogen) with an end (greenhouse gas reduction), Romm explains. This could have harmful consequences, since, he estimates, it will take thirty to fifty years for hydrogen vehicles to have a significant impact on greenhouse gases. A recent National Academy of Sciences study seconds this point, stating, "In the best-case scenario, the transition to a hydrogen economy would take many decades, and any reductions in oil imports and carbon-dioxide emissions are likely to be minor during the next twenty-five years."

"If the goal is to reduce greenhouse gases," Romm argues, "then there are technologies available right now that can have a more immediate effect"—hybrid vehicles, for instance. And diverting existing (and limited) natural-gas supplies to create hydrogen for vehicles "would make that fuel less available where its use could result in a more immediate reduction in greenhouse-gas emissions—in replacing existing oil and coal-burning electric-power plants in the nation's energy grid with cleaner natural-gas power plants."

### As a fuel, hydrogen is "simply a better mousetrap."

In fact, some hydrogen-technology companies have back-burnered research and development on transportation applications. "The horizons for fuel-cell vehicles keep getting pushed out further and further, and it's unlikely that somebody's going to license and

# As GE Goes, So Goes the Nation?

Fuel cells will probably not be a viable market until a company like General Electric gets into the business in a big way, say critics of the hydrogen economy.

GE is indeed researching fuel cells, albeit cautiously, in keeping with its approach to most other energy markets. "We do have an investment in fuel cells," says Jon Ebacher, the company's vice president of power-systems technology. "I don't know if it will ever get to the dimensions where it will work at the huge volumes that were once forecast, but I think it's quite viable in niche markets." Right now, he sees a possible market in "industrial facilities that have isolated power needs, where you have a maintenance crew that deals with heating, ventilating, and air-conditioning." But, he says, "there's a distance between where they are today and the huge potential in consumer markets that was forecast at one time."

GE has also invested substantially in researching a type of fuel-cell system that would employ a gas turbine and hydrogen system working as a combined cycle. Right now, Ebacher says, GE is considering creating a power plant based on this system by 2013. It could be sooner, depending on external factors, including political developments around both fuel and the environment, the price of fuel, and the types of fuel that are available. But there are still unresolved technical challenges that could push that back.

Natural-gas prices in particular are an important barometer. "During the California energy crisis, the price of gas spiked up to $7 per million BTUs," Ebacher says. Higher gas prices, he says, "could spark some other research efforts that may come in front of fuel cells—coal, for instance. It's possible to run a combined-cycle system on coal. You put a chemical plant beside a combined-cycle power plant to process the coal into a gaseous fuel to run the electrical plant." But right now, Ebacher says, "the capital cost of doing that doesn't cross the goal line. However, if the price of natural gas or its availability gets in a bad place, all of a sudden the capital cost of doing that might not look so bad.

"It all revolves around availability, economics, and the environment—where the pressures are, what are the levers. But if you talk about running out of hydrocarbon fuels, then you would have to say that hydrogen had better be in the cards."

—B. K

commit to a uniform, standardized hydrogen technology for at least ten to fifteen years," says Stephen Tang, an industry consultant and former president and CEO of Eatontown, NJ.-based Millennium Cell, which makes a system called Hydrogen on Demand that supplies hydrogen to fuel cells.

To pay the bills in the meantime, Tang says, "Millennium Cell has targeted markets that it believes can tolerate the price of hydrogen and fuel cells, such as consumer electronic devices, standby power, and military portables. In all of those markets, you're competing with an incumbent technology that is rather expensive in its own right and also has some limitations in performance. In these markets, then, we can focus on hydrogen as a performance fuel and not focus so much on the environmental benefits or the energy-independence benefits—attributes that buyers have difficulty valuing. It's simply a better mousetrap: Hydrogen allows you to run your cell phone much longer, or your laptop much longer, without being a slave to the energy grid or inferior batteries."

## Who Will Lead?

Despite the limitations, there is growing momentum for hydrogen vehicles. Hybrid vehicles may be a "bridging technology" toward the hydrogen age, but it's one that "doesn't at all curb the nation's appetite for oil," says Chris Borroni-Bird, GM's director of design-technology fusion. Therefore, the automaker directs about a third of its R&D—over $1 billion thus far and involving more than six hundred people—toward fuel cells. The company insists that it will have a commercially viable fuel-cell vehicle available by the end of the decade.

In other business sectors, investment in hydrogen technology is slowly returning after the boom in hydrogen technology stocks in 1999–2000 and the subsequent bust that lasted until last year. "Behind a lot of the hype, there was tremendous capital inflow in the mid-1990s going into 2000," Tang says. Unfortunately, the number of commercial products—and the resulting revenue—in the industry have been "underwhelming" relative to investment dollars. That has made the investment community more cautious so far, but things are changing. "Right now there is a much more realistic view of the possibilities," Tang says. "The investor today is looking more toward interesting niche strategies and early market penetration rather than the hope of the mass market, the home run where fifty million cars are going to be sold with your product in it."

With the investment community poised and the technology issues coming together, says ABI Research's Ozbek, "Everything is feeding into this giant equation—you can consider it a giant chemical reaction—and once everything has been fed in and the equation solved, it's going to change the whole energy infrastructure." Federal support and direction will be especially important. While Ozbek considers President Bush's $1.7 billion State of the Union pledge for energy research a good start, he would like the government to provide such research incentives as Japan and the European Union have in recent years.

And though the president disappointed many hydrogen proponents by making no specific mention of hydrogen-energy R&D in his 2004 address, his proposed 2005 budget did increase funding for hydrogen research. The federal government, Ozbek argues, should provide enhanced tax credits for buyers of fuel-cell vehicles and fuel credits for energy companies and other investing in building a hydrogen infrastructure. Jeremy Rifkin agrees, urging the federal government to take the lead by establishing benchmarks—mandating tougher fuel-efficiency standards and requiring a greater use of renewable energy sources by power companies—as the European Union currently does.

## California's Hydrogen Highway

One state isn't waiting for action from companies or the federal government. In California, the new Schwarzenegger administration has committed to an energy plan that aims to create a

# Is GM's Hy-Wire the Car of the Future?

General Motors has had a reputation for being rather conservative when it comes to both new technological developments and vehicle design, but it seems to have leapt ahead of other carmakers with its concept car, the Hy-wire.

The idea, says Chris Borroni-Bird, director of design-technology fusion for GM, is that "if you design a vehicle around the fuel cell and hydrogen tanks, you might be able to create a better vehicle than if you just put those same systems in a car designed for an internal-combustion engine."

The Hy-wire design puts the fuel cell and hydrogen storage tanks into a skateboard-like chassis that allows for greater flexibility and interchangeability of body types. Customized car bodies are then effectively "docked into" the uniform chassis.

And because the fuel cell can provide much greater electrical output than today's batteries, GM's designers have replaced mechanical and hydraulic systems for steering and braking with an electronically controlled one. "This system provides more design freedom, because those electrical wires can be routed in numerous ways, replacing a fixed steering column," Borroni-Bird says.

The Hy-wire prototype has no gas engine, no brake pedals, and no instrument panel. The fuel cell enables you to operate everything by wire. The electronic controls are included in a compact handgrip console that extends from the floor from between the front seats of the vehicle. Drivers can steer, brake, or accelerate with the controls built into the handgrips.

Because GM puts the hydrogen directly on board the vehicle, there is no need for the car to convert fossil fuels or other renewable sources into hydrogen. As a result, it can claim to offer a zero-emission vehicle and market the car to be compatible with a network of hydrogen fueling stations.

To that end, GM "applauds any hydrogen infrastructure projects, anywhere in the world," says Tim Vail, GM's director of business development for fuel-cell activities. Yet it will take a lot of applause to get the government to invest the estimated $11 billion to get a sufficient mass of hydrogen refueling stations to support 1 million vehicles, in proximity to 70 percent of the nation's population. "But," says the optimistic Vail, "$11 billion is nothing compared to past infrastructure projects such as the highways or the railroads. So it's not that big an issue to overcome. You just have to have the will to do it."

—B. K.

"hydrogen highway" in the state by 2010. The ambitious plan proposes the construction of hydrogen fueling stations every twenty miles along the state's twenty-one major interstate highways. By taking this step to break the chicken-and-egg dilemma (which comes first, the vehicle or the fueling infrastructure?) and by continuing to impose strict mandates on automakers for fuel efficiencies, California could jumpstart the hydrogen economy.

"The pieces are all on the table," says Terry Tamminen, secretary of California's Environmental Protection Agency, "and there have been demonstration projects, but they have not been pulled together into any kind of unified vision, something that average people can use and where we can more fully commercialize the technology. So we're taking a lot of this work that's already been done, bringing it together, adding some timetables and leadership, and then of course asking for some federal money to help with the pieces that aren't paid for by private industry or other investments."

## California could jumpstart the hydrogen economy.

California already has several hydrogen fueling stations, serving research projects and some municipal fleets, and about a dozen more are in the works. For instance, SunLine Transit Agency, a local public-transit company, now operates a hydrogen fueling station that it uses to test its hydrogen-powered buses. And AC Transit, which provides public transportation in the San Francisco Bay area, expects to have three fuel-cell-powered buses later this year.

The state's goal is to provide an infrastructure of fueling stations to support a consumer market for fuel-cell vehicles. "If we can deliver such a network by a certain date," Tamminen explains, "we can then ask car companies to deliver on their promises to start delivering cars to showrooms."

One of the things driving California's plan is a California Energy Commission report that, Tamminen says, "includes credible evidence that in three to five years we are going to have serious shortages of refined fuels in the state. Not because there's not enough petroleum under the sands of Iraq but, rather, because we don't have enough refinery capacity in the state—or in the country—to keep up with the demand created by longer commutes, poorer fuel economy, and a growing population. The report predicts a likelihood of $3 to $5 per gallon gasoline prices and periodic shortages.

"During the oil embargo of the mid-1970s, we had twenty-four thousand retail gasoline outlets in the state, compared to ten thousand today. If there are shortages, not only will there be gas lines—they will be twice as long."

Consequently, it's not a question of if but when we move toward a hydrogen economy. Even Romm, who is dubious about short-term prospects for hydrogen, concludes: "The longer we wait to act, and the more inefficient, carbon-emitting infrastructure that we lock into place, the more expensive and the more onerous will be the burden on all segments of society when we finally do act."

BILL KEENAN is a freelance business writer and former editor of *Selling* magazine.

# Sunrise for Renewable Energy?

**Energy: Renewable energy may not appear to be competitive with oil and gas at the moment, but the gap is closing.**

A spectacular sea snake has been spotted slithering around Scotland's northern waters. Though it is fiery red in colour, and some 100 metres in length, the writhing beastie has not sent the locals of Orkney running for the hills. That is because it is actually an innovative new device designed to produce electricity by capturing energy from the ocean's waves. Pelamis, manufactured by Ocean Power Delivery, a British firm, is at the vanguard of the next energy revolution. Or at least that is what proponents of renewable energy would have you believe. Orkney is home to the European Union's main marine-energy test centre, and local politicians and academics like to boast that Scotland, ideally suited to wave and wind-power projects, will become "the Saudi Arabia of renewable energy".

If such claims sound a bit over the top, they are entirely in keeping with the euphoria now sweeping through the renewable-energy sector. Money is pouring in as venture-capital firms, including many not previously interested in renewable energy, throw money at renewables. BP and Royal Dutch/Shell, two oil giants, have big renewables divisions. GE has unveiled "Eco-magination", an initiative focused on clean energy. High oil prices, environmental concerns, a desire for greater energy security and improved technologies "are combining to create the best investing environment ever for renewable power", observed Terry Pratt, a credit analyst at Standard & Poor's, in a report published in October. The International Energy Agency (IEA), a quasi-governmental agency not known for excessive greenery, forecasts that over $1 trillion will be invested in non-hydro renewable technologies worldwide by 2030. By then, the IEA predicts, such technologies will triple their share of the world's power generation to 6%. In some regions, such as western Europe and California, the share could top 20%.

Yet such predictions are met with scepticism by those who remember what happened after the oil shocks of the 1970s. Back then, high oil prices and concerns over scarcity led many firms to bet heavily on alternative-energy technologies. Most of them lost those bets when oil and gas prices fell in the late 1980s. One of the biggest losers was Exxon. Its current boss, Lee Raymond, has vowed not to spend another penny of his shareholders' money on renewables, which he calls "a complete waste of money".

The chief drawback of renewables is their cost compared with conventional energy sources. The cost of generating electricity from wind turbines is at least 5 cents per kilowatt hour (kWh), for example. Solar or wave power cost at least 18 or 20 cents per kWh. The cost of electricity from conventional sources, in contrast, is typically much lower—as little as 3 to 5 cents per kWh. Barring some dramatic breakthrough, renewable sources cannot, on the face of it, possibly compete.

## Changing the Rules

But look beyond the headline figures and a different picture emerges. Renewable energy has regulatory, commercial and technological trends on its side, all of which are working to close the cost gap with conventional sources. Taken together, they promise a far more sustainable, market-driven basis for investment in renewables than yesterday's faith in high oil prices—and suggest that renewable energy's cheerleaders could be on to something after all.

First, consider regulatory and policy trends. Critics have long complained that renewables have survived only because of government subsidies. They are right—but every form of energy is subsidised. America's huge Energy Act, signed into law by President Bush in August, hands most of its $80 billion or so of largesse not to wind or solar, but to well-entrenched industries such as oil, coal and nuclear. Germany and Spain handed out cash to their coal industries even as they subsidised windmills.

> **"Pricing schemes that favour renewable energy are also being made possible by the arrival of new technologies such as 'smart' meters."**

Yet many governments, striving to reduce carbon emissions, are now embracing policies that promise more enduring and politically palatable support for renewable energy than subsidies: "externalities" pricing. In some countries, especially in Europe, action has come in the form of direct taxes on carbon emissions—which, of course, greatly benefit renewable energy. Japan is phasing out its solar subsidies altogether next year. Tax is a four-letter word in America, so policymakers there have instead adopted a mix of regulations, rather than a carbon tax, to boost clean energy. These include such measures as tax credits and "renewable portfolio standards" that require a certain proportion of energy production within a particular state to come from renewables.

Second, these policy measures are being accompanied by the arrival of innovative business models built around renewables. A good example is Actus Lend Lease, an American firm, which is developing the world's largest solar-powered residential community in Hawaii to provide housing for American soldiers. "This is a business decision—there is no subsidy," says Chris Sherwood of Actus. Lenders were worried about the volatility of electricity prices, since Hawaii generates most of its electricity by burning imported oil, and the community's residents will pay a fixed rent, including utility bills, that is set by the army and adjusted only once a year. A sudden spike in the electricity price might have meant that the firm running the project would have been unable to make its debt repayments. Solar panels, in contrast, produce electricity at a known price for the lifetime of the panels. Reducing the uncertainty over energy costs, says Mr Sherwood, made it possible for the developers to borrow more.

Similarly, Sun Edison, an American start-up backed by Goldman Sachs and BP, has devised a clever new business model that overcomes a number of the real-world obstacles that have hitherto stymied renewable-energy projects. Simply put, it offers big retailers (such as Whole Foods and Staples) long-term, fixed-price electricity contracts in return for being able to set up solar panels on their rooftops. The retailers benefit from stable power prices, but do not have to buy or run the panels themselves; Goldman Sachs, which finances the panels, benefits from the associated tax credits and other offsets; BP sells more solar panels; and solar power has a better chance of taking off. Meanwhile, other ventures are looking to wind energy for a hedge. Several firms are putting together hybrid financial products that combine the output of wind farms in America's mid-west with that of natural gas-fired plants—thus hedging the volatility of both.

Pricing schemes that favour renewable energy are also being made possible by the arrival of new technologies such as "smart" meters, which allow for hour-by-hour variation in power prices. These make it possible for utilities to charge much more for power during the sweltering midday peak than early in the morning or late at night. Since solar panels produce their greatest power output in the middle of the day—just when prices are at their peak under a variable-pricing regime—Tim Woodward of Nth Power, a venture-capital firm specialising in energy, thinks smart meters with this type of "time of use" or "critical peak" pricing will make solar power far more attractive. "We see a groundswell toward this," he says. Several American states, led by California, are moving towards variable pricing, and the Energy Act encourages utilities to adopt it. Enel, Italy's national energy company, is rolling out smart meters to 30m customers across the country, and there are plans to make smart meters mandatory across the European Union, whenever a meter is installed or replaced.

# Boxing Clever

In the meantime, GridPoint, an American firm, is selling a "black box" at retailers such as Home Depot that its boss, Peter Corsell, claims will "solve the last-mile problem of the stupid grid." Usually, solar panels need a complex tangle of wires, inverters, batteries and other equipment to be installed to make them work. His firm replaces that with a "plug and play" device that also provides backup power. It even uses predictive software and an internet connection to juggle weather forecasts and utility pricing plans to decide when to sell power back on to the grid.

All of this is making renewables more attractive, even without advances in the generating technologies themselves. But those technologies are not standing still either. Wind energy is now a commercially viable business, without subsidies, in a number of places around the world. (The crucial factor is the "wind potential" of the site; even the best sites for wind turbines produce power only 30–40% of the time, and the average across all of Germany's wind turbines, for example, is just 11%.) Of course, government helped the industry get to this point. Denmark, for example, is home to world-class turbine manufacturers, such as NEG Micron and Vestas, thanks to early state aid. And tax credits and other subsidies help wind operators in Germany and elsewhere.

The key to wind's success in becoming commercially viable has been technologies that have allowed turbine size to grow from an average of 10 metres in diameter in the mid-1970s to over 80 metres today. To build and run such monstrous turbines, companies have devised new composites for the blades, variable-pitch blades that catch the slightest of breezes, variable-speed drive motors and other advances. A doubling of wind speed means about an eight-fold gain in a windmill's energy output, so making windmills taller makes sense, as winds tend to be stronger and more stable higher off the ground. Of course, there are practical limits: make a turbine too big and you cannot deliver it to a field or a windy mountain-top. But offshore, where turbines can be moved by ship, that is not a constraint. Experts expect offshore wind to take off dramatically, especially in Europe, which has both plenty of wind and lots of protesters who object to land-based turbines. Robert Kleiburg of Shell muses that the industry may need to rethink turbine design for offshore environments, however.

**Talisman, an oil company, has decided to put up two windmills on top of its gas platforms.**

The prospects are also good for improvements in solar power. Ever since Bell Labs patented its design for a photovoltaic cell in 1954, crystalline silicon—the same stuff that is used to make computer chips—has been the dominant technology for such cells, thanks to its high reliability and conversion efficiency (at least compared with rival technologies). Silicon-based systems typically convert about 15% of the sun's energy into useful electricity. That may seem low, but since the fuel is free, the efficiency of conversion matters less than the overall cost per kilowatt of power delivered.

Alas, silicon photovoltaic cells are now victims of their own success. The solar industry has sucked up so much crystalline silicon that there is a global shortage, and prices have shot up. But crisis breeds invention. "In the old days, we'd get the garbage after the IT industry got the good stuff," says Rhone Resch of America's Solar Industries Association. But now half a dozen silicon-wafer plants are going up around the world dedicated solely to providing silicon for solar energy. "This is a watershed for the silicon industry," says Christopher O'Brien of Sharp Solar.

One firm hoping to capitalise on the silicon shortage is Evergreen Solar. It uses conventional crystalline silicon, but in an unusually frugal fashion. From crucibles of molten silicon, ribbons of the stuff are continuously pulled out. This "string-pulling" uses 30% less silicon than the usual sawing-and-etching method does, with further improvements in sight. But others are betting on a rival technology: thin films. Rather than etch wafers, various firms are creating solar panels on rolls of stainless steel (ECD Ovonics), plate glass (GE's Astropower division), and other materials amenable to continuous manufacturing processes. That means costs can be greatly reduced once full-scale plants are built and perfected, which would compensate for thin films' lower conversion efficiency.

"I'm betting against silicon," says Arno Penzias, a Nobel-winning scientist who is now with NEA, a venture-capital firm. Instead, he favours a flavour of thin-film solar technology known as "CIGS"—a sandwich of thin layers of copper, indium and gallium selenide pioneered at America's National Renewable Energy Laboratory (NREL). His firm invests in HelioVolt, which is trying to commercialise this technology; the firm claims that it can already achieve efficiencies close to those of silicon in the laboratory but using just one-hundredth the material. Billy Stanbery, HelioVolt's boss, thinks this technology could allow solar panels to be built into roofing materials, rather than installed on top. Shell's solar division, which is developing a thin film similar to CIGS, thinks it could reduce the cost of solar panels by more than 50% by 2012.

Another promising, but tricky, approach is organic solar panels. Konarka, whose founder won a Nobel prize for pioneering organic solar cells, is leading the charge in this area—but even one insider admits that commercialisation of its optical organic PV cells "is a long way off". Other researchers are applying nanotechnology and molecular chemistry to solar power, with the aim of mimicking photosynthesis. Most pundits think that is a long way off too. But a paper published by a team from the NREL in May raises a tantalising possibility: it found that tiny nanocrystals known as "quantum dots" could, in theory, make possible solar cells with around 70% efficiency. So the future for solar power could be bright indeed.

# Follow the Money

But what is most striking is that figures compiled by Shell Renewables in April 2004, when the oil price stood at $40 a barrel—it is currently closer to $60—found that wind turbines and solar panels could close the cost gap with conventional energy sources. Provided they are large enough and are sited in suitable locations, the most efficient modern wind turbines can produce electricity at a wholesale price (the price at which electricity producers buy and sell power on the grid) competitive with non-renewable sources.

Solar panels cannot produce power at such low cost, but comparing their cost-per-kWh with wholesale prices is arguably not the most relevant comparison. That is because in general, solar panels are used not by electricity producers selling power to the grid at wholesale prices, but by consumers who use solar power to supplement or replace power bought from utility companies at retail prices (typically 8 to 20 cents per kWh). So solar power need only match these higher retail prices in order for homeowners and businesses to start to consider it as a viable alternative. And it turns out that the most efficient of today's solar panels do indeed match the retail price of electricity in some parts of the world with high retail prices, such as Japan (which is now phasing out its solar subsidies).

Renewables' growing competitiveness is not, in short, simply the result of sky-high oil prices. And that explains why Wall Street is at last getting interested. Not long ago, America's renewable-energy industry held a finance conference in New York at the Waldorf Astoria hotel. Brian Daly, a financier with the Trust Company of the West, stood up to make a presentation in the bejewelled grand ballroom. He observed: "When I made my first presentations in this industry, there were ten guys with ponytails and I had to flip charts myself." Now, he observed, the Waldorf ballroom was packed with besuited bankers—and his slides appeared on a high-tech screen.

If you still need persuading that something big and exciting is happening in renewable energy, head back to the frothy waters of the North Sea off Scotland. There, you will find the energy equivalent of beating swords into ploughshares: the planting of windmills on oil platforms. Talisman, an independent oil company, has decided to put up two windmills on top of one of its gas platforms. Building stable platforms accounts for around a third of the cost of offshore wind farms. But the oil and gas industry in the North Sea, now in decline, has plenty of platforms sitting around.

A Talisman official explains that, for the moment, the energy will be used only to power the platform's operations, but in future it may serve as a generating station, and send power ashore. "This will be the greenest platform in the world," he says. If even hardened oilmen can look to the winds for inspiration, perhaps the time really has come for renewable energy after all.

---

# UNIT 4

# Biosphere: Endangered Species

## Unit Selections

## Key Points to Consider

- Are there ways to assess the value or worth of living organisms other than those from whom we derive direct benefits (our domesticated plant and animals species)? What are the relationships between economic assessments of the biosphere and moral or value judgments on the preservation of species?

- Why is the spread of invasive species through the global trading network so difficult to control and what kinds of damage do invasive species produce? What suggestions would you make to remedy the problem?

- What kinds of changes are taking place in the chemistry, physics, and biology of the ocean—the world's largest ecosystem? Explain some of the interconnections between such things as changing ocean temperatures, and the population of phytoplankton and the marine species that feed upon them.

## Student Web Site

www.mhcls.com/online

## Internet References

Further information regarding these Web sites may be found in this book's preface or online.

**Endangered Species**
> http://www.endangeredspecie.com

**Friends of the Earth**
> http://www.foe.co.uk/index.html

**Natural Resources Defense Council**
> http://nrdc.org

**Smithsonian Institution Web Site**
> http://www.si.edu

**World Wildlife Federation (WWF)**
> http://www.wwf.org

Tragically, the modern conservation movement began too late to save many species of terrestrial and aquatic plants and animals from extinction. In fact, even after concern for the biosphere developed among resource managers, their effectiveness in halting the decline of herds and flocks, packs and schools, or groves and grasslands and coral reefs has been limited by the ruthlessness and efficiency of the competition. Wild plants and animals compete directly with human beings and their domesticated livestock and crop plants for living space and for other resources such as sunlight, air, water, and soil. As the historical record of this competition in North America and other areas attests, since the seventeenth century human settlement has been responsible—either directly or indirectly—for the demise of many plant and wildlife species. It should be noted that extinction is a natural process, part of the evolutionary cycle, and not always created by human activity; but human actions have the capacity to accelerate a natural process that might otherwise take a millennia.

The first selection in the unit addresses one of the most pervasive biological issues throughout human history—that of invasive species—and one that has contributed significantly to the extinction rates that Wilson discusses. In "Strangers in Our Midst: The Problem of Invasive Alien Species," scientist Jeffrey McNeely of the World Conservation Union notes that "invasive" or non-native species that become established in environments and spread rapidly because they lack natural enemies are not only very damaging to human economic and other interests but loom large as one of the world's most serious biological threats. While the migration of people, plants, and animals has been a feature of human history for a millennia, as globalization and an increasing emphasis on international trade has dominated the late 20th and early 21st centuries, natural barriers to the ready movement of plants and animals without human aid have disappeared. But no nation on earth has, as yet, developed an effective coordinated strategy to deal with this more pervasive problem.

In "Markets for Biodiversity Services: Potential Roles and Challenges," authors Michael Jenkins, Sara Scherr, and Mira Inbar (all with non-profit NGOs) argue that while historically it may have been the role of government to deal with such problems as invasive species and the loss of biodiversity, an alternative based in the market exists. Public sector financing for protection of the biosphere is facing increasingly severe challenges, particularly in countries where the prevailing political philosophies may deny that problems exist. In such situations, the authors contend, the lower cost approach to protecting and preserving biodiversity may be to pay landowners to manage their lands in ways that will preserve natural systems and conserve native species. At least some of this payment could come from conservation organizations such as the Nature Conservancy, but other logical sources include private corporations, research institutes, and private individuals. What is essential is to find market-based mechanisms such as eco-tourism that will convince resource owners and managers that good stewardship creates economic value.

# Strangers in Our Midst

## *The Problem of Invasive Alien Species*

JEFFREY A. MCNEELY

Invasive alien species—non-native species that become established in a new environment then proliferate and spread in ways that damage human interests—are now recognized as one of the greatest biological threats to our planet's environmental and economic well-being.[1]

Most nations are already grappling with complex and costly invasive-species problems: Zebra mussels (*Dreissena polymorpha*) from the Caspian and Black Sea region affect fisheries, mollusk diversity, and electric-power generation in Canada and the United States; water hyacinth (*Eichornia crassipes*) from the Amazon chokes African and Asian waterways; rats originally carried by the first Polynesians exterminate native birds on Pacific islands; and deadly new disease organisms (such as the viruses causing SARS, HIV/AIDS, and West Nile fever) attack human, animal, and plant populations in temperate and tropical countries. For all animal extinctions where the cause is known, invasive alien species are the leading culprits, contributing to the demise of 39 percent of species that have become extinct since 1600.[2] The 2000 IUCN Red List of Threatened Species reported that invasive alien species harmed 30 percent of threatened birds and 15 percent of threatened plants.[3] Addressing the problem of these invasive alien species is urgent because the threat is growing daily and the economic and environmental impacts are severe.

A key question is whether the global reach of modern human society can be matched by an appropriate sense of responsibility. One critical element of this question is the definition of "native," a concept with challenging spatial and temporal dimensions. While every species is native to a particular geographic area, this is just a snapshot in time, because species are constantly expanding and contracting their ranges, sometimes with human help. For example, Britain has nearly 40 more species of birds today than were recorded 200 years ago. About a third of these are deliberate introductions, such as the Little Owl (*Athene noctua*), while the others are natural colonizations that may be taking advantage of climate change.[4]

**An invasive alien species is not a "bad" species but rather one "behaving badly" in a particular context.**

According to one view, local biological "enrichment" by non-native species always harms native species at some level, so any introduction should be regarded, at least in principle, as undesirable. An opposing view is that because species are constantly expanding or contracting their range, new species—especially those that are beneficial to people, such as crops, ornamental plants, and pets—should be welcomed as "increasing biodiversity" unless they are clearly harmful. According to this perspective, in the case of British birds noted above, only those introduced by people and that are causing ecological or economic damage, such as pigeons, are considered to be invasive.

All continental areas have suffered from invasions of alien species, losing biological diversity as a result, but the problem is especially acute on islands in general and for small islands in particular. The physical isolation of islands over millions of years has favored the evolution of unique species and ecosystems, so islands often have a high proportion of endemic species. The evolutionary processes associated with isolation have also meant that island species are especially vulnerable to predators, pathogens, and parasites from other areas. More than 90 percent of the 115 birds known to have become extinct over the past 400 years were endemic to islands.[5] Most of these evolved in the absence of mammalian predators, so the arrival of rats and cats carried by people has had a devastating impact.

Island plants are also affected. For example, the tree *Miconia calvescens* replaced the forest canopy on more than 70 percent of the island of Tahiti over a 50-year time span, starting with a few trees in two botanical gardens. Some 40–50 of the 107 plant species endemic to the island of Tahiti are believed to be on the verge of extinction primarily due to this invasion.[6] Introduced animals also can affect plants. For example, goats introduced on St. Clemente Island, California, have caused the extinction of eight endemic species of plants and have endangered eight others.[7]

An invasive alien species is not a "bad" species but rather one "behaving badly" in a particular context, usually due to inappropriate human agency or intervention. A species may be so threatened in its natural range that it is given legal protection, yet it may generate massive ecological and other damage elsewhere.

The degradation of natural habitats, ecosystems, and agricultural lands (through loss of vegetation and soil and pollution of land and waterways) that has occurred throughout the world has made it easier for non-native species to become invasive, opening up new possibilities for them. For all of these reasons, and others that will become apparent below, the issue of invasive alien species is receiving growing international attention.

# The Vectors: How Species Move Around the World

The natural barriers of oceans, mountains, rivers, and deserts have provided the isolation that has enabled unique species and ecosystems to evolve. But in just a few hundred years, these barriers have been overcome by technological changes that helped people move species vast distances to new habitats, where some of them became invasive. The growth in the volume of international trade, from US$192 billion in 1960 to almost $6 trillion in 2003,[8] provides more opportunities than ever for species to be spread either accidentally or deliberately.

Some movement seems accidental, or at least incidental, in that transporting the species was not the purpose of the transporter. For example, ballast water is now regarded as the most important vector for transoceanic movements of shallow-water coastal organisms, dispersing fish, crabs, worms, mollusks, and microorganisms from one ocean to another. Enclosed water bodies like San Francisco Bay are especially vulnerable. The bay already has at least 234 invasive alien species, causing significant economic damage. California has one of the toughest ballast water laws in the nation, requiring ships from foreign ports to exchange their ballast water 200 miles from the California coastline, but enforcement remains spotty at best.

Ballast water may also be important in the epidemiology of waterborne diseases affecting plants and animals. One study measured the concentration of the bacteria *Vibrio cholerae*—which cause human epidemic cholera—in the ballast water of vessels arriving to the Chesapeake Bay from foreign ports, finding the bacteria in plankton samples from all ships.[9]

Other invasives are hitchhikers on global trade. For example, the Asian long-horned beetle (*Anoplophora glabripennis*) is one of the newest and most harmful invasive species in the United States. Originating in northeastern Asia, it finds its way to the United States through packing crates made of low-quality timber (that which is too infested for other uses). The number of insects found in materials imported from China increased from 1 percent of all interceptions in 1987 to 20 percent in 1996.[10] Outbreaks were reported in and around Chicago as early as 1992, in Brooklyn in August 1996, and in California in 1997. The beetle finds a congenial home among native maples, elders, elms, horse chestnuts, and others. The U.S. Department of Agriculture predicted that if the beetle becomes established, it could denude Main Street, USA, of shade trees, affect lumber and maple sugar production, threaten tourism in infested areas, and reduce biological diversity in forests.[11]

Another dangerous trade-related species for North America is the Asian gypsy moth (*Lymantria dispar*), which was first reported in the United States in 1991, entering as egg masses attached to ships or cargo from eastern Siberia. The caterpillars of this species are known to feed on more than 600 species of trees, and as moths, the females can disperse themselves over long distances. Scientists fear that this species could cause vastly more damage than the European gypsy moth, which already defoliates 1.5 million hectares of forest per year in North America.

With almost 700 million people crossing international borders as tourists each year, the opportunities for them to carry potential invasive species, either knowingly or unknowingly, are profound and increasing. Many tourists return with living plants that may become invasive, or carry exotic fruits that may be infested with invasive insects that can plague agriculture back home. Travelers may also carry diseases between countries, as apparently happened with the SARS virus. Tourism is considered an especially efficient pathway for invasive alien species on subAntarctic islands such as South Georgia. Visitors to the island reached 15,000 in 1999. Part of the problem is that many tourists are visiting similar islands on the same trip, increasing the chances of a seed, fruit, or insect being carried, more than would be expected from a single landing of a few people who spend an extended time on one island.[12]

Many species are introduced on purpose but have unintended consequences. One example of purposeful introduction gone wrong is the extensive stocking program that introduced African tilapia *Oreochromis* into Lake Nicaragua in the 1980s, resulting in the decline of native populations of fish and the imminent collapse of one of the world's most distinctive freshwater ecosystems. The alteration of Lake Nicaragua's ecosystem is likely to have effects on the planktonic community and primary productivity of the entire lake—Central America's largest—destroying native fish populations and likely leading to unanticipated consequences.[13]

Sport fishers have also had an influence, importing their favorite game fish into new river systems, where they can have significant negative impacts on native species. For example, the northern pike (*Esox lucius*) has invaded rivers in Alaska and is replacing native species of salmon. While the northern pike occurs naturally in some parts of Alaska, it was introduced to the salmon-rich south-central area in the 1950s, probably by a fisherman who brought it to Bulchitna Lake. Flooding in the 1980s subsequently spread the pike into the streams of the Susitna and Matanuska river basins. Pike have now occupied at least a dozen lakes and four rivers in some of the richest salmon and trout habitat in the Pacific Northwest. Rainbow trout are an even greater threat. Originating in western North America, they have been introduced into 80 new countries, often with devastating impacts on native fish.

Pets are also a problem. Domestic cats can plunder ecosystems that they did not previously inhabit. On Marion Island in the sub-Antarctic Indian Ocean, cats were estimated to kill about 450,000 seabirds annually.[14] Exotic pets may escape—or be released when they have outlived their novelty—and become established in their new home. Stories of crocodiles in the Manhattan's sewer system are probably fanciful, but many former pets are becoming established in the wild. For example, Monk parakeets (*Myiopsitta monachus*), descended from former pets that were released possibly in the 1960s, have invaded some 76 localities in 15 U.S. states.[15] Native to southern South America, they are the only parrots that build their own nests, some of which support several hundred individuals and have separate families living in different chambers. Some believe that they soon will become widespread throughout the lower 48 states, posing a significant threat to at least some agricultural lands by feeding on ripening crops. And Burmese pythons (*Python molurus*) have become established in Everglades National Park, where they reach a very large size and prey on many native species, even alligators.

Pet stores often advertise invasive species that are legally controlled. For example, the July 2000 issue of the magazine *Tropical Fish Hobbyist* recommended several species of the genus *Salvinia* as aquarium plants, even though they are considered noxious weeds in the United States and prohibited by Australian quarantine laws.

The globalization of trade and the power of the Internet offer new challenges, as sales of seeds and other organisms by mail order or over the Internet pose new and very serious risks to the ecological security of all nations. Controls on harvest and export of species are required as part of a more responsible attitude of governments toward the potential of spreading genetic pollution through invasive species. Further, all receiving countries want to ensure that they are able to control what is being imported. Virtually all countries in the world have serious problems in this regard, an issue that some countries are calling "biosecurity."

# The Science of Understanding Invasions

Biodiversity is dynamic, and the movement of species around the world is a continuing process that is accelerating through expanding global trade. By trying to identify which species are especially likely to become invasive, and hence harmful to people, ecologists are improving the quality of invasion biology as a predictive science so that people can continue to benefit from global biodiversity without paying the costs resulting from species that later become harmful.

Previous examples indicate the characteristics that can make a species invasive. For instance, coastal ecosystems are frequently invaded by microorganisms from ballast water for three main reasons. First, concentrations of bacteria and viruses exceed those reported for other taxonomic groups in ballast water by 6 to 8 orders of magnitude, and the probability of successful invasion increases with inoculation concentration. Second, the biology of many microorganisms combines a high capacity for increase, asexual reproduction, and the ability to form dormant resting stages. Such flexibility in life history can broaden the opportunity for successful colonization, allowing rapid population growth when suitable environmental conditions occur. And third, many microorganisms can tolerate a broad range of environmental conditions, such as salinity or temperature, so many sites may be suitable for colonization.[16] Insects are a major problem because they can lay dormant or travel as egg masses and are difficult to detect. The African tilapia introduced to Lake Nicaragua adapted well, because they are able to grow rapidly; feed on a wide range of plants, fish, and other organisms; and form large schools that can migrate long distances. Further, they are maternal mouth brooders, so a single female can colonize a new environment by carrying her young in her mouth.[17] Rapid growth, generalized diet, ability to move large distances, and prolific breeding are all characteristics of successful invaders.

It is not always simple, however, to distinguish a beneficial non-native species from one at significant risk of becoming invasive. A non-native species that is useful in one part of a landscape may invade other parts of the landscape where its presence is undesirable, and some species may behave well for decades before suddenly erupting into invasive status. The Nile Perch (*Lates niloticus*), for example, was introduced to Lake Victoria in the 1950s but did not become a problem until the 1980s, when it was a key factor in the extinction of as many as half of the lake's 500 species of endemic fish, attractive prey for the perch.[18] That said, ecologists over the past several decades have agreed on some broad principles for guiding risk assessment. First, the probability of a successful invasion increases with the initial population size and with the number of attempts at introduction. While it is possible for a species to invade with a single gravid female or fertile spore, the odds of doing so are very low. Second, among plants, the longer a non-native plant has been recorded in a country and the greater the number of seeds or other propagules that it produces, the more likely it will become invasive. Third, species that are successful invaders in one situation are likely to be successful in other situations; rats, water hyacinth, microorganisms, and many others fall into this category. Fourth, intentionally introduced species may be more likely to become established than are unintentionally introduced species, at least partly because the vast majority of these have been selected for their ability to survive in the environment where they are introduced. Fifth, plant invaders of croplands and other highly disturbed areas are concentrated in herbaceous families with rapid growth and a wide range of environmental tolerances, while invaders of undisturbed natural areas are usually from woody families, especially nitrogen-fixing species that can live in nitrogen-poor soils.[19] And sixth, fire, like disturbance in general, increases invasion by introduced species. So ecosystems that are naturally prone to fire, such as the fynbos of South Africa, coastal chaparral in California, and maquis in the Mediterranean,[20] can be heavily invaded if fire-liberated seeds of invasive species are available. (These are all shrub communities adapted to cool, wet winters and hot, dry summers, where fire is a regular phenomenon. They are also rich in species: Fynbos have about 8,500 species that include many endemic *Proteaceae*; chaparral have about 5,000 species; and maquis have 25,000—of which about 60 percent are endemic to the Mediterranean region.[21])

Other ecological factors that may favor nonindigenous species include a lack of controlling natural enemies, the ability of an alien parasite to switch to a new host, an ability to be an effective predator in the new ecosystem, the availability of artificial or disturbed habitats that provide an ecosystem the aliens can easily invade, and high adaptability to novel conditions.[22]

# Invasive Alien Species and Protected Areas

Protected areas are widely perceived as being devoted to conserving natural ecosystems. Ironically, protected areas are in fact heavily damaged by invasive alien species, and many protected-area managers consider this their biggest problem. Some examples:

- Galapagos National Park, a World Heritage site, is being affected by numerous invasive alien species, including pigs, goats, feral cats, fire ants, and mosquitoes.
- Kruger National Park, South Africa's largest, has recorded 363 alien plant species, including water weeds that pose a serious threat to the park's rivers.
- In the Wadden Sea, a biosphere reserve and Ramsar site protected by the Netherlands, Germany, and Denmark, the Pacific oyster has invaded, having escaped captive management. It is disrupting tourism because of its sharp shells. It has also carried with it numerous other invasive alien species.
- The Wet Tropic World Heritage Area of North Queensland, Australia, is infested by numerous invasive alien species, of which the worst is the pond apple from Florida, which has invaded creeks and riverbanks, wetlands, melaleuca swamps, and mangrove communities. Feral pigs, another invasive species, help to spread the species. The pond apple is now rare in its native range in the Florida Everglades.
- Everglades National Park in Florida, another World Heritage site, is threatened by the invasion of melaleuca from Queensland, demonstrating that species that may behave well in their natural habitat can be a serious problem when they invade somewhere else.
- Tongariro National Park, New Zealand, is also a World Heritage site, but a third of its territory has been infested by heather, a European plant deliberately introduced into New Zealand by an early park warden in 1912 in an attempt to reproduce the moors of Scotland.

These are just a few examples among many that could be cited that demonstrate that even the most strictly protected areas can be extremely vulnerable to invasion by non-native species.

---

It is sometimes argued that systems with great species diversity are more resistant to new species invading. However, a study in a California riparian system found that the most diverse natural assemblages are in fact the most invaded by non-native plants, and protected areas worldwide are heavily invaded by non-native plants and animals.[23] Dalmatian toadflax (*Linaria dalmatica*) is invading relatively undisturbed shrub-steppe habitat in the Pacific Northwest, wetland nightshade (*Solanum tampicense*) is invading cypress wetlands in central and south Florida, and garlic mustard (*Allilaria officinalis*) is often found in relatively undisturbed systems in the northern parts of North America.

This work helps resolve the controversy over the relationship between biodiversity and invasions, suggesting that the scale of investigation its a critical factor. Theory suggests that non-native species should have a more difficult time invading a diverse ecosystem, because the web of species interactions should be more efficient in using resources such as nutrients, light, and water than would fewer species, leaving fewer resources available for the nonnative species. But even in well-protected landscapes such as national parks, invaders often seem to be more successful in diverse ecosystems. Even though diversity does matter in fending off invasives, its effects are negated by other factors at larger scales. The most diverse ecosystems might be at the greatest risk of invasion, while losses of species, if they affect community-scale diversity, may erode invasion resistance.[24]

# The Economic Impacts of Invasion

One reason invasive alien species are attracting more attention is that they are having substantial negative impacts on numerous economic sectors, even beyond the obvious impacts on agriculture (weeds), forestry (pests), and health (diseases or disease vectors). The probability that any one introduced species will become invasive may be low, but the damage costs and costs of control of the species that do become invasive can be extremely high (such as the recent invasion of eastern Canada by the European brown spruce longhorn beetle (*Tetropium fuscum*), which threatens the Canadian timber industry).

Estimates of the economic costs of invasive alien species include considerable uncertainty, but the costs are profound—and growing (see Table 1).

Most of these examples come from the industrialized world, but developing countries are experiencing similar, and perhaps proportionally greater, damage. Invasive alien insect pests—such as the white cassava mealybug (*Phenacoccus herreni*) and larger grain borer (*Prostephanus truncates*) in Africa—pose direct threats to food security. Alien weeds constrain efforts to restore degraded land, regenerate forests, and improve utilization of water for irrigation and fisheries. Water hyacinth and other alien water weeds that choke waterways currently cost developing countries in Africa and Asia more than US$100 million annually. Invasive alien species pose a threat to more than $13 billion of current and planned World Bank funding to projects in the irrigation, drainage, water supply, sanitation, and power sectors.[25] And a study of three developing nations (South Africa, India, and Brazil) found annual losses to introduced pests of $138 billion per year.[26]

In addition to the direct costs of managing invasives, the economic costs also include their indirect environmental consequences and other nonmarket values. For example, invasives may cause changes in ecological services by disturbing the operation of the hydrological cycle, including flood control and water supply, waste assimilation, recycling of nutrients,

# Table 1   Indicative Costs of Some Invasive Alien Species (in U.S. Dollars)

| Species | Economic Variable | Economic Impact |
|---|---|---|
| Introduced disease organisms | Annual cost to human, plant, and animal health in the United States | $41 billion per year[a] |
| A sample of alien species of plants and animals | Economic costs of damage in the United States | $137 billion per year[b] |
| Salt cedar | Value of ecosystem services lost in western United States | $7–16 billion over 55 years[c] |
| Knapweed and leafy spurge | Impact on economy in three U.S. states | Direct costs of $40.5 million per year; indirect costs of $89 million[d] |
| Zebra mussel | Damages to U.S. industry | Damage of more than $2.5 billion to the Great Lakes fishery between 1998–2000;[e] $5 billion to U.S. industry by 2000[f] |
| Most serious invasive alien plant species | Costs 1983–1992 of herbicide control in England | $344 million per year for 12 species[g] |
| Six weed species | Costs in Australia agroecosystems | $105 million per year[h] |
| Pinus, Hakeas, and Acacia | Costs on South African floral kingdom to restore to pristine state | $2 billion total for impacts felt over several decades[i] |
| Water hyacinth | Costs in seven African countries | $20–50 million per year[j] |
| Rabbits | Costs in Australia | $373 million per year (agricultural losses)[k] |
| Varroa mite | Economic cost to beekeeping in New Zealand | An estimated $267–602 million over the next 35 years[l] |

[a] P. Daszak, A. Cunningham, and A. D. Hyatt, "Emerging Infectious Diseases of Wildlife: Threats to Biodiversity and Human Health," *Science,* 21 January 2000, 443–49.

[b] D. Pimentel, L. Lach, R. Zuniga, and D. Morrison, "Environmental and Economic Costs of Non-indigenous Species in the United States," *BioScience* 50 (2000): 53–65.

[c] E. Zavaleta, "Valuing Ecosystem Services Lost to Tamarix Invasion in the United States," in H. A. Mooney and R. J. Hobbs, eds., *Invasive Species in a Changing World* (Washington DC: Island Press, 2000).

[d] D. A. Bangsund, F. L. Leistritz, and J. A. Leitch, "Assessing Economic Impacts of Biological Control of Weeds: The Case of Leafy Spurge in the Northern Great Plains of the United States," *Journal of Environmental Management* 56 (1999): 35–43; and D. A. Bangsund, S. A. Hirsch, and J. A. Leitch, *The Impact of Knapweed on Montana's Economy* (Fargo, ND: Department of Agricultural Economics, North Dakota State University, 1996).

[e] P. C. Focazio, "Coordinated Issue Area: Aquatic Nuisance, Non-Indigenous, and Invasive Species," *Coastlines* 30, no.1 (2001): 4–5.

[f] "Coatings to Repel Zebra Mussels," U.S. Army Construction Engineering Research Laboratory fact sheet, http://www.cecer.army.mil/facts/sheets/FL10.html.

[g] M. Williamson, "Measuring the Impact of Plant Invaders in Britain," in S. Starfinger, K. Edwards, I. Kowarik, and M. Williamson, eds., *Plant Invasions: Ecological Mechanisms and Human Responses* (Leiden, Netherlands: Backhuys, 2000), 57–70.

[h] A. Watkinson, R. Freckleton, and P. Dowling, "Weed Invasions of Australian Farming Systems: From Ecology to Economics," in C. Perrings, M. Williamson, and S. Dalmazzone, eds., *The Economics of Biological Invasions* (Cheltenham, UK: Edward Elgar, 2000), 94–116.

[i] J. Turpie and B. Heydenrych, "Economic Consequences of Alien Infestation of the Cape Floral Kingdom's Fynbos Vegetation," in Perrings, Williamson, and Dalmazzone, ibid., pages 152–82.

[j] S. Joffe and S. Cook, *Management of the Water Hyacinth and Other Aquatic Weeds: Issues for the World Bank* (Cambridge, UK: Commonwealth Agriculture Bureau International (CABI) Bioscience, 1997).

[k] P. White and G. Newton-Cross, "An Introduced Disease in an Invasive Host: The Ecology and Economics of Rabbit Carcivirus Disease (RCD) in Rabbits in Australia," in Perrings, Williamson, and Dalmazzone, note h above, pages 117–37.

[l] R. Wittenberg and M. J. W. Cock, eds., *Invasive Alien Species: A Tool Kit of Best Prevention and Management Practices* (Wallingford, UK: Global Invasive Species Programme and CABI, 2001).

SOURCE: J. A. McNeely.

conservation and regeneration of soils, pollination of crops, and seed dispersal. Such services have current-use value and option value (the potential value of such services in the future). In the South African fynbos, for example, the establishment of invasive tree species—which use more water than do native species—has decreased water supplies for nearby communities and increased fire hazards, justifying government expenditures equivalent to US$40 million per year for both manual and chemical control.[27]

**Customs and quarantine practices, developed in an earlier time, are inadequate safeguards against the rising tide of species that threaten native biodiversity.**

Many people in today's globalized economy are driven especially by economic motivations. Those who are importing non-native species are usually doing so with a profit motive and often seek to avoid paying for possible associated negative impacts if those species become invasive. The fact that these negative impacts might take several decades to appear make it all the easier for the negative economic impacts to be ignored. Similarly, those who are ultimately responsible for such "accidental" introductions (for example, through infestation of packing materials or organisms carried in ballast water) seek to avoid paying the economic costs that would be required to prevent these "accidental," but predictable, invasions. In both cases, the potential costs are externalized to the larger society, and to future generations.

# Responses

Customs and quarantine practices, developed in an earlier time to guard against diseases and pests of economic importance, are inadequate safeguards against the rising tide of species that threaten native biodiversity. Globally, about 165 million 6-meter-long, sealed containers are being shipped around the world at any given time. This number is far larger than custom officers can reasonably be expected to examine in detail. In the United States, some 1,300 quarantine officers are responsible for inspecting 410,000 planes and more than 50,000 ships, with each ship carrying hundreds of containers. While they intercept alien species nearly 50,000 times a year, it is highly likely that at least tens of thousands more enter the country uninspected each year. In Europe, inspection at the port of entry is also desperately overextended, and once a container enters the European Union, no further border inspections are done. This is a recipe for disaster.

Instead, a different set of strategies is now needed to deal with invasive species. These include prevention (certainly the most preferable), early eradication, special containment, or integrated management (often based on biological control). Mechanical, biological, and chemical means are available for controlling invasive species of plants and animals once they have arrived. Early warning, quarantine, and various other health measures are involved to halt the spread of pathogens.[28]

The international community has responded to the problem of invasive alien species through more than 40 conventions or programs, and many more are awaiting finalization or ratification.[29] The most comprehensive is the 1992 Convention on Biological Diversity, which calls on its 188 parties to "prevent the introduction of, control, or eradicate those alien species which threaten ecosystems, habitats, or species" (Article 8h).[30] A much older instrument, one that is virtually universally applied, is the 1952 International Plant Protection Convention, which applies primarily to plant pests, based on a system of phytosanitary certificates. Regional agreements further strengthen this convention. Other instruments deal with invasive alien species in specific regions (such as Antarctica), sectors (such as fishing in the Danube River), or vectors (such as invasive species in ballast water, through the International Maritime Organization). The fact that the problem continues to worsen indicates that the international response to date has been inadequate.

On the national level, some legal measures can offer very straightforward methods of preventing or managing invasions. For example, to deal with the problem of Asian beetle invasions, the United States now requires that all solidwood packing material from China must be certified free of bark (under which insects may lurk) and heat-treated, fumigated, or treated with preservatives. China might reasonably issue a reciprocal regulation, as North American beetles are a hazard there.

The nursery industry is by far the largest intentional importer of new plant taxa. Issuing permits for imported species is a good way for the agencies responsible for managing such invasions to keep track of what is being traded and moved around the country. Some people believe that it is impossible to issue a regulation containing a list of permitted and prohibited species, at least partly because the ornamental horticulture industry is always seeking new species. But the Florida Nurserymen and Growers Association recently identified 24 marketed species on a black list drawn up by Florida's Exotic Pest Plant Council and decided to discourage trade in 11 of the species (the least promising sellers in any case).[31]

Sometimes nature itself can fight back against invasive alien species, at least when they reach plague proportions. For example, the zebra mussels that have invaded the North American Great Lakes with disastrous effects are now declining because a native sponge (*Eunapius fragilis*) is growing on the mussels, preventing them from opening their shells to feed or to breathe. The sponge has become abundant in some areas, while the zebra mussel population has fallen by up to 40 percent, although it is not yet clear whether the sponges will be effective in controlling the invasive mussels in the long term.[32]

Biological control—the intentional use of natural enemies to control an invasive species—is an important tool for managers. Some early efforts at biological control agents had disastrous effects, such as South American cane toads (*Bufo marinus*) in Australia, Indian common mynahs (*Acridotheres tristis*) in Hawaii, and Asian mongooses (*Herpestes javanicus*) in the Caribbean. Not only did these species not deal with the problem species upon which they were expected to prey, but they ended up causing havoc to native species and ecosystems. On the other hand, biological control programs are now much more carefully considered and in many cases are the most efficient, most effective, cheapest, and least damaging to the environment of any of the options for dealing with invasives that have already arrived.[33] Examples include the use of a weevil (*Cyrtobagous salviniae*) to control salvinia fern (*Salvinia molesta*), another weevil (*Neohydronomus affinis*) to control water lettuce (*Pistia stratiotes*), and a predatory beetle (*Hyperaspis pantherina*) to control orthezia scale (*Orthezia insignis*) that threatened the endemic national tree of Saint Helena (*Commidendrum robustum*).[34]

Those seeking to use viruses or other disease organisms to control an invasive species need to understand ecological links. When millions of rabbits died after the intentional introduction of the myxomatosis virus in the United Kingdom, for example, populations of their predators, including stoats, buzzards, and owls, declined sharply. The impact affected other species indirectly, leading to local extinction of the endangered large blue butterfly (*Maculina arion*) because of reduced grazing by rabbits on heathlands, which removed the habitat for an ant species that assists developing butterfly larvae.[35] But the use of the myxoma virus in conjunction with 1080 poison on the Phillip Island in the South Pacific successfully eradicated invasive rabbits, allowing the recovery of the island's vegetation (including the endemic *Hibiscus insularis*).[36]

At small scales of less than one hectare, it appears possible with current technology to eradicate invasive species of plants through use of herbicides, fire, physical removal, or a combination of these, but the costs of eradication rise quickly as the area covered increases. With the right approach and technology, invasive alien mammals can be eradicated from islands of thousands of hectares in size. Rat eradication from islands of larger than 2,000 hectares has been successful, and large mammals have been removed from much bigger ones than that, primarily by hunting and trapping.

Environmentally sensitive eradication also requires the restoration of the community or ecosystem following the removal of the invasive. For example, the eradication of Norway rats from Mokoia Island in New Zealand was followed by greatly increased densities of mice, also alien species. Similarly, the removal of Pacific rats (*Rattus exulans*) from Motupao Island, New Zealand, to protect a native snail led to increases of an exotic snail to the detriment of the natives. And on Motunau Island, New Zealand, the exotic box-thorn (*Lycium ferocissimum*) increased after the control of rabbits. On Santa Cruz Island, off the west coast of California, removing goats led to dramatic increases in the abundance of fennel (*Foeniculum vulgare*) and other alien species of weeds. Thus reversing the changes to native communities caused by non-native species will often require a sophisticated understanding of ecological relationships. It is now well recognized that eradication programs are only the first step in a long process of restoration.[37] Sometimes native species become dependent on invasive ones, causing dilemmas for managers. For example, giant kangaroo rats (*Dipodomys ingens*) in the American West continually modify their burrow precincts by digging tunnels, clipping plants, and other activities. This chronic disturbance to soil and vegetation sometimes promotes the establishment of invasive species of plants that were originally imported as ornamentals from the Mediterranean so that they constitute a very large proportion of the vegetation on giant kangaroo rat territories. They have significantly larger seeds than do native species so are favored by the grain-eating kangaroo rats.[38] Because the kangaroo rats depend on non-native plant species for food and the non-native plant species depend upon the kangaroo rats to disturb their habitat continually, the relationship is mutualistic. This strong relationship may also inhibit population growth of native grassland plants that occupy disturbed habitats but have difficulty competing with nonnative weeds for resources. This mutualism presents an intractable conservation management dilemma, suggesting that it may be impossible to restore valley grasslands occupied by endangered kangaroo rats to conditions where native species dominate.

High-tech management measures are also being tried. For example, Australian scientists are planning to insert a gene known as "daughterless" into invasive male carp (*Cyprinus carpio*) in the Murray-Darling River, the country's longest, thereby ensuring that their offspring are male. The objective is to release them into the wild, sending wild carp populations into a decline and making room for the native species that are being threatened by the invasive European carp.[39] Using genetic modification can help eradicate an invasive alien species, but if the detrimental gene is released into nature and starts to flourish, many other species could be negatively affected. Thus the precautionary approach needs to be applied to control techniques as well as to introductions.

The problems of invasive alien species are so serious that actions must be taken even before we can be "certain" of all of their effects. However, mechanical removal, biocontrol, chemical control, shooting, or any other approach to controlling alien invasive species needs to be carefully considered prior to use to ensure that the implications have been fully and carefully considered, including impacts on human health, other species, and so forth. A public information program is also needed to ensure that the proposed measures are likely to be effective as well as socially and politically acceptable. Many animal-rights groups oppose the killing of any species of wildlife, for instance, even if they are causing harm to native species of plants and animals. The recent controversy surrounding the population of mute swans in the Chesapeake Bay is a good example.[40]

## Conclusions

Ecosystems have been significantly influenced by people in virtually all parts of the world; some have even called these "engineered ecologies." Thus, a much more conscious and better-informed management of ecosystems—one that deals with non-native species—is critical.

In just a few hundred years, major global forces have rendered natural barriers ineffective, allowing non-native species to travel vast distances to new habitats and become invasive alien species. The globalization and growth in the volume of trade and tourism, coupled with the emphasis on free trade, provide more opportunities than ever for species to be spread accidentally or deliberately. This inadvertent ending of millions of years of biological isolation has created major ongoing environmental problems that affect developed and developing countries, with profound economic and ecological implications.

Because of the potential for economic and ecological damage when an alien species becomes invasive, every alien species needs to be treated for management purposes as if it is potentially invasive, unless and until convincing evidence indicates that it is harmless in the new range. This view calls for urgent action by a wide range of governmental, intergovernmental, private sector, and civil institutions.

A comprehensive solution for dealing with invasive alien species has been developed by the Global Invasive Species Programme.[41] It includes 10 key elements:

- *An effective national capacity to deal with invasive alien species.* Building national capacity could include designing and establishing a "rapid response mechanism" to detect and respond immediately to the presence of potentially invasive species as soon as they appear, with sufficient funding and regulatory support; as well as implementing appropriate training and education programs to enhance individual capacity, including customs officials, field staff, managers, and policymakers. It could also include developing institutions at national or regional levels that bring together biodiversity specialists with agricultural quarantine specialists. Building basic border control and quarantine capacity and ensuring that agricultural quarantine, customs, and food inspection officers are aware of the elements of the Biosafety Protocol are other ways to deal with invasive alien species on a national level.

- *Fundamental and applied research at local, national, and global levels.* Research is required on taxonomy, invasion pathways, management measures, and effective monitoring. Further understanding on how and why species become established can lead to improved prediction on which species have the potential to become invasive; improved understanding of lag times between first introduction and establishment of invasive alien species; and better methods for excluding or removing alien species from traded goods, packaging material, ballast water, personal luggage, and other methods of transport.

## The problems of invasive alien species are so serious that actions must be taken even before we can be "certain" of all their effects.

- *Effective technical communications.* An accessible knowledge base, a planned system for review of proposed introductions, and an informed public are needed within countries and between countries. Already, numerous major sources of information on invasive species are accessible electronically and more could also be developed and promoted, along with other forms of media.
- *Appropriate economic policies.* While prevention, eradication, control, mitigation, and adaptation all yield economic benefits, they are likely to be undersupplied, because it is difficult for policymakers to identify specific beneficiaries who should pay for the benefits received. New or adapted economic instruments can help ensure that the costs of addressing invasive alien species are better reflected in market prices. Economic principles relevant to national strategies include ensuring that those responsible for the introduction of economically harmful invasive species are liable for the costs they impose; ensuring that use rights to natural or environmental resources include an obligation to prevent the spread of potential invasive alien species; and requiring importers of such potential species to have liability insurance to cover the unanticipated costs of introductions.

- *Effective national, regional, and international legal and institutional frameworks.* Coordination and cooperation between the relevant institutions are necessary to address possible gaps, weaknesses, and inconsistencies and to promote greater mutual support among the many international instruments dealing with invasive alien species.
- *A system of environmental risk analysis.* Such a system could be based on existing environmental impact assessment procedures that have been developed in many countries. Risk analysis measures should be used to identify and evaluate the relevant risks of a proposed activity regarding alien species and determine the appropriate measures that should be adopted to manage the risks. This would also include developing criteria to measure and classify impacts of alien species on natural ecosystems, including detailed protocols for assessing the likelihood of invasion in specific habitats or ecosystems.
- *Public awareness and engagement.* If management of invasive species is to be successful, the general public must be involved. A vigorous public awareness program would involve the key stakeholders who are actively engaged in issues relevant to invasive alien species, including botanic gardens, nurseries, agricultural suppliers, and others. The public can also be involved as volunteers in eradication programs of certain nonnative species, such as woody invasives of national parks for suggested actions that individuals can take.
- *National strategies and plans.* The many elements of controlling invasive alien species need to be well coordinated, ensuring that they are not simply passed on to the Ministry of Environment or a natural resource management department. A national strategy should promote cooperation among the many sectors whose activities have the greatest potential to introduce them, including military, forestry, agriculture, aquaculture, transport, tourism, health, and water-supply sectors. The government agencies with responsibility for human health, animal health, plant health, and other relevant fields need to ensure that they are all working toward the same broad objective of sustainable development in accordance to national and international legislation. Such national strategies and plans can also encourage collaboration between different scientific disciplines and approaches that can seek new approaches to dealing with problems caused by invasive alien species.
- *Invasive alien species issues built into global change initiatives.* Global change issues relevant to invasives begin with climate change but also include changes in nitrogen cycles, economic development, land use, and other fundamental changes that might enhance the possibilities of these species becoming established. Further, responses to global change issues, such as sequestering carbon, generating biomass energy, and recovering degraded lands, should be designed in ways that use native species and do not increase the risk of the spread of non-native invasives.

# What Can an Individual Do?

While the problem of invasive alien species seems daunting, an individual can make an important contribution to the problem, and if thousands of individuals work toward reducing the spread of invasive aliens, real progress can be made. Here are some steps that can be taken:

- Become informed about the issue.
- Grow native plants, keep native pets, and avoid releasing non-natives into the wild.
- Avoid carrying any living materials when traveling.
- Never release plants, fish, or other animals into a body of water unless they came out of that body of water.
- Clean boats before moving them from one body of water to another, and avoid using non-native species as bait.
- Support the work of organizations that are addressing the problem of invasive alien species.

- *Promotion of international cooperation.* The problem of invasive alien species is fundamentally international, so international cooperation is essential to develop the necessary range of approaches, strategies, models, tools, and potential partners to ensure that the problems of such species are effectively addressed. Elements that would foster better international cooperation could include developing an international vocabulary, widely agreed upon and adopted; cross-sector collaboration among international organizations involved in agriculture, trade, tourism, health, and transport; and improved linkages among the international institutions dealing with phytosanitary, biosafety, and biodiversity issues and supporting these by strong linkages to coordinated national programs.

Because the diverse ecosystems of our planet have become connected through numerous trade routes, the problems caused by invasive alien species are certain to continue. As with maintaining and enhancing health, education, and security, perpetual investments will be required to manage the challenge they present. These 10 elements will ensure that the clear and present danger of invasive species is addressed in ways that build the capacity to address any future problems arising from expanding international trade.

# Notes

1. H. A. Mooney, J. A. McNeely, L. E. Neville, P. J. Schei, and J. K. Waage, eds., *Invasive Alien Species: Searching for Solutions* (Washington, DC: Island Press, 2004).

2. B. Groombridge, ed., *Global Biodiversity: Status of the Earth's Living Resources* (Cambridge, UK: World Conservation Monitoring Centre, 1992).

3. C. Hilton-Taylor, IUCN *Red List of Threatened Species* (Gland, Switzerland: IUCN–The World Conservation Union (IUCN). 2000).

4. R. May, "British Birds by Number," *Nature,* 6 April 2000, 559–60.

5. Ibid., and Groombridge, note 2 above.

6. J-Y. Meyer, "Tahiti's Native Flora Endangered by the Invasion of *Miconia calvesens,*" *Journal of Geography* 23 (1997): 775–81.

7. D. Pimentel, L. Lach, R. Zuniga, and D. Morrison, "Environmental and Economic Costs of Nonindigenous Species in the United States," *BioScience* 50 (2000): 53–65.

8. World Trade Organization (WTO), *International Trade Statistics 2003* (Geneva: WTO, 2004).

9. G. M. Ruiz et al., "Global Spread of Microorganisms by Ships," *Nature,* 2 November 2000, 49–50.

10. J. E. Pasek, "Assessing Risk of Foreign Pest Introduction via the Solid Wood Packing Material Pathway," presentation made at the North American Plant Protection Organization Symposium on Pet Risk Analysis, Puerto Vallerta, Mexico, 18–21 March 2002.

11. U.S. Department of Agriculture, *Agricultural Research Service Research to Combat Invasive Species,* www.invasivespecies.gov/toolkit/arsisresearch.doc (accessed 27 April 2004).

12. S. L. Chown and K. J. Gaston, "Island-Hopping Invaders Hitch a Ride with Tourists in South Georgia," *Nature,* 7 December 2000, 637.

13. K. R. McKaye et al., "African Tilapia in Lake Nicaragua," *BioScience* 45 (1995): 406–11.

14. L. Winter, "Cats Indoors!" *Earth Island Journal,* Summer 1999, 25–26.

15. G. Zorpette, "Parrots and Plunder," *Scientific American,* July 1997, 15–17.

16. Ruiz et al., note 9 above.

17 McKaye et al., note 13 above.

18. A. J. Ribbink, "African Lakes and Their Fishes: Conservation Scenarios and Suggestions," *Environmental Biology of Fishes* 19 (1987): 3–26; T. Goldschmit, F. Witte, and J. Wanink, "Cascading Effects of the Introduced Nile Perch on the Detritivorousphytoplanktivorous Species in the Sublittoral Areas of Lake Victoria," *Conservation Biology* 7 (1993): 686–700; and R. Ogutu-Ohwayo, "Nile Perch in Lake Victoria: The Balance between Benefits and Negative Impacts of Aliens," in O. T. Sandlund, P. J. Schei, and A. Viken, eds., *Invasive Species and Biodiversity Management* (Dordrecht, Netherlands: Kluwer Academic Publishers, 1999), 47–64.

19. M. L. McKinney and J. L. Lockwood, "Biotic Homogenization: A Few Winners Replacing Many Losers in the Next Mass Extinction," *Tree* 14 (1999): 450–53.

20. C. M. D'Antonio, T. L. Dudley, and M. Mack, "Disturbance and Biological Invasions: Direct Effects and Feedbacks," in L. Locker, ed., *Ecosystems of Disturbed Ground* (Amsterdam: Elziveer, 1999).

21. B. Groombridge and M. D. Jenkins, *World Atlas of Biodiversity: Earth's Living Resources in the 21st Century* (Berkeley, CA: University of California Press, 2002)

22. Pimentel, Lach, Zuniga, and Morrison, note 7 above.

23. N. L. Larson, P. J. Anderson, and W. Newton, "Alien Plant Invasion in Mixed-Grass Prairie: Effects of Vegetation Type and Anthropogenic Disturbance," *Ecological Applications* 11 (2001):

128–41; and J. M. Levine, "Species Diversity and Biological Invasions: Relating Local Process to Community Pattern," *Science,* 5 May 2000, 852–54.

24. Ibid.

25. S. Noemdoe, "Putting People First in an Invasive Alien Clearing Programme: Working for Water Programme," in J. A. McNeely, ed., *The Great Reshuffling: Human Dimensions of Invasive Alien Species,* (Gland, Switzerland: IUCN, 2001), 121–26

26. D. Pimentel et al., "Economic and Environmental Threats of Alien Plant, Animal, and Microbe Invasions," *Agriculture, Ecosystems and Environment* 84 (2001): 1–20.

27. Ibid.

28. J. Kaiser, "Stemming the Tide of Invading Species," *Science,* 17 September 2000, 836–841.

29. C. Shine, N. Williams, and L. Gündling, *A Guide to Designing Legal and Institutional Frameworks on Alien Invasive Species* (Bonn, Germany: IUCN, 2000).

30. L. Glowka, F. Burhenne-Guilmin, and H. Synge, *A Guide to the Convention on Biological Diversity* (Gland, Switzerland: IUCN, 1994). See also K. Raustiala and D. G. Victor, "Biodiversity Since Rio: The Future of the Convention on Biological Diversity," *Environment,* May 1996, 16–20, 37–45.

31. Kaiser, note 28 above.

32. A. Ricciardi, F. Sneider, D. Kelch, and H. Reiswig, "Lethal and Sub-Lethal Effects of Sponge Overgrowth on Introduced Dreissenide Mussels in the Great Lakes—St. Lawrence River System," *Canadian Journal of Fisheries and Aquatic Sciences* 52: 2695–703.

33. M. S. Hoddle, "Restoring Balance: Using Exotic Species to Control Invasive Exotic Species," *Conservation Biology* 18 (2004): 38–49; S. M. Louda and P. Stiling, "The Double-Edged Sword of Biological Control in Conservation and Restoration," *Conservation Biology* 18 (2004): 50–53; and R. Wittenberg and M. J. W. Cock, eds., *Invasive Alien Species: A Tool Kit of Best Prevention and Management Practices* (Wallingford, UK: Global Invasive Species Programme and Commonwealth Agricultural Bureau International, 2001).

34. Wittenberg and Cock, ibid.

35. P. Daszak, A. Cunningham, and A. D. Hyatt. "Emerging Infectious Diseases of Wildlife: Threats to Biodiversity and Human Health," *Science,* 21 January 2000, 443–49.

36. P. Coyne, "Rabbit Eradication on Phillip Island," in Wittenberg and Cock, note 33 above, page 176.

37. R. C. Klinger, P. Schuyler, and J. D. Sterner, "The Response of Herbaceous Vegetation and Endemic Plant Species to the Removal of Feral Sheep from Santa Cruz Island, California," in C. R. Veitch and M. N. Klout, eds., *Turning the Tide: the Eradication of Invasive Species* (Gland, Switzerland, and Cambridge, UK: IUCN, 2002), 14 1–54.

38. P. Schiffman, "Promotion of Exotic Weed Establishment by Endangered Giant Kangaroo Rats (*Dipodomys ingens*) in a California Grassland," *Biodiversity and Conservation* 3 (1994): 524–37.

39. R. Nowak, "Gene Warfare: One Small Tweak and a Whole Species Will Be Wiped Out," *New Scientist,* 11 May 2002, 6.

40. See B. Engle, "No Swansong in the Chesapeake Bay," *Environment,* December 2003, 7.

41. The Global Invasive Species Programme (GISP) was established in 1997 as a consortium of the Scientific Committee on Problems of the Environment (SCOPE), CABI, and IUCN, in partnership with the United Nation Environment Programme and with funding from the Global Environment Facility (GEF). See J. A. McNeely, H. A. Mooney, L. E. Neville, P. J. Schei, and J. K. Waage, eds., *Global Strategy on Invasive Alien Species* (Gland, Switzerland: IUCN, 2001).

**JEFFREY A. MCNEELY** is chief scientist at IUCN-The World Conservation Union in Gland, Switzerland. His research focuses on a broad range of topics relating to conservation and sustainable use of biodiversity, with a particular focus in recent years on the relationship between agriculture and wild biodiversity, the relationship between biodiversity and human health, and the impacts of war on biodiversity. McNeely has written or edited more than 30 books, from *Mammals of Thailand* (Association for the Conservation of Wildlife, 1975), his first, to *Ecoagriculture: Strategies to Feed the World and Save Wild Biodiversity* (Island Press, 2003). He has also published extensively on biodiversity, protected areas, and cultural aspects of conservation. He may be reached at jam@iucn.org.

From *Environment,* July/August 2004, pp. 17–31. Reprinted by permission of the Helen Dwight Reid Educational Foundation. Published by Heldref Publications, 1319 Eighteenth St., NW, Washington, DC 20036-1802. Copyright © 2004. www.heldref.org

# Markets for Biodiversity Services

## *Potential Roles and Challenges*

MICHAEL JENKINS, SARA J. SCHERR, AND MIRA INBAR

H istorically, it has been the responsibility of governments to ensure biodiversity protection and provision of ecosystem services. The main instruments to achieve such objectives have been

- direct resource ownership and management by government agencies;
- public regulation of private resource use;
- technical assistance programs to encourage improved private management; and
- targeted taxes and subsidies to modify private incentives.

But in recent decades, several factors have stimulated those concerned with biodiversity conservation services to begin exploring new market-based instruments. The model of public finance for forest and biodiversity conservation is facing a crisis as the main sources of finance have stagnated, despite the recognition that much larger areas require protection. At the same time, increasing recognition of the roles that ecosystem services play in poverty reduction and rural development is highlighting the importance of conservation in the 90 percent of land outside protected areas. It is thus urgent to find new means to finance the provision of ecosystem services, yet under current conditions private actors lack financial incentives to do so.

## Crisis in Biodiversity Conservation Finance

Financing and management of natural protected areas has historically been perceived as the responsibility of the public sector. According to the United Nations Environment Programme, there are presently 102,102 protected areas worldwide, covering an area of 18.8 million square kilometers. Seventeen million square kilometers of these areas—11.5 percent of the Earth's terrestrial surface—are forests. Two-thirds of these have been assigned to one of the six protected-area management categories designated by the World Conservation Union (IUCN).

However, over the last few decades, severe cutbacks in the availability of public resources have undermined the effectiveness of such strategies. Protected areas in the tropics are increasingly dependent on international public or private donors for financing. Yet budgets for government protection and management of forest ecosystem services

are declining, as are international sources from overseas development assistance. Land acquisition for protected areas and compensation for lost resource-based livelihoods are often prohibitively expensive. For example, it has been estimated that $1.3 billion would be required to fully compensate inhabitants in just nine central African parks.[1] The donation-driven model is often unsustainable, both economically and environmentally. Sovereignty is also an issue: About 30 percent of private forest concessions in Latin America and the Caribbean and 23 percent in Africa are already foreign owned. At the same time, public responsibility for nature protection is shifting with processes of devolution and decentralization, and new sources of financing for local governments to take on biodiversity and ecosystem service protection have not been forthcoming.

Moreover, scientific studies increasingly indicate that biodiversity cannot be conserved by a small number of strictly protected areas.[2] Conservation must be conceived in a landscape or ecosystem strategy that links protected areas within a broader matrix of land uses that are compatible with and support biodiversity conservation in situ. To achieve such outcomes, it will be essential to engage private actors in conservation finance on a large scale. Yet the markets for products from natural areas and forests face at least three serious challenges: declining commodity prices for traditionally important products, such as timber; competition from illegal sources; and poorly functioning, overregulated markets. Thus, private forest owners and landowners need to find new revenue streams to justify retaining forests on the landscapes and to manage them well in the context of declining commodity prices and competition in natural forests from illegal sources of timber.

## Rural Development, Poverty Reduction, and Biodiversity

The vast majority of biodiversity resources in the world are found in populated landscapes, and it can be argued that the biodiversity that underpins ecosystem services critical to human health and livelihoods should have high priority in conservation efforts. An estimated 240 million rural people live in the world's high-canopy forest landscapes. In Latin America, for example, 80 percent of all forests are located in areas of medium to high human population density.[3] Population growth in the world's remaining "tropical wilderness areas" is twice the global average. More than a billion people live in the 25 biodiversity "hotspots" identified by Conservation International; in 16 of these hotspots, population growth is higher than the world average.[4] While species

**Table 1**  Estimated Financial Flows for Forest Conservation (in millions, U.S. dollars)

| Sources of Finance | SFM (early 1990s) | SFM (early 2000) | PAS (early 1990s) | PAS (early 2000) |
|---|---|---|---|---|
| Official development assistance | $2,000–$2,200 | $1,000–$1,200 | $700–$770 | $350–$420 |
| Public expenditure | NA | $1,600 | NA | $598 |
| Philanthropy[a] | $85.6 | $150 | NA | NA |
| Communities[b] | $365–$730 | $1,300–$2,600 | NA | NA |
| Private companies | NA | NA | NA | NA |

[a]Underestimates self-financing and in-kind nongovernmental organization contributions.
[b]Self-financing and in-kind contributions from indigenous and other local communities.
Note: In 1990, there were an estimated 100 million hectares of community-managed forests worldwide. SFM is "sustainable forest management." PAS stands for "protected area system."
Source: A. Molnar, S. J. Scherr, and A. Khare, *Current Status and Future Potential of Markets for Ecosystem Services of Tropical Forests: An Overview* (Washington, DC: Forest Trends, 2004).

richness is lower in drylands and other ecosystems not represented among the "hot spots," the species that play functional ecosystem roles are all the more important and difficult to replace.

Poor rural communities are especially dependent upon natural biodiversity. Low-income rural people rely heavily on the direct consumption of wild foods, medicines, and fuels, especially for meeting micronutrient and protein needs, and during "hungry" periods. An estimated 350 million poor people rely on forests as safety nets or for supplemental income. Farmers earn as much as 10 to 25 percent of household income from nontimber forest products. Bushmeat is the main source of animal protein in West Africa. The poor often harvest, process, and sell wild plants and animals to buy food. Sixty million poor people depend on herding in semiarid rangelands that they share with large mammals and other wildlife. Thirty million low-income people earn their livelihoods primarily as fishers, twice the number of 30 years ago. The depletion of fisheries has serious impacts on food security. Wild plants are used in farming systems for fodder, fertilizer, packaging, fencing, and genetic materials. Farmers rely on soil microorganisms to maintain soil fertility and structure for crop production, and they also rely on wild species in natural ecological communities for crop pollination and pest and predator control. Wild relatives of domesticated crop species provide the genetic diversity used in crop improvement. The rural poor rely directly on ecosystem services for clean and reliable local water supplies. Ecosystem degradation results in less water for people, crops, and livestock; lower crop, livestock, and tree yields; and higher risks of natural disasters.

**More than a billion people live in the 25 biodiversity "hotspots" identified by Conservation International; in 16 of these hotspots, population growth is higher than the world average.**

Three-quarters of the world's people living on less than $1 per day are rural. Strategies to meet the United Nations Millennium Development Goals in rural areas—to reduce hunger and poverty and to conserve biodiversity—must find ways to do so in the same landscapes. Crop and planted pasture production—mostly in low-productivity systems—dominate at least half the world's temperate, subtropical, and tropical forest areas; a far larger area is used for grazing livestock.[5]

Food insecurity threatens biodiversity when it leads to overexploitation of wild plants and animals. Low farm productivity leads to depletion of soil and water resources and increases the pressure to clear additional land that serves as wildlife habitat. Some 40 percent of cropland in developing countries is degraded. Of more than 17,000 major protected areas, 45 percent (accounting for one-fifth of total protected areas) are heavily used for agriculture, while many of the rest are islands in a sea of farms, pastures, and production forests that are managed in ways incompatible for long-term species and ecosystem survival.[6]

Despite this high level of dependence by the poor on biodiversity, the dominant model of conservation seeks to exclude people from natural habitats. In India, for example, 30 million people are targeted for resettlement from protected areas.[7] From the perspective of poverty reduction and rural development, it is thus urgent to identify alternative conservation systems that respect the rights of forest dwellers and owners and address conservation objectives in the 90 percent of forests outside public protected areas. Markets for ecosystem services potentially offer a more efficient and lower-cost approach to forest conservation.[8]

# Need for Financial Incentives to Provide Ecosystem Services

There is growing recognition that regulatory and protected area approaches, while critical, are insufficient to adequately conserve biodiversity. A fundamental problem is financial, especially for resources that lie outside protected areas. For these to be conserved, they need to be more valuable than the alternative uses of the land. And for such resources to be well managed, good stewardship needs to be more profitable than bad stewardship. The failure of forest owners and producers to capture financial benefits from conserving ecosystem benefits leads to overexploitation of forest resources and undersupply of ecosystem services.

This reality is hard for many people to accept, because most ecosystem services are considered "public goods." The "polluter pays" principle has argued that the right of the public to these services trumps the private rights of the landowner or manager. Yet good management has a cost. While the individual who manages his or her resources to protect biodiversity produces public benefits, the costs incurred are private. Under current institutions, those who benefit from these services have no incentive to compensate suppliers for these services. In most of the world, forest ecosystem services are not traded and have no "price." Thus, where the opportunity costs of forest land for agricultural enterprises, infrastructure, and human settlements are higher than the use or income value of timber and nontimber forest products (NTFPs), habitats will be cleared and wild species will be allowed to disappear. Because they receive little or no direct benefit from them, resource owners and producers ignore the real economic and noneconomic values of ecosystem services in making decisions about land use and management.

---

**A lower-cost approach to securing conservation is to pay only for the biodiversity services themselves, by paying landowners to manage their assets so as to achieve biodiversity or species conservation.**

---

Mechanisms are needed by which resource owners are rewarded for their role as stewards in providing biodiversity and ecosystem services. Anticipation of such income flows would enhance the value of natural assets and thus encourage their conservation. Compared to previous approaches to forest conservation, market-based mechanisms promise increased efficiency and effectiveness, at least in some situations. Experience with market-based instruments in other sectors has shown that such mechanisms, if carefully designed and implemented, can achieve environmental goals at significantly less cost than conventional "command-and-control" approaches, while creating positive incentives for continual innovation and improvement. Markets for ecosystem services could potentially contribute to rural development and poverty reduction by providing financial benefits from the sale of ecosystem services, improving human capital through associated training and education, and strengthening social capital through investment in local cooperative institutions.

# New Market Solutions to Conserve Biodiversity

The market for biodiversity protection can be characterized as a nascent market. Many approaches are emerging to financially remunerate the owners and managers of land and resources for their good stewardship of biodiversity (see Table 2). Market mechanisms to pay for other ecosystem services—watershed services, carbon sequestration or storage, landscape beauty, and salinity control, for example—can be designed to conserve biodiversity as well. However, in general, biodiversity services are the most demanding to protect because of the need to conserve many different elements essential for diverse, interdependent species to thrive. Figure 1 illustrates potential market solutions and some of the complexities involved.

## Land Markets for High-Biodiversity-Value Habitat

National governments (in the form of public parks and protected areas), NGO conservation organizations (for example, The Nature Conservancy), and individual conservationists have long paid for the purchase of high-biodiversity-value forest habitats. Direct acquisition can be expensive, as underlying land and use values are also included.

Local sovereignty concerns arise when buyers are from outside the country—or even the local area—or where extending the area of noncommercial real estate reduces the local tax base. New commercial approaches are being developed to encourage the establishment of privately owned conservation areas, such as conservation communities (the purchase of a plot of land by a group of people mainly for recreation or conservation purposes), ecotourism-based land protection projects, and ecologically sound real estate projects being organized in Chile.[9] These build on growing consumer demand for housing and vacation in biodiverse environments.

## Payments for Use or Management

A lower-cost approach to securing conservation is to pay only for the biodiversity services themselves, by paying landowners to manage their assets so as to achieve biodiversity or species conservation. It is likely that the largest-scale payments for land-use or management agreements belong to one of two categories. One encompasses government agroenvironmental payments made to farmers in North America and Europe for reforesting conservation easements. The other category describes management contracts aiming to conserve aquatic and terrestrial wildlife habitat. In Switzerland, "ecological compensation areas," which use farming systems compatible with biodiversity conservation, have expanded to include more than 8 percent of total agricultural land. In the tropics, diverse approaches include nationwide public payments in Costa Rica for forest conservation and in Mexico for forested watershed protection.

Conservation agencies are organizing direct payments systems, such as conservation concessions being negotiated by Conservation International, and forest conservation easements negotiated by the *Cordão de Mata* ("linked forest") project with dairy farmers in Brazil's Atlantic Forest. The dairy farmers in the latter example receive, in exchange, technical assistance and investment resources to raise crop and livestock productivity. Some countries that use land taxes are using tax policies in innovative ways to encourage the expansion of private and public protected areas.

## Payment for Private Access to Species or Habitat

Private sector demand for biodiversity has tended to take the form of payments for access to particular species or habitats that function as "private goods" but in practice serve to cover some or all of the costs of providing broader ecosystem services. Pharmaceutical companies have contracted for bioprospecting rights in tropical forests. Ecotourism companies have paid forest owners for the right to bring tourists into their lands to observe wildlife, while private individuals are willing to pay forest owners for the right to hunt, fish, or gather nontimber forest products.

## Tradable Rights and Credits within a Regulatory Framework

Multifactor markets for ecosystem services have been successfully established, notably for sulfur dioxide emissions, farm nutrient pollutants, and carbon emissions. These create rights or obligations within a broad regulatory framework and allow those with obligations to "buy" compliance from other landowners or users. Developing such markets for biodiversity is more complicated, because specific site conditions matter so much. The United States has operated a wetlands mitigation

## Table 2  Types of Payments for Biodiversity Protection

**Purchase of high-value habitat**

| Type | Mechanism |
|---|---|
| Private land acquisition | Purchase by private buyers or nongovernmental organizations explicitly for biodiversity conservation |
| Public land acquisition | Purchase by government agency explicitly for biodiversity conservation |
| **Payment for access to species or habitat** | |
| Bioprospecting rights | Rights to collect, test, and use genetic material from a designated area |
| Research permits | Right to collect specimens, take measurements in area |
| Hunting, fishing, or gathering permits for wild species | Right to hunt, fish, and gather |
| Ecotourism use | Rights to enter area, observe wildlife, camp, or hike |
| **Payment for biodiversity-conserving management** | |
| Conservation easements | Owner paid to use and manage defined piece of land only for conservation purposes; restrictions are usually in perpetuity and transferable upon sale of the land |
| Conservation land lease | Owner paid to use and manage defined piece of land for conservation purposes for defined period of time |
| Conservation concession | Public forest agency is paid to maintain a defined area under conservation uses only; comparable to a forest logging concession |
| Community concession in public protected areas | Individuals or communities are allocated use rights to a defined area of forest or grassland in return for commitment to protect the area from practices that harm biodiversity |
| Management contracts for habitat or species conservation on private farms, forests, or grazing lands | Contract that details biodiversity management activities and payments linked to the achievement of specified objectives |
| **Tradable rights under cap-and-trade regulations** | |
| Tradable wetland mitigation credits | Credits from wetland conservation or restoration that can be used to offset obligations of developers to maintain a minimum area of natural wetlands in a defined region |
| Tradable development rights | Rights allocated to develop only a limited total area of natural habitat within a defined region |
| Tradable biodiversity credits | Credits representing areas of biodiversity protection or enhancement that can be purchased by developers to ensure they meet a minimum standard of biodiversity protection |
| **Support biodiversity-conserving businesses** | |
| Biodiversity-friendly businesses | Business shares in enterprises that manage for biodiversity conservation |
| Biodiversity-friendly products | Eco-labeling |

Source: S. J. Scherr, A. White, and A. Khare, *Current Status and Future Potential of Markets for Ecosystem Services in Tropical Forests: An Overview* (Washington, DC: Forest Trends, 2003).

program since the early 1980s in which developers seeking to destroy a wetland must offset that by buying wetland banks conserved or developed elsewhere. A similar approach is used for "conservation banking," described in the box on the next page.

A variant of this approach is being designed for conserving forest biodiversity in Brazil by permitting flexible enforcement of that country's "50 percent rule," which requires landholders in Amazon forest areas to maintain half of their land in forest. This rule is also applied in other regions in Brazil, where lesser proportional areas are set aside for forest use. Careful designation of comparable sites is required.

Another approach, biodiversity credits, is under development in Australia. In this system, legislation creates new property rights for private landholders who conserve biodiversity values on their land. These landholders can then sell resulting "credits" to a common pool. The

law also creates obligations for land developers and others to purchase those credits. The approach requires that the "value" of the biodiversity unit can be translated into a dollar value.

## Biodiversity-Conserving Businesses

Conservation values are beginning to inform consumer and investor decisions. Eco-labeling schemes are being developed that advertise or certify that products were produced in ways consistent with biodiversity conservation. The global trade in certified organic agriculture was worth $21 billion worldwide in 2000.[10] International organic standards are expanding to landscape-scale biodiversity impacts. The Rainforest Alliance and the Sustainable Agriculture Network certify coffee, bananas, oranges, and other products grown in and around

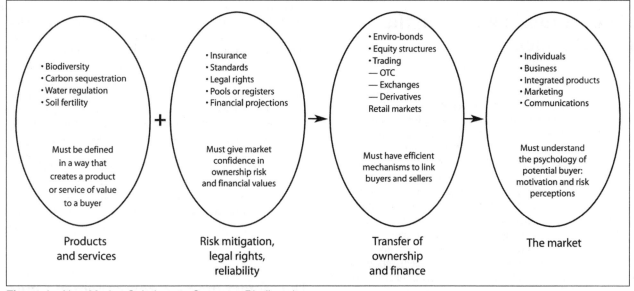

**Figure 1** New Market Solutions to Conserve Biodiversity.

Note: OTC ("over-the-counter") trading involves direct negotiation with buyers and sellers rather than an official stock market.

Source: D. Brand, "Emerging Markets for Forest Services and Implications for Rural Development, Forest Industry, and Government," presentation to the Katoomba Group Meeting, "Developing Markets for Ecosystem Services," Vancouver, October 2000.

## A New Fund to Finance Forest Ecosystem Services

The Mexican government recently announced the creation of a new fund to pay indigenous and other communities for the forest ecosystem services produced by their land.[1] Indigenous and other communities own approximately 80 percent of all forests in Mexico—totaling some 44 million hectares—as collectively held, private land. The Mexican Forestry Fund has been under design since 2002, guided by a consultative group with government, nongovernmental organization, and industry representatives. The purpose of the US$20 million fund is to promote the conservation and sustainable management of natural forests, leverage additional financing, contribute to the competitiveness of the forest sector, and catalyze the development of mechanisms to finance forest ecosystem services. Operational manuals are being prepared, and priority conservation sites have already been identified. The fund proposes to pay $40 per hectare (ha) per year to owners of deciduous forests in critical mountain areas and $30 per ha per year to other forest types.

### Notes

1. Comisión Nacional Forestal (CONAFOR), presentation given at the Mexican Forestry Expo, Guadalajara, Mexico, 8 August 2003.

who are protecting biodiversity. In 2002, more than 100 million hectares of forest were certified (a fourfold increase over 1996), although only 8 percent of the total certified area is in developing countries, and most of that is in temperate forests.

## Current Market Demand

Available information suggests that biodiversity protection services are presently the largest market for ecosystem services. A team from McKinsey & Company, the World Resources Institute, and The Nature Conservancy estimated the annual international finance for the conservation market (conservation defined as protecting land from development) at $2 billion, with the forest component a large share of that.[11] Buyers are predominantly development banks and foundations in the United States and Europe.

A study by the International Institute for Environment and Development (IIED) of 72 cases of markets for forest biodiversity protection services in 33 countries found that the main buyers of biodiversity services (in declining order of prevalence) were private corporations, international NGOs and research institutes, donors, governments, and private individuals.[12] Communities, public agencies, and private individuals predominate as sellers. Most of these cases took place in Latin America and in Asia and the Pacific. Only four cases were found in Europe and Russia and one was found in the United States.

Three-quarters of the cases in the IIED study were international markets, and the rest were distributed among regional, national, and local buyers. International actors—as well as many on the national level—who demand biodiversity protection services tend to focus on the most biodiverse habitats (in terms of species richness) or those perceived to be under the greatest threat globally (for example, places like the Amazon, where there are a high number of endemic species and where habitat area has greatly declined). Most of the private corporations were interested in eco-labeling schemes for crops or timber, investment in biodiversity-friendly companies, horticultural companies concerned with ecosystem services, or pharmaceutical bioprospecting.

high-biodiversity-value areas. The Sustainable Agriculture Initiative is a coalition of multinational commercial food producers (Nestle, Dannon, Unilever, and others) who are seeking to ensure that all of the products they purchase along the supply chain come from producers

## Conservation Banking in the United States

Amendments to the United States Endangered Species Act in 1982 provided for an "incidental take" of enlisted species, if "a landowner provides a long-term commitment to species conservation through development of a Habitat Conservation Plan (HCP)." These amendments have opened the door to a series of market-based transactions, described as conservation banking, which permits land containing a natural resource (such as wetlands, forests, rivers, or watersheds) that is conserved and maintained for specified enlisted species to be used to offset impacts occurring elsewhere to the same natural resource.[1] A private landowner may request an "incidental take" permit and mitigate it by purchasing "species credits" from preestablished conservation banks. Credits are administered according to individuals, breeding pairs, acres, nesting sites, and family units. Conservation banking has maximized the value of underutilized commercial real estate and given private landowners incentive to conserve habitat.

California was the first state to authorize the use of conservation banking and has established 50 conservation banks since 1995. Other states, including Alabama, Colorado, and Indiana, have followed suit. In April 2002, the Indiana Department of Transportation, the Federal Highway Association Indiana Division, and four local government agencies finalized an HCP for the endangered Indiana bat as part of the improvement of transportation facilities around Indianapolis International Airport. These highway improvements will occur in an area of known Indiana bat habitat that is predicted to experience nearly $1.5 billion in economic development during the next ten years. Under the HCP, approximately 3,600 acres will be protected, including 373 acres of existing bat habitat.

### Notes

1. A. Davis, "Conservation Banking," presentation to the Katoomba Group-Lucarno Workshop, Lucarno, Switzerland, November 2003.

---

Such private payments are usually site-specific. Local actors more commonly focus on protecting species or habitats of particular economic, subsistence, or cultural value.

## Projected Growth in Market Demand

The fastest-growing component of future market demand for biodiversity services is likely to be in eco-labeling of crop, livestock, timber, and fish products for export and for urban consumers. In 1999, the value of the organic foods market was US$14.2 billion. Its value is growing at 20–30 percent a year in the industrialized world, as the international organic movement is strengthening standards for biodiversity conservation.[13] Pressures continue to increase on major international trading and food processing companies to source from suppliers who are not degrading ecosystem services. Donor and international NGO conservation will continue to expand as NGOs begin to establish entire research departments aimed at developing new market-based instruments. Voluntary biodiversity offsets are also a promising source of future demand, as many large companies are seeking ways to maintain their "license to operate" in environmentally sensitive areas, and offsets are of increasing interest to them.

The costs of and political resistance to land acquisition are rising. Construction of biological corridors in and around production areas is an increasingly important conservation objective. At the same time, however, many of the most important sites for biodiversity conservation are in more densely populated areas with high opportunity costs for land. Thus we are likely to see a major shift from land acquisition to various types of direct payments for easements, land leases, and management contracts.

---

**The fastest-growing component of future market demand for biodiversity services is likely to be in eco-labeling of crop, livestock, timber, and fish products for export and for urban consumers.**

---

A rough back-of-the-envelope estimate suggests that the current value of international, national, and local direct payments and trading markets for ecosystem services from tropical forests alone could be worth several hundred million dollars per year, while the value of certified forest and tropical tree crop products may reach as much as a billion dollars. While this is a large and significant amount, it represents a small fraction of the value of conventional tropical timber and other forest product markets. For example, by comparison, the total value of tropical timber exports is $8 billion (including only logs, sawnwood, veneer, and plywood), which is a small fraction of the total exports and domestic timber, pulpwood, and fuelwood markets in tropical countries. NTFP markets are far larger still.[14] The total value of international trade for NTFPs is $7.5 billion–$9 billion per year, with another $108 billion in processed medicines and medicinal plants.[15] Domestic markets for NTFPs are many times larger (for example, domestic consumption accounted for 94 percent of the global output of fresh tropical fruits 1995–2000.)[16] Nonetheless, these rough figures are quite interesting when compared with the scale of public and donor forest conservation finance summarized in Table 1.

## Scaling Up Payments for Biodiversity: Next Steps

Markets for ecosystem services are steadily growing and can be expected to grow even more rapidly in the next decade. Yet they predominate as pilot projects. What will it take to transform these markets to impact ecosystem conservation on the global scale? The four most strategic and catalytic areas for policy and action are to

- structure emerging markets to support community-driven conservation;
- mobilize and organize buyers for ecosystem services;

- connect global and national action on climate change to biodiversity conservation; and
- invest in the policy frameworks and institutions required for functioning ecosystem service payment systems.

## Supporting Community-Driven Conservation

The benefits of investments in ecosystem services will be maximized over the long term if markets reward local participation and utilize local knowledge. In community forests and agroforestry landscapes, communities have already established sophisticated conservation strategies. Studies of indigenous timber enterprises document conservation investments on the order of $2 per hectare per year apart from other management activities and investments of community time and labor; this is equal to the average available budget per hectare for protected areas worldwide. Conservation policies must recognize the role that local people are playing in the conservation of forest ecosystems worldwide and support them (either with cash or in-kind support) to continue to be good environmental stewards.

To enable conservation-oriented management to remain or become economically viable, it is important that ecosystem service payments and markets are designed so that they strategically channel financial payments to rural communities. Such payments can be used to develop and invest in new production systems that increase productivity and rural incomes, and enhance biodiversity at a landscape scale—an approach referred to as "ecoagriculture."[17] Ecosystem service payments to poor rural communities that are providing stewardship services of national or international value can help to meet multiple Millennium Development Goals. For any semblance of a sustainable future to be realized, it is crucial that our long-term vision includes biodiversity and natural ecosystems as part of the "natural infrastructure" of a healthy economy and society.

## Mobilizing and Organizing Buyers for Ecosystem Services

Turning beneficiaries into buyers is the driving force of ecosystem service markets. Because beneficiaries are often hesitant to pay for goods previously considered free, "willingness to pay" for ecosystem services must be organized on a greater scale. The private sector must be called upon to engage in responsible corporate behavior in conserving biodiversity. For example, Insight Investment, a major financial firm, has developed a biodiversity policy that uses conservation as a screen for investment. Voluntary payments by consumers, retail firms, and other actors can be encouraged through social advertising. This approach is growing rapidly now for eco-labeling programs (labeling of some personal care products and foods) and voluntary carbon emission offset programs involving investment in reforestation. Stockholder pressure is beginning to influence some firms to avoid investments and activities that harm biodiversity, and this is evolving to positive action. Civil society campaigns can also mobilize willingness to pay for biodiversity offsets and payments to local partners for conservation.

## Connecting Climate Action with Biodiversity Conservation

Far more aggressive action must and will be taken to mitigate and adapt to climate change. Land use and land-use change currently contribute more than 20 percent of carbon emissions and other greenhouse gases. Action to reduce these emissions must be a central part of our response, and it is critical that action to sequester carbon through improved land uses accompanies strategies to reduce industrial emissions. There is thus an unprecedented opportunity at this time to structure our responses to climate change so that actions related to land use are also designed to protect and restore biodiversity. Moreover, such actions can be designed in ways that enhance and protect livelihoods, especially for those most vulnerable to the impacts of climate change. Indeed, it is imperative that they do so.

As a result of the deliberations at the Conference of the Parties of the United Nations Framework Convention on Climate Change last year, payments for forest carbon through the Clean Development Mechanism (CDM) of the Kyoto Protocol can be used to finance forest restoration and regeneration projects that conserve biodiversity while providing an alternative income source for local people.[18] But the scale of forest carbon under CDM is very small—too small to have a major impact on climate, biodiversity, or livelihoods. It is critical that we aim for a much larger program in the second commitment period, and it is crucial that nations affiliated with the Organisation for Economic Co-operation and Development (OECD) create initiatives to utilize carbon markets for biodiversity conservation in their own internal trading programs. It is imperative to develop a new principle of international agreements on climate response and carbon trading, one that builds a system that encourages overlap of the major international environmental agreements and the Millennium Development Goals. This could mobilize demand by creating an international framework for investing in good ecosystem service markets. It is also important that emerging private voluntary markets for carbon (that is, with actors who do not have a regulatory obligation) are encouraged to pursue such biodiversity goals as well. The Climate, Community and Biodiversity Alliance, for example, is seeking to develop guidelines and indicators for private investments in carbon projects that will achieve these multiple goals. The Forest Climate Alliance of The Katoomba Group is seeking to mobilize the international rural development community to advocate for such approaches.[19]

## Investing in Policy Frameworks and Institutions for Biodiversity Markets

Ecosystem service markets are genuinely new—and biodiversity markets are the newest and most challenging. Every market requires basic rules and institutions in order to function, and this is equally true of biodiversity markets. The biodiversity conservation community needs to act quickly and strategically to ensure that as these markets develop, they are effective, equitable, and operational and are used sensibly to complement other conservation approaches.

Policymakers and public agencies play a vital role in creating the legal and legislative frameworks necessary for market tools to operate effectively. This includes establishing regulatory rules, systems of rights over ecosystem services, and mechanisms to enforce contracts and settle ownership disputes. Ecosystem service markets pose profound equity implications, as new rules may fundamentally change the distribution of rights and responsibilities for essential ecosystem services. Forest producers and civil society will need to take a proactive role to ensure that rules support the public interest and create development opportunities.

New institutions will also be needed to provide the business services required in ecosystem service markets. For example, in order for beneficiaries of biodiversity services to become willing to pay for them, better methods of measuring and assessing biodiversity in working landscapes must be developed, as well as the institutional capacity to do so. New institutions must be created to encourage transactions and

# Protecting Brazil's Atlantic Forest: The Guaraqueçaba Climate Action Project

Due to excessive deforestation, the Atlantic Forest of Brazil has been reduced to less than 10 percent of its original size. The Guaraqueçaba Climate Action Project has sought to regenerate and restore natural forest and pastureland.[1] Companies such as American Electric Power Company, General Motors, and Chevron-Texaco have invested US$18.4 million to buy carbon emission offset credits from the approximately 8.4 million metric tons of carbon dioxide that the project is expected to sequester during its lifespan. The project has initiated sustainable development activities both within and outside the project boundary, including ecotourism, organic agriculture, medicinal plant production, and a community craft network. The project has made significant contributions toward enhancing biodiversity in the area, creating economic opportunities for local people (such as jobs), restoring the local watershed, and substantially mitigating climate change.

## Notes

1. The Nature Conservancy (TNC), *Climate Action: The Atlantic Forest in Brazil* (Arlington, VA: TNC, 1999).

reduce transaction costs. Such institutions could include "bundling" biodiversity services provided by large numbers of local producers, as well as investment vehicles that have a diverse portfolio of projects to manage risks. Registers must be established and maintained, to record payments and trades. For example, The Katoomba Group is developing a Web-based "Marketplace" to slash the information and transaction costs for buyers, sellers, and intermediaries in ecosystem service markets.[20]

## Conclusion

Conservation of biodiversity and of the services biodiversity provides to humans and to the ecological health of the planet requires financing on a scale many times larger than is feasible from public and philanthropic sources. It is essential to find new mechanisms by which resource owners and managers can realize the economic values created by good stewardship of biodiversity. Moreover, private consumers, producers, and investors can financially reward that stewardship. New markets and payment systems, strategically shaped to deliver critical public benefits, are showing tremendous potential to move biodiversity conservation objectives to greater scale and significance.

## Notes

1. M. Cernea and K. Schmidt-Soltau, 2003. "Biodiversity Conservation versus Population Resettlement, Risks to Nature and Risks to People," paper presented at CIFOR (Center for International Forestry Research) Rural Livelihoods, Forests and Biodiversity Conference, Bonn, Germany, 19–23 May 2003.

2. See S. Wood, K. Sebastian, and S. J. Scherr, *Pilot Analysis of Global Ecosystems: Agroecosystems* (Washington, DC: International Food Policy Research Institute and the World Resources Institute, 2000), 64; and E. W. Sanderson et al., "The Human Footprint and the Last of the Wild," *Bioscience* 52, no. 10 (2002): 891–904.

3. K. Chomitz, *Forest Cover and Population Density in Latin America,* research notes to the World Bank (Washington, DC: World Bank, 2003).

4. R. P. Cincotta and R. Engelman, *Nature's Place: Human Population and the Future of Biological Diversity* (Washington, DC: Population Action International, 2000).

5. Wood, Sebastian, and Scherr, note 2 above.

6. J. McNeely and S. J. Scherr, *Ecoagriculture: Strategies to Feed the World and Conserve Wild Biodiversity* (Washington, DC: Island Press, 2003).

7. A. Khare et al., *Joint Forest Management: Policy Practice and Prospects* (London: International Institute for Environment and Development, 2000).

8. S. J. Scherr, A. White, and D. Kaimowitz, *A New Agenda for Forest Conservation and Poverty Reduction: Making Markets Work for Low-Income Communities* (Washington, DC: Forest Trends, CIFOR, and IUCN-The World Conservation Union, 2004).

9. E. Corcuera, C. Sepulveda, and G. Geisse, "Conserving Land Privately: Spontaneous Markets for Land Conservation in Chile," in S. Pagiola et al., eds., *Selling Forest Environmental Services: Market-Based Mechanisms for Conservation and Development* (London: Earthscan Publications, 2002).

10. J. W. Clay, *Community-Based Natural Resource Management within the New Global Economy: Challenges and Opportunities,* a report prepared by the Ford Foundation (Washington, DC: World Wildlife Fund, 2002).

11. M. Arnold and M. Jenkins, "The Business Development Facility: A Strategy to Move Sustainable Forest Management and Conservation to Scale," proposal to the International Finance Corporation (IFC) Environmental Opportunity Facility from Forest Trends, Washington, DC, 2003.

12. N. Landell-Mills, and I. Porras. 2002. *Markets for Forest Environmental Services: Silver Bullet or Fool's Gold? Markets for Forest Environmental Services and the Poor, Emerging Issues* (London: International Institute for Environment and Development, 2002).

13. International Federation of Organic Agriculture Movements (IFOAM), "Cultivating Communities," 14th IFOAM Organic World Congress, Victoria, BC, 21–28 August 2002.

14. S. J. Scherr, A. White, and A. Khare, *Current Status and Future Potential of Markets for Ecosystem Services of Tropical Forests: A Report for the International Tropical Timber Organization* (Washington, DC: Forest Trends, 2003).

15. M. Simula, *Trade and Environment Issues in Forest Protection,* Environment Division working paper (Washington, DC: Inter-American Development Bank, 1999).

16. Food and Agricultural Organization of the United Nations (FAO), *FAOSTAT* database for 2000, accessible via http://www.fao.org.

17. For more information on ecoagriculture, see the Ecoagriculture Partners' Web site at http://www.ecoagriculturepartners.org.

18. S. J. Scherr and M. Inbar, *Clean Development Mechanism Forestry for Poverty Reduction and Biodiversity Conservation: Making*

*the CDM Work for Rural Communities* (Washington, DC: Forest Trends, 2003).

19. For more information on this project, see http://www. katoombagroup.org/ Katoomba/forestcarbon.

20. The Katoomba Group is a unique network of experts in forestry and finance companies, environmental policy and research organizations, governmental agencies and influential private, community, and nonprofit groups. It is dedicated to advancing markets for some of the ecosystem services provided by forests, such as watershed protection, biodiversity habitat, and carbon storage. For more information on the Katoomba Group, see http://www.katoombagroup.org. Forest Trends serves as the secretariat for the group. More information on Forest Trends can be found at http://www.forest-trends.org.

**Michael Jenkins** is the founding president of Forest Trends, a nonprofit organization based in Washington, D.C., and created in 1999. Its mission is to maintain and restore forest ecosystems by promoting incentives that diversify trade in the forest sector, moving beyond exclusive focus on lumber and fiber to a broader range of products and services. Previously he worked as a senior forestry advisor to the World Bank (1998), as associate director for the Global Security and Sustainability Program of the MacArthur Foundation (1988–1998), as an agroforester in Haiti with the U.S. Agency for International Development (1983–1986), and as technical advisor with Appropriate Technology International (1981–1982). He has also worked in forestry projects in Brazil and the Dominican Republic and was a Peace Corps volunteer in Paraguay. He speaks Spanish, French, Portuguese, Creole, and Guaraní, and can be contacted by telephone at (202) 298-3000 or via e-mail at mjenkins@forest-trends.org. **Sara J. Scherr** is an agricultural and natural resource economist who specializes in the economics and policy of land and forest management in tropical developing countries. She is presently director of the Ecosystem Services program at Forest Trends, and also director of Ecoagriculture Partners, the secretariat of which is based at Forest Trends. She previously worked as principal researcher at the International Center for Research in Agroforestry, in Nairobi, Kenya, as (senior) research fellow at the International Food Policy Research Institute in Washington, D.C., and as adjunct professor at the Agricultural and Resource Economics Department of the University of Maryland, College Park. Her current work focuses on policies to reduce poverty and restore ecosystems through markets for sustainably grown products and environmental services and on policies to promote ecoagriculture—the joint production of food and environmental services in agricultural landscapes. She also serves as a member of the Board of the World Agroforestry Centre, and as a member of the United Nations Millennium Project Task Force on Hunger. Scherr can be reached by telephone at (202) 298-3000 or via e-mail at sscherr@forest-trends.org. **Mira Inbar** is program associate with Forest Trends. She works in the Ecosystem Services program, supporting efforts to establish frameworks and instruments for emerging transactions in environmental services worldwide. Before joining Forest Trends, she worked with communities in the Urubamba River Valley of Peru to initiate a forest conservation plan. She has worked with the National Fishery Department in Western Samoa, the Marie Selby Botanical Gardens, and Environmental Defense. Inbar may be reached by telephone at (202) 298-3000 or via e-mail at minbar@forest-trends.org. This article is © The Aspen Institute and is published with permission.

# UNIT 5

# Resources: Land and Water

## Unit Selections

## Key Points to Consider

- How has ecosystem management in the drylands region south of the Sahara produced some hopeful mitigation of the trend toward desertification in that area? What lessons can the management techniques developed for this area be applied to other parts of the world?

- How are groundwater reserves in the United States being monitored in terms of the differences between the amount of water extracted and the amount of water returned to groundwater reserves? Does current scientific analysis hold the idea that many of our groundwater supplies are being depleted to the point where they will no longer yield usable water?

- Is there a relationship between water pollution and the cost of water? How can conservation methods alter the price of water for commercial, domestic, industrial, and agricultural needs?

- Is it possible to reach a balance between the demands of commercial timber industries for tropical hardwoods and the utility of the forest environment for agriculture? Has the development of a transportation system in the Amazon improved or worsened the environmental prospects for the world's largest tropical forest area?

## Student Web Site
www.mhcls.com/online

## Internet References
Further information regarding these Web sites may be found in this book's preface or online.

**Global Climate Change**
http://www.puc.state.oh.us/consumer/gcc/index.html

**National Oceanic and Atmospheric Administration (NOAA)**
http://www.noaa.gov

**National Operational Hydrologic Remote Sensing Center (NOHRSC)**
http://www.nohrsc.nws.gov

**Terrestrial Sciences**
http://www.cgd.ucar.edu/tss

The worldwide situations regarding reduction of biodiversity, scarcity of energy resources, and pollution of the environment have received the greatest amount of attention among members of the environmentalist community. But there are a number of other resource issues that demonstrate the interrelated nature of all human activities and the environments in which they occur. One such issue is the declining quality of what has often been mistakenly referred to as "renewable resources"—soils, groundwater reserves, and forests. In the developing world, excessive rural populations have forced the overuse of lands and sparked a shift into marginal areas, and the total availability of new farmland is decreasing at an alarming rate of 2 percent per year. In the developed world, intensive mechanized agriculture has resulted in such a loss of topsoil that some areas are experiencing a decline in food production. Other natural resources, such as minerals and timber, are declining in quantity and quality as well; in some areas they are no longer usable at present levels of technology. The overuse of groundwater reserves has resulted in potential shortages beside which the energy crisis pales in significance. And

the very productivity of Earth's environmental systems—their ability to support human and other life—is being threatened by processes that derive at least in part from energy overuse and inefficiency and from pollution. Many environmentalists believe that both the public and private sectors, including individuals, are continuing to act in a totally irresponsible manner with regard to the natural resources upon which we all depend.

Uppermost in the minds of many who think of the environment in terms of an integrated whole, as evidenced by many of the selections in this unit, is the concept of the threshold or critical limit of human interference with natural systems of land and water. This concept suggests that the environmental systems we occupy have been pushed to the brink of tolerance in terms of stability and that destabilization of environmental systems has consequences that can only be hinted at, rather than predicted. Although the broader issue of system change and instability, along with the lesser issues such as the quantity of agricultural land, the quality of iron ore deposits, the sustained yield of forests, or the availability of fresh water seems to be quite diverse, all are closely tied to a pair of concepts—that of resource marginality and of the globalization of the economy that has made marginality a global rather than a regional problem.

Not all the utilizations of marginal resources turn out quite so badly. In the first article in this unit, independent researcher Michael Mortimore describes recent successes in developing the highly marginal environments of dryland West Africa. In "Dryland Development: Success Stories from West Africa," Mortimore describes the fragility (or marginality) of the African drylands that encompass 40% of the continent's surface. Away from river valleys, dryland soils are easily eroded by wind and water, lose their fertility quickly under cultivation, and are subject to desertification—the conversion of soils and accompanying vegetation from merely a dryland condition to one of absolute desert. While the general trend of human use of drylands for farming has produced degradation, experimental farming techniques begun in Kenya in the 1930s have recently been applied to West African dryland farming regions in Nigeria, Niger, and Senegal. As farmers in these areas continue to live in poverty, their lot has been improved by the new farming techniques adapted specifically to drylands and involving conservation of water and soil. The results have been striking: sustainable ecosystem management, increasing farm investment, stable outputs per areal unit, and increasing farm incomes.

The use of water also figures importantly—but in entirely different ways—in the second selection of this unit. William Alley, Chief of Ground Water at the U.S. Geological Survey, analyzes the future of ground water in the United States in "Tracking U.S. Groundwater: Reserves for the Future?" Alley notes that groundwater, like other so-called "renewable" or "flow" resources, can be depleted. If the rate of extraction of withdrawal exceeds the rate of replacement or recharge, then the resource continues to dwindle in the same way that a bank account will decrease if monthly deposits are less than monthly withdrawals. Throughout the United States, groundwater reserves are being depleted as rates of withdrawal exceed rates of recharge. While the U.S. has technologies for monitoring the recharge vs. withdrawal ratios that are better than nearly anywhere else in the world, even here the ability to predict the speed of groundwater depletion is limited. What can be said is that virtually everywhere in the world groundwater reserves are being drawn down faster than they are being replenished. And, given the degree to which we depend on groundwater, this trend is alarming.

In the third and fourth articles in the unit, the struggle between different stakeholders in the land and water use wars also takes center stage. Journalist Jim Morrison in "How Much Is Clean Water Worth?" asks the essential question in water management: can we put a dollar value not just on the price of water per quart, gallon, or acre-foot, but also on the ecosystems that water sustains? Morrison answers his own question in an affirmative but non-quantitative way. The watershed that supplies the city of New York with its basic water supply is easy to calculate: $1.3 billion per year. But calculating the dollar value of the Catskills in terms of recreation, scenic value, and wildlife habitat is more difficult, although even a cursory look at the region would suggest that habitat and wildlife are powerful economic engines for both the region and for a wider area. If the ecosystem service values could be more accurately calculated, then wiser choices could be made about how to use the Catskills' water supply—is it best to reserve the water for use of residents of New York City or to leave it as a part of an integrated ecosystem that has its own inherent value? A similar set of questions on a larger scale is asked in "Searching for Sustainability: Forest Policies, Smallholders, and the Trans-Amazon Highway." A team of research scientists (Eirivelthon Lima, Frank Merry, Daniel Nepstad, Gregory Amacher, Claudia Azevedo-Ramos, Paul Lefebvre, and Felipe Resque, Jr.) inquires into the relationship between traditional methods of logging in tropical forests, the use of forested lands, and the connections between logging, land use (particularly agriculture), and the presence of the Trans-Amazon Highway, a 3000 kilometer road, mostly dirt and impassable for half the year during the rainy season. They conclude that the traditionally-accepted selective cutting method of harvesting timber from the forest is compatible both with sustained yields and the opening of forest lands to sustainable subsistence or marginally-commercial agriculture. But government mismanagement and corruption combine to make what would be an ecologically sound process an environmentally destructive one.

There are two possible solutions to all these problems posed by the use of increasingly marginal and scarce resources and by the continuing pollution of the global atmosphere. One is to halt the basic cause of the problems—increasing population and consumption. The other is to provide incentives and techniques for the conservation and management of existing resources and for the discovery of alternative resources to eliminate the demand for more marginal resources and the use of heavily polluting ones.

# Dryland Development

## Success Stories from West Africa

Michael Mortimore

The African continent spans a huge range of climates, from equatorial to tropical, desert, and subtropical on both sides of the equator. Much of Africa—around 40 percent—is dryland, receiving less than 1,000 millimeters (mm) of mean annual rainfall in short rainy seasons, the remaining months being relatively or absolutely rainless.[1] High temperatures during the tropical rainy season cause much of the rainfall to be lost in evaporation, and the high intensity of storms may cause much of it to run off in floods.

For securing human livelihoods, the principal constraints of dryland climates are aridity and the variability of the rainfall. In areas receiving less than 250 mm of rainfall during the growing season, farming is more or less impossible without irrigation, and most livelihoods are based on livestock. Yet in some parts of semiarid Africa, with rainfall of only 400–800 mm, very large and dense rural populations (up to 400 people per square kilometer (km$^2$)) are found. The variability of the rainfall compounds the effects of aridity and tends to increase the drier the location. Frequent and unpredictable droughts (and occasional floods) introduce high levels of risk into farming and livestock production. Rainfall variability also occurs over the long term. In the Sahel, Africa's biggest dryland, there was a decline of up to 33 percent in the average rainfall between the periods 1931–1960 and 1961–1990.[2]

Dryland soils—away from river valleys—have low natural fertility, measured in terms of the key plant nutrients.[3] Many soils are derived from desert sands and are incompletely formed. Scarce organic matter reflects the poor natural vegetation that grows with such a low rainfall.[4] Thus, nutrients limit plant growth where rainfall is higher, and aridity limits growth where rainfall is lower.[5] The economic output of cropping systems therefore depends on inputs and management.[6]

The African drylands are home to 268 million people—40 percent of the continent's population. That many of these people are very poor is not in doubt. However, because drylands form subregions within nations, statistics on dryland incomes are rare. Access to health and education services is poor in rural areas; infant, child, and maternal mortality are high; and average life expectation at birth is low. Poor nutrition reduces available energy, and the high temperatures in early summer make farm work arduous. Given the large numbers of people, failure to reduce poverty in the drylands will prejudice the achievement of the Millenium Development Goals (MDGs).[7]

For the past five decades, the threat of desertification has dominated dryland development policy and debate.[8] The term "desertification" describes a set of land degradation processes, which include soil fertility decline, erosion by wind or water, dune formation, hydrological decline, biodiversity loss, deforestation, and declining bioproductivity. Some changes are irreversible and extend beyond the normal climatic oscillations of the desert edge and the effects of random droughts to forms of land degradation that are commonly attributed to the actions of humans.[9]

The mainstream view of dryland management continues to be that widespread and "inappropriate" land use practices require transformation. This view is reflected in the Convention to Combat Desertification, which was formulated after the Rio Earth Summit in 1992. In neo-Malthusian interpretations, unsustainable practices are blamed on population growth, which has driven human and animal populations beyond the rather low carrying capacity of the drylands.[10]

## Challenging Malthus: The Machakos Story

In an article published in *Environment* in 1994,[11] based on a study of long-term change in Machakos District, Kenya, it was shown that degradation is not inevitable in African drylands. As long ago as the 1930s, the Machakos Reserve acquired some infamy among conservationists, who thought they saw "every phase of misuse of the land," leading to soil erosion and deforestation on a large scale, with its inhabitants consequently "rapidly drifting to a state of hopeless and miserable poverty and their land to a parching desert of rocks, stones and sand."[12]

By the 1990s, the district's population had multiplied sixfold, while expanding into previously uninhabited areas (most of them dry and risky). The study found, however, that erosion had been largely brought under control on private farmlands. This was achieved through innumerable small investments in terracing and drainage, advised by the extension services but carried out by voluntary work groups, hired laborers, or the farmers themselves. On some grazing land, significant improvements in management were also taking place. The value of agricultural production per km² increased between 1930 and 1987 by a factor of six and doubled on a per-capita basis.[13] At the same time, a rapid change in agricultural technology occurred, with a switch from an emphasis on livestock production to increasingly intensive farming, close integration of crops with livestock production, and increased marketing of higher value commodities (such as fruit, vegetables, and coffee). A social transformation also occurred with the enthusiastic pursuit of education, giving increased access to employment opportunities outside the district and intensifying rural-urban linkages.[14]

The Machakos story upset the Malthusian scenario for drylands, suggesting in its place the hypothesis that positive linkages between population growth and environmental management may occur under the right conditions.[15] As a widely cited "success story," is it a model for other African drylands?

## What Is a "Success Story"?

Dryland communities live and work on an intersection between managed ecosystems and human systems, which are coevolving as time goes on.[16] Seen this way, such interactions may take negative forms, leading to irreversible damage to the ecosystem, and/or impoverishment in the human system. Such a model means that change in the natural ecosystems should be in harmony with the course of human development. If not, negative impacts such as those characterized as desertification must occur. To guide policy, therefore, it is necessary to define what might be called a "success story." Farmers' and pastoralists' knowledge and achievements are central to success: Unlike outsiders, local peoples possess accumulated experience of specific dryland environments.[17]

Based on the Machakos experience, it is proposed that "success" may be evident in one or more of the following four interconnected domains: ecosystem management (soil and biological resources), land investments, productivity, and personal incomes or welfare (see Table 1). For the conservationist, the first of these may be considered preeminent. However, without a healthy level of investment, ecosystem stability, functions, and services (economic products, biodiversity, hydrological cycles, microclimates, and so on) cannot be sustained under increasing exploitation. Likewise, if productivity fails, new investments cease, and human incomes or welfare stagnate. Conservation cannot be achieved by edict, because it is socially constructed in a given situation. Integrated approaches to understanding and managing ecosystems are seen to be essential for the future relations between societies and nature.[18]

In the Machakos story, long-term evidence suggests success in all four domains but not in all of the indicators (Table 1). A transition was achieved from acute risk of further degradation to a pathway characterized instead by conserving the ecosystem, with a strong narrative of soil and water conservation and terracing—on arable land in particular—that now extends to every corner of the old district.[19] These works present impressive evidence of investments at the farm level, and there is a range of other investments such as water catchments and storage and livestock structures. In the individual household, financial flows occurred between sectors, whereby farm profits financed education and off-farm employment, while off-farm incomes were themselves used to finance capital developments on-farm.[20] Farm output indicators show increased value of output per km², including crop and livestock production. This fed into income effects, with an improving trend in farm incomes per capita until 1987. Rising educational achievement supported off-farm employment and the diversification of livelihoods.

However, all success is relative and incomplete. Thus it is unlikely that any system, including that of Machakos, can score highly in all four domains at one time or perform equally well at all times. Incomes in particular are at the mercy of macroeconomic changes. For example, in Machakos, increasing value of output per capita may have hesitated after 1987, with adverse trends in global markets.[21] The indicators are not all relevant everywhere. This descriptive model, however, suggests key variables in very complex systemic interactions and measurable indicators that are amenable to policies.

## Are There Success Stories in West Africa?

Success stories have been documented from a variety of locations in African drylands.[22] In West Africa, three of these stories are told in studies that were designed specifically to test the Machakos model under different conditions. Findings suggest that success is by no means unqualified.[23] However, evidence has been found of significant achievements in the first three domains—ecosystem management, land investments, and productivity—identified in Table 1. Such evidence, which emerges from analyses of long-term data and from well-supported inferences, runs counter to some current perceptions.

# Table 1 Defining "Success" in Drylands, Based on the Machakos Experience

| Domain | Outcome | Indicators |
|---|---|---|
| Ecosystem management | Stabilization or reversal of degradation | Soil erosion controlled<br>Soil water holding-capacity improved<br>Nutrient losses minimized or compensated<br>Trees managed sustainably[a]<br>Useful biodiversity maintained |
| Land investments | Viability and sustainability in economic and/or social terms[b] | Private farm investments<br>Cross-sectoral financial flows<br>Acceptable economic rate of return on public investments |
| Productivity | Maintenance or increase | Stable or increasing crop yields or livestock production per hectare (ha)<br>Increasing value of output per ha<br>Increasing market participation |
| Incomes and welfare | Maintenance or increase in real terms | Increasing value of output per capita<br>Strengthened access to off-farm incomes<br>Rising achievement in education<br>Asset accumulation on- and off-farm |

[a] The major factor driving forest clearance in dryland Africa is agricultural expansion. In Europe and South and East Asia, it is accepted that the historical benefits of agricultural expansion exceed the value of lost woodland. In African drylands, where deforestation is often declaimed, woodland clearance is often later followed by conservation of trees on farms. M. Mortimore and B. Turner, "Does the Sahelian Smallholder's Management of Woodland, Farm Trees and Rangeland Support the Hypothesis of Human-Induced Desertification?" *Journal of Arid Environments,* forthcoming, 2005. It is not practicable to propose a reversal of agricultural expansion.

[b] Investments are made for social as well as financial benefits (such as houses for retirement, domestic water catchments, and gardens).

## The Kano Close-Settled Zone, Nigeria

The significance of the Kano story lies in the scale and longevity of the area's intensive farming system. As originally delimited,[24] the zone now has a population of more than 5 million, excluding the urban population of the city of Kano, which exceeds 1.5 million.[25] More than 85 percent of the land surface is occupied by farmland. Intensive farming of small holdings under annual cultivation, with less than 0.5 hectares (ha) per person, a dense scattering of trees, and livestock, is centuries old in the zone and is strongly oriented toward the conservation and care of land resources.[26]

It is possible to analyze ecosystem management as follows:

- Soil organic matter is protected by applying manure from domestic animals, dry compost, and waste at rates of 4–6 tons per ha per year. In the dry season, animals roam the fields during the day but are penned at night, where they stay during the growing season, feeding on cut-and-carried fodder. The manure and bedding are mixed and returned to the fields by cart, donkey's back, or even head loading.
- Plant nutrients are recycled through feeding crop residues, tree browse, and weeds to the animals; growing nitrogen-fixing crops such as cowpea or groundnuts that fix nitrogen with the grain (millet or sorghum); protecting leguminous trees (*Faidherbia albida*); and

taking advantage of dust deposition in the dry season and low rates of leaching in the wet. When farm budgets permit, inorganic fertilizers are applied, but in small quantities—only about a third of recommended doses even when they were subsidized in the 1980s. Quantities of nitrogen, phosphorus, potassium, calcium, and magnesium in the soil vary from field to field and from year to year and tend to be low; nevertheless, no evidence was found of a general decline over a period of 13 years.[27]

- Soil moisture control is important where rainfall is scarce but intensive. It is achieved through field ridging (formerly by hand-hoeing but increasingly with ox-ploughing). This conserves runoff in furrows and maximizes infiltration between rainfall events. Competition between crops and weeds for scarce moisture is reduced by three or even four weeding operations during the short growing season (which is only 12–15 weeks long).
- Biodiversity is conserved in a functional sense through protecting or planting a variety of tree and shrub species on farms or along field boundaries and through on-farm selection and breeding of seed for at least 76 cultivars.[28] Biodiversity occurring naturally in the ecosystem is protected for its contribution to food security.[29]

- Erosion by wind or water is controlled on the gentle slopes of the farmlands (usually at grades less than five degrees) by planted field boundaries (including the henna bush, *Lawsonia inermis,* and the grasses *Andropogon gayanus* and *Vetiveria nigritana*), by planting spreading intercrops (cowpea, groundnut) among the grain, and by maintaining densities of mature farm trees at 7–15 per ha.

The result is an intricate, anthropogenic landscape of permanent fields, farmed parkland, and hamlets. A misleadingly sterile appearance in the dry season is confounded by a startling luxuriance of biomass during the wet.

The investments that made this possible in the past were mainly in human labor (manuring, weeding, and animal tending), but during the twentieth century, groundnut exports and increasing opportunities for employment or trading in the cities provided cash. After groundnut production failed in 1975 (on account of rosette disease and drought), farmers switched to selling some of their grain. Monetary transactions are extremely common, and during the 1990s farm investments that were visibly evident, especially near Kano, included ox-ploughs and teams, livestock kept for fattening, and "modern" building materials such as metal roofing sheets.

For long-term yield trends it is necessary to depend on inferences, as compatible and accurate data are not available. At least one-third of households in inner, high-density villages were estimated to be food-insufficient in the 1960s, although whether this was from poverty or from specializing in groundnuts for export is unclear. Poverty was noticeable on highly subdivided holdings in a very densely populated village close to the walls of Kano City in 1970.[30] Yet in the 1990s, except after a drought,[31] farm holdings with only one-third of a hectare per person could still be self-sufficient in basic grains.[32] The population of Kano Emirate, which roughly coincides with the zone, increased at a rate of 1.9 percent per year from 1931 to 1991.[33] Certainly there are food-insecure households today. Nevertheless, the information available does not suggest a general decline in yields but rather a sustained effort to increase the output of food in line with increasing consumption needs.

The livestock population, the integration of which into the farming system is essential for maintaining organic inputs to cultivated soils, increases in density in line with the human population. People like to invest their savings in animals, and fattening the animals for market has become a thriving industry. Sheep and goats, being easier to manage on the crowded farmlands, are increasing at the expense of cattle, the owners of which have to rely on transhumance (grazing away from home during the cropping season).

With their sales of animals, higher-value crops (cowpea; sesame; and new, disease-resistant groundnuts), and labor and skills in the rapidly growing urban markets of Nigeria, farming families are participating in more monetary transactions than ever before. Kano City's markets are supplying a population six times larger than in the 1960s, and although the zone contributes only a small part of its needs, what it does sell is very important to local livelihoods.[34] During the same period, increased retail activity and increasing domestic capital, such as bicycles, motorcycles, radios, clothing, equipment, and furnishings, suggested a slow, incremental growth for better-off families until the 1990s.[35] However, there is much anecdotal evidence of rural poverty.

Long-term change in the zone suggests significant successes in ecosystem management, many small-scale investments, and no evidence of a decline in yields per hectare. But trends in rural incomes depend on impressionistic evidence that is ambivalent. Recently, positive trends have been held ransom by economic recession, inflation, and adjustment policies.[36]

## Diourbel Region, Senegal

According to land-cover maps of the administrative departments of Bambey and Diourbel, the Diourbel Region, like the Kano Close-Settled Zone, has more than 85 percent of its surface under cultivation. But rural population densities are lower (46–150 per km²); therefore, there is more land per capita (0.6–0.7 ha). More than a century of production for export, promoted by strong policy incentives, reduced the priority given to grain production.

Farm expansion, the demarcation of boundaries with perennial grasses or shrubs, the protection of economic farm trees, and the scattering of small settlements have produced a landscape resembling that seen around Kano. Consequently, in the face of repeated cultivation and the export of plant nutrients, maintaining soil fertility is a focal theme in ecosystem management. But the soil fertility equation in Diourbel differs from that in Kano. Less rainfall (500 mm compared with 650 mm), labor, and animal manure mean that fertility, measured in terms of chemical soil properties, can only be sustained on about a fifth of cultivated land, the *champs de case* (the fields nearest to the house).[37] On these, familiar strategies are employed: recycling organic matter and available plant nutrients, erosion control, and biodiversity protection. The remaining fields, the *champs de brousse,* used to be fallowed regularly, but economic pressures to bring them under annual cultivation led to the substitution of inorganic fertilizers for fallowing. This strategy was promoted by fertilizer subsidies under the national agricultural program from 1960 to 1980. When the subsidy was removed, fertilizer use fell, and soil fertility faced a crisis of economic sustainability.

The Senegalese economy depended on groundnut exports from the inception of French rule in the 1880s until the 1980s.[38] In the late 1960s, a million tons per year were produced, making Senegal (with Nigeria) a leading world exporter. Under the French colonial policy and the independent government's *programme agricole,* the state took over responsibility not only for fertilizer promotion and supply but also for credit, groundnut seed supply, technical advice, marketing, processing, and export. From the sales of groundnuts, it deducted credit repayments, thus reducing the producers to a high level of dependency. It even subsidized rice imports so that Senegal could exploit its comparative advantage in growing groundnuts. Food commodity markets developed only slowly and late, as the Senegalese developed a strong preference for eating rice (mostly imported). Because the state was supplying much of the need for farm investments, private funds tended to be channelled into urban investments instead of agriculture.

The system proved to be financially unsustainable, as costs increased and global prices fell during the 1970s. At the same time, the Sahel Drought of 1968–1974 was followed by a persistent decline in annual rainfall until the 1990s. Structural adjustment policies, introduced in 1984, included an ending to credit and subsidies, the withdrawal of some agricultural services, late payments and stagnant, low producer prices. There was a dramatic fall in groundnut production, a diversion of output to local consumption, and agricultural stagnation. The amount of fallow land increased, and some farm tree populations may have declined.[39] In some areas, there was a slowing in the growth of the population and even absolute decline.[40]

However, not all the evidence on productivity is negative.[41] Long-term improvements in millet yields per ha and per mm of rainfall (though not per capita) are visible in official data.[42] Expanding production of cowpeas, which command a good price, and of crops for urban niche markets appear in the statistics or have been widely observed. Livestock populations have increased in response to buoyant meat prices, and fattening animals is a popular sideline for farming families even though they often have to purchase supplementary feed. Rural markets, for many years frustrated by the state's dominant role in commodity movements, are being reinvigorated.

The search for alternative incomes from outside farming is proportionally most important for households that, owing to small holdings or underinvestment, only produce a fraction of their annual cereal needs. On the other hand, households that are food-secure throughout the year can generate five times as much agricultural income from crops and livestock.[43] Income diversification thus plays a critical role in the adjustment to changing conditions.

The evidence of success in Diourbel, therefore, is very different from that in Machakos and Kano. With regard to managing the ecosystem, there has been partial success in stabilizing or reversing degradation, but the collapse of state investment has left farmers unable to compensate, and only a proportion of the soil benefits from organic fertilization. Crop production and farm tree populations in some areas are said to be languishing.[44] But private investment is bouncing back in response to market signals, and the livestock sector is

growing. Trends in rural incomes cannot be established from the data available, but much energy is devoted to widening the household's portfolio—whether from sheer necessity or in response to opportunity. The case for success rests essentially in the resilience that has characterized the responses of farming families, both to the failure of the government's state-led agricultural policy and the concurrent decline in rainfall.

## *Maradi Department, Niger*

The evidence of success so far assembled is found in densely populated regions where, according to Danish economist Ester Boserup's theory of agricultural change, land scarcity drives farmers to use labor-intensive practices and adopt new technologies, thereby increasing their output per hectare.[45] It is, however, in low-density regions that rapid population growth and agricultural expansion often draw attention to large-scale erosion and deforestation. One such region is Maradi Department in Niger, where until the nineteenth century, large areas lay uninhabited and where farming villages were mostly concentrated in the extreme south.

French colonization encouraged rapid migration from the south and from places further afield, taking advantage of relatively good rainfall during the 1950s and extensive cultivable land that stood unclaimed in the department's central *arrondissements.* The pace of change was rapid. The cultivated fraction in four representative village territories averaged 31 percent in 1957, according to aerial photograph interpretation.[46] In 1999, all four were virtually entirely cultivated, and between 1974 and 1996, satellite imagery of the whole department south of the effective limit of rainfed cultivation indicated an increase from 59 percent to 73 percent.[47] This confirms that little uncultivated land (and even less unclaimed land) remained (see Figure 1).

The Sahel Drought exposed the risk associated with farming in such marginal areas, even as a growing population sought to appropriate more land for millet and groundnut production. Deforestation, soil fertility decline, erosion, falling yields, overgrazing, food insecurity, poverty, and increasing dependence on resources outside the area (food aid and urban employment) seemed to threaten not only the environment but also economic sustainability of rural communities.[48] A major rural development project was put in place. The demand for cultivable land was driven by demographic, market, and institutional forces, as well as a threat of declining productivity.[49]

This was not an auspicious setting for a success story. However, there are some surprises to be found in long-term data and recent field investigations.[50] Two changes in the management of farmland have been detected. The first of these is a growing use of animal manure—especially in the southern arrondissements where land is most scarce—as well as inorganic fertilizer when it can be afforded. The second is the practice, called *défrichement amélioré,* of protecting

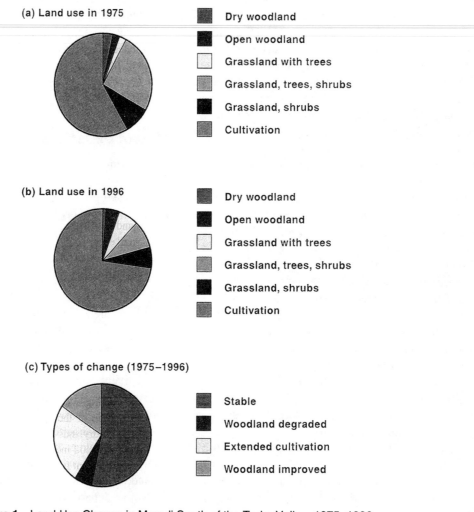

**Figure 1**   Land-Use Change in Maradi South of the Tarka Valley, 1975–1996.
Source: A. Mahamane, *Usages des Terres et Évolutions Végétales dans le Département de Maradi* (Land Use and Vegetation Change in Maradi Department), Drylands Working Paper 27 (Crewkerne, UK: Drylands Research, 2001), Tables 1, 2, and 3.

trees that are regenerating naturally and that are considered to have economic value. Promoted by the development program, this indigenous practice is now firmly established, creating an increasingly wooded appearance in the once bare farmland around the villages. Individuals rather than patrilineal families are tending to take charge of the natural resources, and conservationist attitudes toward biodiversity resources are being reasserted.[51]

Farm investments are thus already significant, and permanent fields and vegetated boundaries are appearing as private rights to the use of land are confirmed.[52] Livestock numbers in the department trended upward throughout the 1990s. Especially in southern areas, ox-drawn ploughs, seeders, and carts are becoming commonplace.

Crop yield data are available for each arrondissement and when compared, these suggest a positive trend for millet in the south (in response to the need to intensify farming practice) while in the north, where land is relatively abundant,

they are stagnant.[53] In the department as a whole, cereal production kept ahead of the nutritional requirements of a rapidly growing population from 1964 until 1998, except in drought years.[54]

There are no long-term data on incomes and welfare, and recent surveys confirmed the continuing existence of poverty in Maradi.[55] Not too much should be claimed, therefore, in terms of success. But Niger as a whole is now less dependent on food aid than it was in the 1980s, despite its larger population. Maradi Department lies across the border from Katsina and Kano in Nigeria and is increasingly incorporated into buoyant transborder marketing chains in food commodities.[56]

## Policies to Promote Success

Other regions in West Africa have also yielded counter-intuitive evidence of positive trends in environmental management and productivity.[57] From the studies that have been carried out in all these regions, we can learn that the following policy-dependent agents have had positive impacts:

- *Agricultural product markets.* Without effective demand, no dryland farmer can generate an income from agriculture with which to cross-subsidize other sectors of the household economy (such as education fees and trading capital). Without a prospect of profit, no dryland farmer will use off-farm income to capitalize agriculture (such as soil or water conservation and purchased inputs). It is remarkable that there is no dryland region of any size that lacks linkages with one or more markets, however distant. The relevant policies concern price stabilization, the control of competition from imports, and deregulation that is properly balanced with the needs of local producers.

- *Physical infrastructure.* Increased market participation is unavoidable for dryland producers; they need to sell surpluses, buy their way out of deficit, achieve food security, and access labor and commodity markets. The provision of roads and transport systems reduces marketing costs and thereby increases producer prices, increases interaction, opens up new income-earning opportunities and educational access. With technological change, electronic communications infrastructure is also critical for accessing market information. The relevant policies are public investments in roads and telephones and regulatory frameworks that facilitate private sector investments in communications systems.

- *Institutional infrastructures.* Given a withdrawal of the state from many areas (including produce buying and processing and agricultural extension support), a failure of private sector enterprises to adequately fill the gap, and failures of local governance, attention is shifting to building the capacity of decentralized and autonomous institutions. These can restore local ownership of natural resources, negotiate on behalf of local communities with higher levels of government, control access to endangered resources, defend the common interests of local producers in wider markets, and supply needed information of all kinds to communities. The relevant policies concern genuine decentralization of government powers, recognition and support of producers' or traders' associations, and the facilitation of new or adapted institutions for ecosystem management (such as common resources) and of information networks. An area offering major potential for poverty reduction is institutional frameworks to manage risk, which go to the heart of the uncertainty of living in dryland environments.[58]

- *Knowledge management.* Useful technologies, rather than being spread by promotional blueprint, tend to be taken up in response to users' particular circumstances. Discriminating users in rainfed farming systems need a variety of options from which to choose. There have been no "miracle" technologies so far that resemble the high-yielding crop varieties used in humid or irrigated South and East Asia. Farmers' own knowledge, exchanges, and experimentation have been as important to the process as experimental agriculture. The relevant policies concern education provision and partnering local producers in extension systems. As knowledge is a public good, the state still has a role to play in putting new knowledge into the public domain and facilitating access to it by poor rural people.

- *Investment incentives.* The assumption that all dryland families are too poor to invest is shown to be false in the cumulative growth of small on- and off-farm investments in the longer-settled and successful systems. Meanwhile, many of the problems of recently occupied, sparsely settled, and highly dynamic systems stem from a lack of investment. Attention was drawn to this priority in Maradi as long ago as 1980,[59] and the evidence suggests that the capitalization of the landscape using local peoples' own resources does indeed take decades rather than years. Clearly, secure title to use of and benefits from ecosystem resources is a necessary condition for investing scanty, private financial resources. The remarkable landscape of Machakos is a witness to what can be achieved in a half-century of incremental growth based on inputs of personal finance and family labor. The relevant policies, therefore, concern an enlightened adjustment by central government of variable incentives (such as taxation and exchange rates) and a prioritization of enabling incentives (such as secure resource tenure, dispute settlement, and conflict prevention) and efficient markets.[60]

- *Income diversification incentives.* Rather than posing a threat to urban areas, seasonal migrants from rural areas provide a range of services at minimal cost, in a dynamic and adaptive informal sector (which tends to defy government regulation). Urban informal sectors offer low average incomes, insecurity, squalor and sometimes ill-health. Paradoxically, perhaps, having such options available provides some of the flexibility necessary in rural drylands to cope with risk and recurrent food insecurity. They also provide investment funds for raising productivity and incomes in drylands, thereby helping to stabilize rural populations. Dryland peoples not only help themselves, they also help to integrate the national economy through consumption and investment streams. The relevant policies are relaxing controls on the movement of people, commodities, and capital around the country and ensuring the personal security of migrants. In an economic union such as the Economic Community of West African States (ECOWAS), these considerations apply to international movements, as drylands are interdependent with more humid or urbanized regions.

Policies, of course, evolve in response to articulated interests. To plead a special case for drylands, even if evidence-based, may have little impact on such a process. However, given the recent advances in democratic government in many countries of Africa, there are new opportunities to empower dryland people to plead their own cause with local and central government. This is especially true where evidence is available from targeted research. This type of empowerment relies on alliances between local people, researchers, and the private and voluntary sectors and on resources to support interactive debate and negotiation. Villagers can express their demands effectively when barriers of language and isolation are removed, as shown by experience in Kenya, Senegal, Niger, and northern Nigeria.[61]

# Conclusion

Is "success" a justified term in dryland environments where risk, low productive potentials, and high levels of poverty are endemic? Some areas can show evidence of achievements that run counter to the expectations generated by some orthodox models of desertification. There are four components of success: a movement toward more sustainable ecosystem management, evidence of increasing investment both on and off the farm, stable or improving output or output value per hectare, and evidence of improving incomes and welfare. Not all of these components are evident in every story, and the last in particular is difficult to pin down in the absence of data. However, a positive trajectory sustained over a timescale of decades is evidence too important to ignore, even though it may conflict with "expert" opinions based on shallow timeframes. It is remarkable that many African smallholders have sustained such achievements against policy failure and urban bias.[62]

The dryland cases show that rural development should be seen in a broader context than that of agriculture alone. Currently, efforts are being made to reemphasize the role of science and technology for improving productivity and food security in Africa. But it is acknowledged that if the potential of technology is to be fully realized, a market-led strategy is required to raise productivity.[63] The dryland success stories support the view that productivity trends respond to economic incentives. The capacities of resource-poor farmers to invest in on-farm improvements should not be underestimated, notwithstanding many constraints.

Basing dryland development policy on evidence of poor peoples' achievements offers a radical alternative to the doomsday scenarios that have a profound though not always admitted influence on policymakers. In starting with a perceived failure in ecosystem management—as implied in the idea of desertification—governments or donors drive themselves into a blind alley. Because the diagnosed "mismanagement" is defined and assessed in quantitative, biophysical terms, it calls for technical solutions that have to be imposed through an assertion of outsider knowledge over insider experience, devaluing the indigenous resource. Participatory rhetoric cannot on its own correct this distortion. On the other hand, knowledge partnerships, constructed on local peoples' achievements, imply equality between the agents of intervention and the intended beneficiaries, an equality that is too often denied in authoritarian governmental structures. This dilemma is widely recognized. It is especially acute in drylands, where the "knowledge gap" between insiders and outsiders is all the greater: The outsider is often unfamiliar with the harsh realities of managing relatively unproductive natural resources at high levels of risk.

By providing evidence of real achievements and internal potentials, success stories can therefore point the way toward laying a new foundation for evidence-led policies for dryland development: Rather than aiming to transform "inappropriate" local practices, such policies instead aim to build on local experience, suggesting a more organic model for development.

# Notes

1. The United Nations (UN) Food and Agricultural Organization (FAO) defines drylands by the average length of the growing period, while the UN Environment Programme (UNEP) defines drylands by an index of aridity based on precipitation over potential evapotranspiration. See FAO, *Land, Food and People,* FAO Economic and Social Development Series 30 (Rome: FAO, 1984); and UNEP, *World Atlas of Desertification* (Nairobi: Arnold for UNEP, 1992). According to the first, drylands have fewer than 180 growing days per year, and according to the second, an aridity index between 0.05 and 0.65. M. Mortimore, *Roots in the African Dust: Sustaining the Sub-Saharan Drylands* (Cambridge, UK: Cambridge University Press, 1998), 12.

2. M. Hulme, "Rainfall Changes in Africa: 1931–1960 to 1961–1990," *International Journal of Climatology* 12 (1992): 685–99.

3. For example, in Maradi Department, Niger, levels of carbon are less than 0.2 percent, nitrogen less than 0.02 percent, and available phosphorus less than 4 parts per million (ppm) on uncultivated soils. M. Issaka, *Évolution à Long Terme de la Fertilité de la Sol dans la Région de Maradi* (Long-Term Change in Soil Fertility in the Maradi Region), Drylands Research Working Paper 30 (Crewkerne, UK: Drylands Research, 2001). These values for phosphorus are exceptionally low; in Kano, Nigeria, and Machakos, Kenya, 11–16 ppm are found. F. Harris and M. A. Yusuf, "Manure Management by Smallholder Farmers in the Kano Close-Settled Zone, Nigeria," *Experimental Agriculture* 37 (2001): 319–32; and J. P. Mbuvi, *Makueni District Profile: Soil Fertility Management,* Drylands Research Working Paper 6 (Crewkerne, UK: Drylands Research, 2000).

4. Organic soils comprise less than 0.4 percent of Maradi soils (see Issaka, ibid.).

5. F. W. T. Penning de Vries and M. A. Djiteye, eds., *La Productivité des Pâturages Sahèliens. Une Etude des Sols, des Végétations et l'Exploitation de Cette Ressource Naturelle* (The Productivity of Sahelian Pastures. A Study of Soils, Vegetation, and the Exploitation of this Natural Resource) (Wageningen, Netherlands: Pudoc, 1991).

6. Issaka, note 3 above; and F. M. A. Harris, "Farm-level Assessment of the Nutrient Balance in Northern Nigeria," *Agriculture, Ecosystems and Environment* 71 (1998): 201–14.

7. The Millenium Development Goals are: halving the proportion of people in extreme poverty and suffering hunger between 1990 and 2015; achieving universal primary education by 2015; eliminating gender disparity in education by 2015; reducing the under-five child mortality ratio by two thirds by 2015; reducing the maternal mortality ratio by three quarters by 2015; halting and beginning to reverse the spread of HIV/AIDS and the incidence of malaria and other diseases by 2015; ensuring environmental stability; and creating a global partnership for development.

8. The most widely used characterization is the revised definition used for the 1992 UN Rio Earth Summit: "Desertification means land degradation in arid, semi-arid and sub-humid areas resulting from various factors, including climatic variations and human activities." UN, *UN Earth Summit. Convention on Desertification,* UN Conference on Environment and Development, Rio de Janeiro, Brazil, 3–14 June 1992, DPI/SD/1576 (New York: United Nations, 1994). Definition is not, however, straightforward: More than 100 attempts have been recorded. M. Glantz and N. Orlovsky, "Desertification: A Review of the Concept," *Desertification Control Bulletin* no. 9 (1983): 15–20.

9. J. F. Reynolds and D. M. Stafford Smith, eds., *Global Desertification. Do Humans Cause Deserts?* (Berlin: Dahlem University Press, 2002).

10. Thomas Malthus argued, with respect to eighteenth-century England, that human populations outgrow their capacity to feed themselves. In his own time, his theory was answered by the industrial revolution, which financed the English in importing food from a growing world market and led to technical innovations that increased agricultural productivity. Since World War II, the inability of many poor countries to follow the same path led to a revival of "neo-Malthusian" thinking. Evidence of land degradation under pressure from rapidly growing farming populations, considered as a system closed to the larger economy, appears to offer only impoverishment and starvation. A summary is provided in M. Mortimore, "Technological Change and Population Growth," in P. Demeny and G. McNicoll, eds., *Encyclopedia of Population* (New York: Macmillan Reference, 2003), 932–35; and for a new discussion of some issues, see Q. Gausset, M. Whyte, and T. Birch Thomsen, eds., *Beyond Territory and Scarcity: Social, Cultural and Political Aspects of Conflicts on Natural Resource Management* (Uppsala, Sweden: Nordic African Institute, 2004).

11. M. Mortimore and M. Tiffen, "Population Growth and a Sustainable Environment: The Machakos Story," *Environment,* October 1994, 10–20, 28–32.

12. C. Maher, *Soil Erosion and Land Utilisation in the Ukamba Reserve (Machakos),* Report to the Department of Agriculture, Mss, Afr.S.755, Rhodes House Library, Oxford (Nairobi: Department of Agriculture, 1937).

13. M. Tiffen, M. Mortimore, and F. Gichuki, *More People Less Erosion: Environmental Recovery in Kenya* (Chichester, UK: John Wiley & Sons, 1994), 93–96.

14. The Machakos story is told in great detail elsewhere: ibid.; J. English, M. Tiffen, and M. Mortimore, *Land Resource Management in Machakos District, Kenya: 1930–1990,* World Bank Environment Paper No. 5, (Washington, DC: World Bank, 1994); and M. Tiffen and M. Mortimore, "Malthus Controverted: The Role of Capital and Technology in Growth and Environment Recovery in Kenya," *World Development* 22, no. 7 (1994): 997–1010.

15. E. Boserup, *The Conditions of Agricultural Growth: The Economics of Agricultural Change under Population Pressure* (London: Allen and Unwin, 1965); and E. Boserup, *Economic and Demographic Relationships in Development* (Baltimore, MD: Johns Hopkins University Press, 1990).

16. P. F. Robbins et al., "Desertification at the Community Scale. Sustaining Dynamic Human-Environment Systems," in Reynolds and Stafford Smith, note 9 above, pages 326–56.

17. M. Mortimore, *Adapting to Drought, Farmers, Famines and Desertification in West Africa* (Cambridge, UK: Cambridge University Press, 1989); and Mortimore, note 1 above.

18. R. D. Smith and E. Maltby, *Using the Ecosystem Approach to Implement the Convention on Biological Diversity. Key Issues and Case Studies,* Ecosystem Management Series No. 2 (Gland, Switzerland: IUCN-The World Conservation Union, 2003).

19. The former Machakos District is now divided between Machakos and Makueni Districts. F. N. Gichuki, S. G. Mbogoh, M. Tiffen, and M. Mortimore, *District Profile: Synthesis,* Drylands Research Working Paper 11 (Crewkerne, UK, Drylands Research, 2000).

20. J. Nelson, *Makueni District Profile: Income Diversification and Farm Investment, 1989–1999,* Drylands Research Working Paper 10 (Crewkerne, UK: Drylands Research, 2000).

21. Tiffin, Mortimore, and Gichuki, note 13 above, 93–95.

22. Global Mechanism of the Convention to Combat Desertification (GM-CCD), *Why Invest in Drylands? A Study Carried out for the Global Mechanism of the Convention to Combat Desertification* (Rome: GM-CCD, 2004); and C. Reij and D. Steeds, *Success Stories in Africa's Drylands: Supporting Advocates and Answering Skeptics* (Amsterdam: CIS/Centre for International Cooperation, Vrije Universiteit Amsterdam, 2003).

23. The studies are reported in Drylands Research Working Papers 1–41, available at www.drylandsresearch.org.uk. See also M. Tiffen and M. Mortimore, "Questioning Desertification in Dryland Sub-Saharan Africa," *Natural Resources Forum* 26, no. 3 (2002): 218–33.

24. The Kano Close-Settled Zone was defined by Mortimore as having local population densities of 350 per square mile (147 per square kilometer (km²) or more, according to the census of 1962. M. Mortimore, "Land and Population Pressure in the Kano Close-Settled Zone, Northern Nigeria," *The Advancement of Science* 23 (1967): 677–88. Over an area of about

10,000 km², densities decrease regularly outward to distances of 40–100 km from Kano City. This threshold density is now exceeded over a wider area. In the innermost periurban districts, densities of more than 800 per km² are found. M. Tiffen, *Profile of Demographic Change in the KanoMaradi Region, 1960–2000,* Drylands Research Working Paper 24 (Crewkerne, UK: Drylands Research, 2001).

25. Tiffin, ibid.

26. Harris, note 6 above; and M. Mortimore, "The Intensification of Peri-Urban Agriculture: The Kano Close-Settled Zone, 1964–86," in B. L. Turner II, R. W. Kates, and H. L. Hyden, eds., *Population Growth and Agricultural Change in Africa* (Gainesville, FL: University Press of Florida, 1993), 358–400.

27. The exception was potassium, which declined significantly in the 59 samples analyzed. M. Mortimore, "Northern Nigeria: Land Transformation under Agricultural Intensification," in C. L. Jolly and B. B. Torrey, eds., *Population and Land Use in Developing Countries. Report of a Workshop* (Washington, DC: National Academy Press, 1993), 42–69.

28. M. Mortimore and W. Adams, *Working the Sahel: Environment and Society in Northern Nigeria* (London: Routledge, 1999), 78.

29. F. M. A. Harris and S. Mohammed, "Relying on Nature: Wild Foods in Northern Nigeria," *Ambio,* 32, no. 1 (2003): 24–29.

30. P. Hill, *Population, Prosperity and Poverty. Rural Kano, 1900 and 1970* (Cambridge, UK: Cambridge University Press, 1977).

31. Droughts have occurred unpredictably throughout the history of Kano and recently in the great Sahel Drought that hit Kano in 1972–1974, 10 years later in 1982–1984, and in several individual years during the 1990s (M. Mortimore, *Adapting to Drought: Farmers, Famines and Desertification in West Africa* (Cambridge, UK: Cambridge University Press, 2001); and M. Mortimore, *Profile of Rainfall Change in the Kano-Maradi Region, 1960–2000,* Drylands Research Working Paper 25 (Crewkerne, UK: Drylands Research, 2000)).

32. Grain yields ranged between 0.5 and 3 tons per hectare (depending on rainfall, fertilization, and weeding) on a small but accurately measured sample located 40 km from Kano. Some output might be sold, lent, or given away, rendering the household food-insecure. F. Harris, *Intensification of Agriculture in Semi-arid Areas: Lessons from the Kano Close-Settled Zone, Nigeria,* Gatekeeper Series No. 59 (London: International Institute for Environment and Development, 1996).

33. Tiffen, note 24 above.

34. J. A. Ariyo, J. P. Voh, and B. Ahmed, *Long-Term Change in Food Provisioning and Marketing in the Kano Region,* Drylands Research Working Paper 34 (Crewkerne, UK: Drylands Research, 2001).

35. An absence of long-term compatible data on real income is frustrating the analysis of income and welfare trends. Meanwhile, general price inflation (linked to a depreciation of the national currency by 200 percent since the 1970s) makes people strongly aware of hardship, especially since structural adjustment policies were introduced in 1986. A. R. Mustapha and K. Meagher, *Agrarian Production, Public Policy and the State in Kano Region, 1900–2000,* Drylands Research Working Paper 35 (Crewkerne, United Kingdom, Drylands Research, 2000).

36. Ibid.

37. A. N. Badiane, M. Khouma, and M. Sène, *Région de Diourbel: Gestion des Sols* (Diourbel Region: Soil Management), Drylands Research Working Paper 15 (Crewkerne, UK: Drylands Research, 2000); and P. Garin, B. Guigou, and A. Lericollais, "Les Pratiques Paysannes dans le Sine" (Farming Practices in Sine), in A. Lericollais, ed., *Paysans Sereer: Dynamiques Agraires et Mobilité au Sénégal* (Sereer Farmers: Agrarian Dynamics and Mobility in Senegal) (Paris: Editions Institut de Recherches en Développement, 1999), 211–98.

38. M. Gaye, *Région de Diourbel: Politiques Nationales Affectant l'Investissement chez les Petits Exploitants* (Diourbel Region: National Policies Affecting Small Farmers' Investments), Drylands Research Working Paper 12 (Crewkerne, UK: Drylands Research, 2000).

39. Farm tree populations are said to be under threat in the well-studied village of Sob, just outside the boundary of Diourbel Region. (See Garin, Guigou, and Lericollais, note 37 above.) Data gathered in 1999 in three villages indicated that the status of farmed parkland is more ambiguous than in Kano, with large variations between villages, abundant regeneration in places, but decline in some species and scarcities of middle-aged trees. S. Sadio, M. Dione, and S. Ngom, *Région de Diourbel: Gestion des Ressources Forestières et de l'Arbre* (Diourbel Region: Management of Forest Resources and Trees), Drylands Research Working Paper 17 (Crewkerne, UK: Drylands Research, 2000).

40. A. Barry, S. Ndiaye, F. Ndiaye, and M. Tiffen, *Région de Diourbel: Les Aspects Démographiques* (Diourbel Region: Demographic Aspects), Drylands Research Working Paper 13 (Crewkerne, UK: Drylands Research, 2000).

41. A. Faye, A. Fall, M. Mortimore, M. Tiffen, and J. Nelson, *Région de Diourbel: Synthesis,* Drylands Research Working Paper 23e (Crewkerne, UK: Drylands Research, 2001).

42. Official data at region and department level were used in the long-term analysis (1960–2000). They are based on a mixture of censuses and estimates. Small trends were only considered potentially significant if sustained over many years.

43. Such a difference is too large to be fully explained by the size of household, which averaged l0 members, 13 and 12 in types 1, 2, and 3 shown in Figure 2.

44. Garin, Guigou, and Lericollais, note 37 above; and Sadio, Dione, and Ngom, note 39 above.

45. See Boserup 1965 and 1990, note 15 above.

46. C. Raynaut, *Recherches Multidisciplinaires sur la Région de Maradi: Rapport de Synthèse* (Multidisciplinary Research in the Maradi Region: Synthesis) (Bordeaux: Maradi Region Research Programme, Université de Bordeaux II, 1980), 10.

47. A. Mahamane, *Usages des Terres et Évolutions Végétales dans le Département de Maradi* (Land Use and Vegetation Change in Maradi Department), Drylands Working Paper 27 (Crewkerne, UK: Drylands Research, 2001).

48. Raynaut, note 46 above; C. Raynaut, "Le Cas De La Région de Maradi (Niger)" (The Maradi Region Case, Niger), in J. Copans, ed., *Sécheresses et Famines du Sahel. Tome II. Paysans et Nomads* (Droughts and Famines in the Sahel. Farmers and Nomads) (Paris: Librairie Francoix Maspero, 1975), 5–43;

and C. Raynaut, J. Koechlin, C. Cheung, and M. Stigliano, *Le Dévelopement Rural de la Région au Village—Analyser et Comprendre la Diversité* (Rural Development from Region to Village—Analysing and Understanding Diversity) (Bordeaux, France: Maradi Region Research Programme, Université de Bordeaux II, 1988).

49. The rural population of Maradi Department grew at more than 3.0 percent per year between 1960 and 1988 (Tiffin, note 24 above). Maradi Department was a breadbasket of Niger: It produced the staple food crop, bulrush millet and was the main center of groundnut production until 1975. It was also a cotton-producing area. During the 1990s, the development of Niger's Code rural (which was based on the principle of "land to the tiller") motivated farmers to claim as much free land as they could plant.

50. M. Mortimore, M. Tiffen, Y. Boubacar, and J. Nelson, *Department of Maradi: Synthesis,* Drylands Research Working Paper 39e (Crewkerne, UK: Drylands Research, 2001).

51. A. Luxereau, "Usages, Représentations, Évolutions de la Biodiversité Végétales chex les Haoussa du Niger" (Use, Perceptions, and Change in Plant Biodiversity among the Hausa of Niger), *Journal d'Agriculture et de Botanique Appliqué NS* (Journal of Agriculture and Applied Botany NS) 36, no. 2 (1994): 67–85; and A. Luxereau and B. Roussel, *Changements Écologiques et Sociaux au Niger* (Ecological and Social Change in Niger) (Paris: L'Harmattan, 1997).

52. Y. Boubacar, *Évolution des Régimes de Propriété et d'utilisation des Ressources Naturelles dans la Région de Maradi* (Changes in the Tenure and Use of Natural Resources in the Maradi Region), Drylands Research Working Paper 29 (Crewkerne, UK: Drylands Research, 2000).

53. Tiffin and Mortimore, note 23 above.

54. The nutritional requirements are estimated at 200 kilograms per person per year. This finding does not exclude the possibility that cereals were exported from the department, leaving some families food-insecure.

55. Cooperative for Assistance and Relief Everywhere (CARE), *Evaluation de la Sécurité des Conditions de Vie dans le Département de Maradi* (Assessment of Living Conditions in Maradi Department) (Niamey, Niger: CARE Niger, 1997).

56. Ariyo, Voh, and Ahmed, note 34 above; and K. Meagher, *Current Trends in Cross-Border Grain Trade between Nigeria and Niger* (Paris: IRAM, 1997).

57. V. Mazzucato and D. Niemeijer, *Rethinking Soil and Water Conservation in a Changing Society,* Tropical Resource Management Papers, No. 32 (Wageningen, Netherlands: Wageningen University and Research Centre, 2000); D. Niemeijer and V. Mazzucato, "Soil Degradation in the West African Sahel: How Serious Is It?" *Environment,* March 2002, 20–31; and C. Reij and T. Thiombiano, *Développement Rural et Environnement au Burkina Faso: La Réhabilitation de la Capacité Productive*

*des Terroirs sur la Partie Nord du Plateau Central entre 1980 et 2000* (Rural Development and the Environment in Burkina, Faso: The Restoration of the Productive Capacity of Village Lands in the North of the Central Plateau between 1980 and 2000) (Amsterdam: CIS, Vrije Universiteit Amsterdam, 2003).

58. Grain banks and other attempts to counter the insecurity of living under drought risk have enjoyed only chequered success. It is arguable that this problem will be found as challenging as achieving sustainable natural resource management. Drought management will be the focus of a new initiative by UNDP's Dryland Development Centre, Nairobi.

59. Raynaut, note 46 above, page 66.

60. D. Knowler, G. Acharya, and T. van Rensburg, *Incentive Systems for Natural Resources Management,* (Rome: FAO Investment Centre, 1998).

61. M. Mortimore and M. Tiffen, "Introducing Research into Policy: Lessons from District Studies of Dryland Development in Sub-Saharan Africa," *Development Policy Review* 22, no. 3 (2004): 259–86.

62. Drylands Research, *Livelihood Transformations in Semi-Arid Africa 1960–2000: Proceedings of a Workshop Arranged by the ODI with Drylands Research and the ESRC,* in the series "Transformations in African Agriculture," Drylands Research Working Paper 40 (Crewkerne, UK: Drylands Research, 2001).

63. InterAcademy Council, *Realizing the Promise and Potential of African Agriculture. Science and Technology Strategies for Improving Agricultural Productivity and Food Security in Africa* (Amsterdam: InterAcademy Council, 2004.

---

**MICHAEL MORTIMORE** is an independent researcher with Drylands Research, Crewkerne, United Kingdom. His research focuses on policy studies of environmental management by small-scale farmers in Africa's drylands. Previously, he carried out research studies as a senior research associate in the Department of Geography, Cambridge University, the Overseas Development Institute, and as an honorary fellow of the Centre of West African Studies, University of Birmingham. Mortimore is the author or coauthor of several books, including *Working the Sahel: Environment and Society in Northern Nigeria* (Routledge, 1999), *Roots in the African Dust: Sustaining the Sub-Saharan Drylands* (Cambridge University Press, 1998), *More People, Less Erosion: Environmental Recovery in Kenya* (John Wiley, 1994), and *Adapting to Drought: Farmers, Famines and Desertification in West Africa* (Cambridge University Press, 1989). He may be reached at (44) (0) 1963-250-198 or via e-mail at mikemortimore@compuserve.com. The author wishes to thank Bill Adams, Abdou Fall, Adama Faye, Francis Gichuki, Mary Tiffen, and Boubacar Yamba, who, together with other colleagues in Africa and the United Kingdom, made valuable contributions to the ideas expressed in this article. The article is based on work mainly funded by the U.K. Department for International Development.

# Tracking U.S. Groundwater

## *Reserves for the Future?*

WILLIAM M. ALLEY

During the past 50 years, groundwater depletion has spread from isolated pockets to large areas in many countries throughout the world. Groundwater occurs almost everywhere beneath the land surface. Its widespread occurrence is a major reason it is used as a source of water supply worldwide. Moreover, groundwater plays a crucial role in sustaining streamflow between precipitation events and especially during protracted dry periods. In addition to human uses, many plants and aquatic animals are dependent upon groundwater discharge to streams, lakes, and wetlands.

A growing awareness of groundwater as a critical natural resource leads to some basic questions. How much groundwater do we have left? Are we running out? Where are groundwater resources most stressed? Where are they most available for future supply? To address these basic and seemingly simple questions requires consideration of several complexities of defining groundwater availability and a review of how one monitors groundwater reserves.

The term "groundwater reserves" is used to emphasize the fact that groundwater, like other limited natural resources, can be depleted. This potential for depletion is a key concept, despite the fact that unlike nonrenewable resources such as mineral deposits, most groundwater resources are replenished. On the other hand, some "fossil" groundwaters in arid and semiarid areas have accumulated over tens of thousand of years (often under cooler, wetter climatic conditions) and are effectively nonrenewable except by artificial recharge of surface water or treated wastewater.

Groundwater management decisions in the United States are made at a local level, which may be a state, municipality, or special district formed for groundwater management. Thus, monitoring of groundwater reserves should be designed to provide the information needed by these entities as a primary consideration. The issues to be addressed are varied and occur at many scales from preservation of a small spring fed by a nearby water source to the management of groundwater development throughout a large aquifer system or river basin.

The nation's groundwater reserve is not a single vast pool of underground water, but rather is contained within a variety of aquifer systems.[1] In general, the locations of the nation's aquifers are known, so much of the current research focuses on characterizing aquifer systems and how they respond to human activities.[2]

Many aquifers cross political divides, including county, state, and international boundaries. This characteristic (as well as the specialized nature of the science of groundwater hydrology) drives the need for a federal role and multijurisdictional collaboration in groundwater monitoring. Concerns about groundwater reserves have become more regional, national, and even global in scale in recent years, as exemplified by interstate and international conflicts over the salinization, contamination, or overexploitation of groundwater.[3] The effects of groundwater development may require many years to become evident. Thus, there is an unfortunate tendency to forgo the data collection and analysis that is needed to support informed decisionmaking until after problems materialize.

## How Much Groundwater Is Available?

The volume of water stored as groundwater is often compared to other major global pools of water within the Earth's hydrological cycle. For example, if one ignores water frozen in glaciers and polar ice, groundwater comprises more than 95 percent of the world's freshwater resources. This statistic illustrates the considerable value of groundwater, but it is also misleading, as it misses the variation in quantity, quality, and availability from location to location. The volume of groundwater in storage, its quality, and the yield to wells vary greatly across the planet. Typically, groundwater is used locally, so the effects of localized pumping on a given region are the primary concern of hydrogeologists.

Estimates of the volume of groundwater are poorly known relative to other pools of water. For example, the volume of the Earth's oceans has been well known for many years, whereas global estimates for groundwater storage vary by orders of magnitude (see Table 1). In part, this variability is due to different considerations of depth and salinity in defining the global groundwater pool. In addition, the variability reflects less knowledge about groundwater than other global pools of water. Early estimates of the global groundwater pool greatly underestimated its volume. It was not until after development began in earnest in the mid-twentieth century that an appreciation of the large storage volume of groundwater emerged universally. More recently, scientists have viewed this resource as an important component of the world water cycle and have expressed increasing interest in quantifying its role.[4]

As a practical matter, it is virtually impossible to remove all water from storage with pumping wells. However, the volume of

## Table 1 Volume of Water Attributed to Oceans and Groundwater Over Time

| | Cubic kilometers of water (in thousands) | |
|---|---|---|
| Date | Oceans | Groundwater |
| 1945 | 1,372,000 | 250 |
| 1967 | 1,320,000 | 8,350 |
| 1978 | 1,338,000 | 10,530–23,400 |
| 1979 | 1,370,000 | 4,000–60,000 |
| 1997 | 1,350,000 | 15,300 |

Note: These data come from different studies of the world water balance. Significant figures are largely retained from original sources.

Source: W. M. Alley, J. W. LaBaugh, and T. E. Reilly, "Groundwater as an Element in the Hydrological Cycle," in M. Anderson, ed., *The Encyclopedia of Hydrological Sciences* (Chichester, UK: John Wiley and Sons Ltd., 2005), 2215–28.

recoverable groundwater in storage for a particular area or aquifer can be estimated as the product of the area, saturated thickness, and specific yield (accounting as appropriate for differences in the estimates of saturated thickness and specific yield among multiple layers or zones).[5] To assess the value and limitations of estimates of groundwater in storage, it is helpful to first consider how aquifers are drained and then look at their dynamic links to the surface environment.

### Aquifer Drainage

The mechanism of aquifer drainage depends on whether an aquifer is unconfined or confined (see Figure 1). In an unconfined aquifer, the upper surface of the saturated zone (water table) is

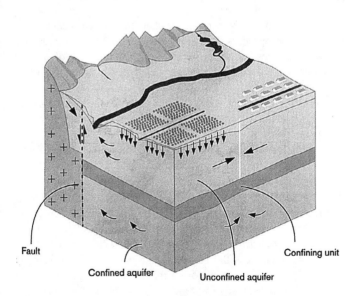

**Figure 1** Hypothetical Basin-Fill Aquifer System.

Note: The arrows show the direction of groundwater flow. Among the features shown are an unconfined aquifer overlying a confining unit and confined aquifer, a gaining stream, recharge from irrigated agriculture, and mountain-front recharge.

Source: Modified from S. A. Leake, *Modeling Ground-Water Flow with MOD-FLOW and Related programs*, U.S. Geological Survey Fact Sheet, 121–97 (Washington, DC. 1997).

free to rise and decline. The principal source of water from pumping an unconfined aquifer is the dewatering of the aquifer material by gravity drainage. The volume of water that is usable in practice is limited by the aquifer's permeability (how easily water moves through a rock unit), water quality, cost of drilling wells, and design of the well and pump.

Consider as an example the unconfined High Plains aquifer, which underlies an area stretching from southern South Dakota to the end of the panhandle of Texas and is the most heavily pumped aquifer in the United States. Depletion of aquifer storage from pumping has had substantial effects on irrigated agriculture in the High Plains, particularly in the southern half, where more than 50 percent of the saturated thickness has been dewatered in some areas. In Kansas, scientists have estimated the lifespan of the aquifer by projecting past trends into the future until the saturated thickness of the aquifer reaches a level at which groundwater pumping for irrigation becomes impractical.[6] The results suggest that many areas in western Kansas have less than 50 years of usable groundwater remaining. Thirty feet of saturated thickness was the critical level in this study, although the researchers noted additional studies that suggest that 30 feet is not enough saturated thickness to provide sufficient well yields for irrigation.

Changes in groundwater levels throughout the High Plains aquifer are tracked annually through the cooperative effort of the U.S. Geological Survey and state and local agencies in the High Plains region (see Figure 2a). Despite the considerable effects of storage depletion in much of the High Plains, only 6 percent of the volume of water in the High Plains aquifer has been depleted since pumping began (see Figure 2b), illustrating how aggregated information about storage depletion over large areas can mask significant local effects.

Confined aquifers, which underlie low permeability confining systems, are filled by water under pressure and respond to pumping differently. The water for pumping is derived not from pore drainage but rather from aquifer compression and water expansion as the hydraulic pressure is reduced. Pumping from confined aquifers results in more rapid water-level declines covering much larger areas when compared to pumping the same quantity of water from unconfined aquifers. If water levels in an area are reduced to the point where an aquifer changes from a confined to an unconfined condition (becomes dewatered), the source of water becomes gravity drainage as in an unconfined aquifer. A major complication arises, however, because the drawdowns in the confined aquifer will induce leakage from adjacent confining units. Slow leakage over large areas can result in the confining unit supplying much, if not most, of the water derived from pumping.[7] Therefore, it is particularly difficult to relate estimates of the volume of groundwater in storage to the usable volume of groundwater in confined aquifers.

Further complications may arise for those aquifers with silt and clay layers that can permanently compact as a result of pumping. Consider, for example, the Central Valley aquifer in California, the nation's second-most-pumped aquifer. By 1977, about 28 percent of the decrease in aquifer storage of 60 million acre-feet was the result of permanent reduction of pore space by compaction, resulting in land subsidence throughout much of the area.[8] Farmers in the Central Valley have drawn on imported surface water as a major source of irrigation water to reduce groundwater depletion and associated subsidence. The decrease in aquifer storage of 60

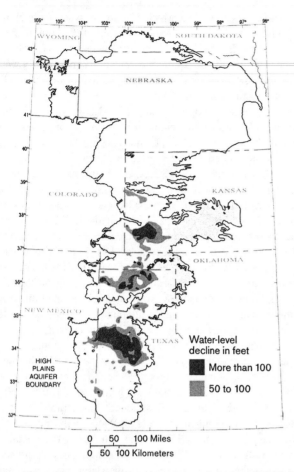

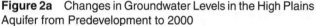

**Figure 2a** Changes in Groundwater Levels in the High Plains Aquifer from Predevelopment to 2000
Source: V. L. McGuire et al., *Water in Storage and Approaches to Ground-Water Management, High Plains Aquifer, 2000*, U.S. Geological Survey Circular 1243 (Reston, VA, 2003).

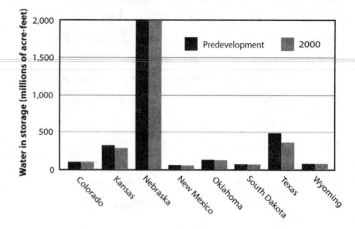

**Figure 2b** Comparison of Predevelopment and 2000 Ground-Water in Storage by State.
Source: V. L. McGuire et al., *Water in Storage and Approaches to Ground-Water Management, High Plains Aquifer, 2000*, U.S. Geological Survey Circular 1243 (Reston, VA, 2003).

## Interactions with Surface Water

Groundwater flows from areas of recharge to areas of discharge. Recharge includes water that naturally enters a groundwater system and water that enters the system at artificial recharge facilities or as a consequence of human activities such as irrigation and waste disposal. Discharge may occur to the atmosphere by transpiration; to streams, lakes, and other surface-water bodies; or through a pumping well. The balance between groundwater recharge and discharge controls groundwater levels and storage in a manner analogous to how deposits and withdrawals control savings in a bank account. If recharge exceeds discharge for some period, groundwater levels and storage will increase. Conversely, groundwater levels and storage will decline during periods when discharge exceeds recharge.

A common misperception is that the development of a groundwater system is "safe" if the average rate of groundwater withdrawal does not exceed the average annual rate of natural recharge. People sometimes make the erroneous assumption that natural recharge is equivalent to the basin sustainable yield.[10] Even further misinterpretations suggest that pumping at less than the recharge rate will not cause water levels and groundwater storage to decline.

To understand the fallacy inherent in these conclusions, one needs to consider how groundwater systems respond to pumping. Under natural conditions, a groundwater system is in long-term equilibrium. That is, averaged over some period (and in the absence of climate change), the amount of water recharging the system is approximately equal to the amount of water leaving (discharging from) the system. Withdrawal of groundwater by pumping changes the natural flow system, and the water is supplied by some combination of increased recharge, decreased discharge, and removal of water that was stored in the system.

Initially, water levels in pumping wells will decline to induce the flow of water to these wells, and water is removed from storage. Subsequently, the groundwater system readjusts to the pumping stress by "capturing" recharge or discharge. Also, the storage contribution to the water budget decreases with time for any given

million acre-feet, although very large, represented only a small part of the more than 800 million acre-feet of freshwater stored in the upper 1,000 feet of sediments in the Central Valley. [9]

Similarly, subsidence caused by groundwater pumping in the low-lying coastal environment of Houston, Texas, has increased its vulnerability to flooding and tidal surges. As a result, Houston has undertaken an expensive shift from sole reliance on its vast groundwater resource to partial reliance on surface water for its water supply.

Several key points arise from the examples and discussion thus far. First, measurement of storage depletion should be placed in the context of individual aquifer systems. For example, in evaluating water-level declines, one has to distinguish carefully between confined and unconfined aquifers, as the two respond very differently to pumping. Second, aquifer-wide estimates of recoverable water in storage have limited utility without considering the distribution of water-level changes and their effects. Finally, depletion of a small part of the total volume of water in storage can have substantial effects that become the limiting factors to development of the groundwater resource. These issues are further reinforced when one considers the response of surface-water bodies to groundwater pumping.

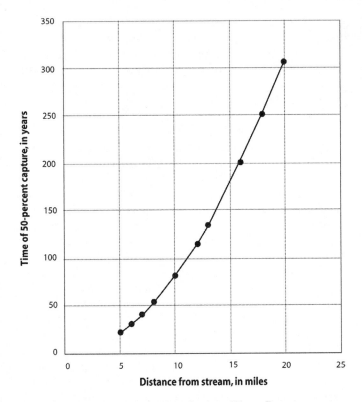

**Figure 3** Effect of Pumping on Surface-Water Resources.
Note: Time of 50-percent capture is the number of years until 50 percent of the pumping rate is accounted for as reduced groundwater discharge to the stream. The relation is for a fully penetrating stream in an aquifer having a transmissivity to storage ratio of 110,000 square feet per day.

Source: C. Fillippone and S. A. Leake, "Time Scales in the Sustainable Management of Water Resources," *Southwest Hydrology* 4, no. 1 (2005): 17. © SAHRA–University of Arizona.

withdrawal. If the system can come to a new equilibrium, the changes in storage will cease (at a new reduced level of groundwater storage), and inflows will again balance outflows. Thus, the long-term source of water to discharging wells becomes a change in the inflow and outflow from the groundwater system.

The amount of groundwater available for use depends upon how the changes in inflow and outflow affect the surrounding environment and upon the extent to which society is willing to accept the resultant environmental changes. Consequences include reduced availability of water to riparian and aquatic ecosystems and reduced availability of surface water for use by humans. Further complicating matters, the effects of pumping on surface-water resources can be spread out over a long period of time, as illustrated by the alluvial aquifer example in Figure 3.

In many areas, the effects of groundwater pumping on surface-water resources, and importantly, the large uncertainties associated with these effects, become the limiting factors to groundwater development. For example, University of Arizona water law and policy expert Robert Glennon in his popular book *Water Follies* describes controversial situations from throughout the United States where groundwater pumping affects streams and lakes.[11] The effects on surface water can occur with relatively little depletion of the total amount of groundwater in storage. One of the areas Glennon discusses is the Upper San Pedro River Basin in southeastern Arizona, where concerns about stream-flow depletion have caused conflicts between development and

environment interests in this ecologically diverse riparian system. The health of the riparian system is dependent on the groundwater level and hydraulic gradient near the stream. A key question is how pumping in the basin affects these components of riparian system health. Congressionally mandated efforts are under way to reduce the annual storage depletion (overdraft) in the Sierra Vista area—a subwatershed of the Upper San Pedro Basin.[12] Current overdraft in the Sierra Vista subwatershed is about 10,000 acre-feet per year, which is small, relative to estimates ranging from 20 to 26 million acre-feet of total groundwater storage.[13] A monitoring plan is an important element for verifying the effectiveness of management measures in reducing overdraft in the Sierra Vista subwatershed, with the ultimate goal of mitigating impacts on the riparian system.

## *Water-Quality Limitations*

Groundwater contamination from human activities clearly places constraints on groundwater availability. Likewise, water-quality constraints on groundwater availability can result from pumping. Perhaps best known are the many cases of saltwater intrusion from pumping groundwater along coastal areas. Groundwater pumping also can induce movement of saline water from underlying aquifers in inland areas. Likewise, shallow polluted groundwater may be induced or accelerated downward and throughout an aquifer by prolonged pumping, such that contaminated groundwater penetrates further and more quickly than otherwise anticipated. The removal of water from storage also changes the quality of the remaining groundwater because good quality water commonly is withdrawn first, and the residual often includes poorer quality groundwater from elsewhere in the aquifer or groundwater that has leaked into the aquifer from adjacent units in response to declining water levels. All these and other possible changes in water quality need to be considered in conjunction with information about changes in water levels and water in storage in evaluating the availability of groundwater. In some cases, the quality of groundwater will be suitable for some uses but not others. Water treatment may be necessary to meet some needs.

# Groundwater Use

An average of 85 billion gallons of groundwater are withdrawn daily in the United States. More than 90 percent of these withdrawals are used for irrigation, public supply (deliveries to homes, businesses, industry), and self-supplied industrial uses. Irrigation is the largest use, accounting for about two-thirds of the amount. The percentage of total irrigation withdrawals provided by groundwater increased from 23 percent in 1950 to 42 percent in 2000. Groundwater provides about half the nation's drinking water with nearly all those in rural areas reliant upon groundwater.[14]

The importance of groundwater withdrawals in the United States is similar to that in the rest of the world, with some variations from country to country. Rapid expansion in groundwater use occurred between 1950 and 1975 in many industrial nations and subsequently in much of the developing world. The intensive use of groundwater for irrigation in arid and semi-arid countries has been called a "silent revolution" as millions of independent farmers worldwide have chosen to become increasingly dependent on the reliability of groundwater resources, reaping abundant social and economic benefits but with limited management controls by government water agencies.[15] Perhaps as many as two billion people worldwide depend directly upon groundwater for drinking water. The dependence on groundwater for drinking water is particularly high in Europe, where about 75 percent of the drinking-water supply is obtained from groundwater.[16]

Water-use data, when coupled with a scientific understanding of how aquifers respond to withdrawals, are crucial for water planning. Yet information on groundwater use is spotty and often inaccurate within the United States and worldwide. In the United States, practices for collecting water-use data vary significantly from state to state and from one water-use category to another, in response to laws regulating water use and interest in water-use data as an input for water management. Programs to collect water-use data in each state are summarized in a review by the National Research Council.[17]

Some water-use data, such as withdrawals for drinking water and other household uses and withdrawals by some industrial users are obtained by direct measurement, and some may be estimated as the amount reported or allowed by permit. Many uses, such as for self-supplied domestic use, agriculture, and some industries, are often estimated using coefficients relating water use to another characteristic, such as number of employees, number of units manufactured, irrigated acreage, or number of livestock. For example, self-supplied domestic water withdrawals are typically determined by multiplying an estimate of the self-supplied population by a per-capita use coefficient. Likewise, water use for a particular type of industry might be estimated using information on employment or production and estimates of gallons per day per employee or per unit of product. Ideally, coefficients used for water-use estimation are grounded in representative data records. In practice, they are often derived empirically or developed using data that are sparsely sampled in time and space and perhaps extrapolated beyond the climatic, technological, and economic conditions for which they were originally developed. Other complications arise in these calculations because it may be difficult to separate surface-water and groundwater withdrawals without site-specific data and because small-scale use may be excluded from official statistics.

In determining the effects of pumping, it is important to recognize that not all the water pumped is necessarily consumed. For example, some of the water pumped for irrigation is lost to evapotranspiration, and some of the water returns to the groundwater system by infiltration, canal leakage, and other paths of irrigation return flow. Of course, water that is not used for consumption can undergo substantial changes in quality between withdrawal and recharge. Ideally, information on groundwater use includes estimates of consumptive use and return flow as well as withdrawals.

## Groundwater Sustainability and Management

Achieving an acceptable tradeoff between groundwater use and the long-term effects of that use is a central theme in the evolving concept of groundwater sustainability.[18] Initially, people viewed groundwater as a convenient resource for general use, and they focused their attention on the economic aspects of groundwater development. Sustainability concerns, emerging in the early 1980s, have brought environmental viewpoints and an intergenerational perspective to the forefront in discussions about groundwater availability.

Groundwater sustainability is commonly defined in a broad context as the development and use of groundwater resources in a manner that can be maintained for an indefinite amount of time without causing unacceptable environmental, economic, or social consequences. The amount of time it takes for the effects of pumping to be manifested elsewhere in the environment reinforces the importance of sustainability as a concept for groundwater management but also makes sustainable solutions difficult to apply in practice. Application of sustainability concepts to water resources requires that the effects of many different human activities on water resources and the overall environment be understood and quantified to the greatest extent possible over the long term. Thus, sustainability likely requires an iterative process of continued monitoring, analysis, application of management practices, and revision. For some cases, particularly in arid areas, the groundwater resource is treated as nonsustainable.[19]

The tradeoff between the water used for consumption and the effects of groundwater withdrawals—on maintenance of instream-flow requirements for fish and other aquatic species, the health of riparian and wetland areas, and other environmental needs—is the driving force behind discussions about the sustainability of many groundwater systems. Considerable scientific uncertainty is associated with disputes over whether pumping will have a specific impact on a particular river or spring. Further complicating matters is the fact that although they are linked through the hydrologic cycle, groundwater and surface water are typically managed separately under different laws and administrative bodies.

Groundwater management strategies are composed of a small number of general approaches:[20]

- use of sources of water other than local groundwater, by shifting the local source of water (either completely or in part) from groundwater to surface water or importing water from outside the local water-system boundaries (the California Central Valley and Houston have implemented these approaches);

- changing rates or spatial patterns of groundwater pumping to minimize existing or potential unwanted effects (examples include moving well fields inland to avoid saltwater intrusion, shifting from deep to shallow groundwater or vice versa, and maintaining sufficient distances between wells to avoid excessive drawdown);

- control or regulation of groundwater pumping through implementation of guidelines, policies, taxes, or regulations by water management authorities (these imposed actions may include restrictions on some types of water use, limits on withdrawal volumes, or establishment of critical levels for aquifer hydraulic heads);

- artificial recharge through the deliberate introduction of local or imported surface water—whether potable, reclaimed, or waste-stream discharge—into the subsurface for purposes of augmenting or restoring the quantity of water stored in developed aquifers (options include infiltration from engineered impoundments, direct-well injection, and pumping designed to induce inflow of freshwater from surface waterways);

- use of groundwater and surface water through the coordinated and integrated use of the two sources to ensure optimum long-term economic and social benefits;
- conservation practices, techniques, and technologies that improve the efficiency of water use, often accompanied by public education programs on water conservation;
- reuse of wastewater (gray water) and treated wastewater (reclaimed water) for non-potable purposes such as irrigation of crops, lawns, and golf courses;
- desalination of brackish groundwater or treatment of otherwise impaired groundwater to reduce dependency on fresh groundwater sources.

These general approaches are not mutually exclusive; that is, the various approaches overlap, or the implementation of one approach will inevitably involve or cause the implementation of another. For example, many approaches involve combinations of surface water, groundwater, and artificial recharge. During periods of excess surface-water runoff, and when surface-water impoundments are at or near capacity, surplus surface water can be stored in aquifer systems through artificial recharge. Conversely, during droughts, increased groundwater pumping can be used to offset shortfalls in surface-water supplies. Depleted aquifer systems can be seen as potential subsurface reservoirs for storing surplus imported or local surface water.

It is important to frame the hydrologic implications of various alternative development strategies in such a way that their long-term implications can be properly evaluated, including effects on the water budget. For example, changing the rates or patterns of groundwater pumping will lead to changes in the spatial patterns of recharge to or discharge from groundwater systems. As another example, in some areas of extensive use of artificial recharge, such as parts of southern California, water from artificial recharge may have replaced much of the native groundwater.

# Monitoring Groundwater Reserves

Water-level measurements in observation wells provide the primary source of information about groundwater reserves. Water-level data collected over periods of days to months are useful for determining an aquifer's hydraulic properties; however, data collected over years to decades are required to monitor the long-term effects of aquifer development and management.

The amount of effort in collecting long-term water-level data varies greatly from state to state, and many long-term monitoring wells are clustered in certain areas.[21] Although they are difficult to track, the number of long-term observation wells appears to be declining because of limitations in funding and human resources. For example, the number of long-term observation wells monitored by the U.S. Geological Survey (USGS) declined by about half from the 1980s to 2000.

For many decades, hydrogeologists and others have been making periodic calls for a nationwide program to obtain more systematic and comprehensive records of water levels in observation wells. O. E. Meinzer, a longtime chief of the USGS Ground Water Division and considered by many to be the father of the science of hydrogeology, described the characteristics of such a program about 70 years ago:

> The program should cover the water-bearing formations in all sections of the country; it should include beds with water-table conditions, deep artesian aquifers, and intermediate sources; moreover, it should include areas of heavy withdrawal by pumping or artesian flow, areas which are not affected by heavy withdrawal but in which the natural conditions of intake and discharge have been affected by deforestation or breaking up of prairie land, and, so far as possible, areas that still have primeval conditions. This nation-wide program should furnish a reliable basis for periodic inventories of the ground-water resources, in order that adequate provision may be made for our future water supplies. [22]

More recently, the Heinz Center report *The State of the Nation's Ecosystems* indicated that data on groundwater levels and rates of change are "not adequate for national reporting."[23] This report advocated supplementing existing networks to develop a national indicator of trends in groundwater levels. The U.S. Government Accountability Office noted that no federal agencies are collecting groundwater data on a national scale and only the USGS and National Park Service are collecting water-level data on a regional scale.[24]

Historically, water-level measurements were simply tabulated, recorded in a paper file, and possibly published in reports. Today, many agencies use the Internet to enhance users' access to current and historical monitoring data. Furthermore, continuous collection, processing, and transmission of water-level data on the Internet in "real time" (typically updated every few hours) is becoming more of a standard procedure. Real-time groundwater data are useful in formulating drought warnings, as they suggest potential effects on water levels in shallow domestic wells. Real-time capability can lead to improved data quality (from continual review of the data) as well as to increased interest in groundwater conditions on the part of the general public.

In addition to water-level monitoring, certain geophysical techniques can enhance the delineation and interpretation of water-level changes over a region. For example, microgravity methods can be used to measure the small gravitational changes that result from changes in groundwater storage (including water stored in the unsaturated zone) over an area (see Figure 4). More recently, researchers have proposed satellite-based measurements of gravity to measure changes over areas the size of a large part of the High Plains aquifer.[25] Meeting the majority of needs for water-level information requires much finer detail than that which satellite measurements can provide, but future technologies may improve this technique. Land- and satellite-based gravity measurements provide area-wide information on changes in the volume of water in storage but do not provide information on vertical changes in heads (water levels) in aquifer systems.

A second geophysical technique, Interferometric Synthetic Aperture Radar (InSAR), uses repeated radar signals from space to measure land-surface uplifts or subsidence at high degrees of

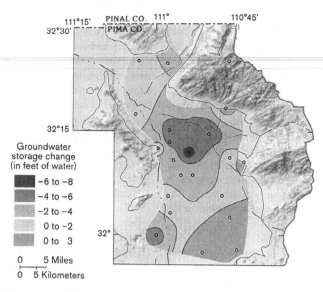

**Figure 4** Change in Groundwater Storage in the Tucson Basin, 1989–1998.

Note: Change in storage was estimated using microgravity measurements.

Source: D. R. Pool, D. Winster, and K. C. Cole, *Land Subsidence and Ground-Water Storage Monitoring in the Tucson Active Management Area, Arizona,* U.S. Geological Survey Fact Sheet 084–00 (Reston, VA, 2000).

measurement resolution and spatial detail. [26] Like gravity methods, InSAR has the advantage of being able to make measurements over large areas and between monitoring wells. The InSAR information can provide additional insights into the areal extent of groundwater depletion where it is linked to subsidence, and can even detect uplift from artificial recharge. InSAR has been found to be particularly useful in identifying faults and geologic structures that may impede groundwater flow and affect the response of an aquifer system to pumping.

## *Integrated Monitoring and Assessment of Groundwater Reserves*

As previously noted, the desire for a national network for monitoring groundwater levels has been discussed since the early 1900s but remains unfulfilled. Meanwhile, groundwater issues have evolved beyond early concerns focused on the hydraulics of individual wells and well field development to encompass many aspects of groundwater, including quantity, quality, and interactions with surface water. Technological advances have been made in sensors, communications, and electronic control systems to monitor groundwater, and computer modeling has become widely used to evaluate groundwater systems. From today's perspective, what might an ideal national program involve?

First and foremost, a national water-level monitoring program should be a collaborative process that involves discourse among local, state, and federal governmental agencies, nongovernmental organizations, and the public.[27] Ideally, data collected would serve double-duty by contributing to the larger regional and national picture while meeting local needs. There should be sufficient consistency in approach to describe the status of groundwater reserves across the country and to show how different constraints affect

utilization of the nation's aquifers. A major early goal would be to identify critical gaps in existing coverage.[28]

Many of the primary issues affecting groundwater availability require analysis at the scale of aquifers to achieve a meaningful perspective. To that end, monitoring programs should be designed in the context of the specific characteristics of each aquifer system. A comprehensive national monitoring program should track major aquifers that are affected by groundwater pumping, areas of future groundwater development, and areas of groundwater recharge. Water levels should be measured in wells open to different depths and in the context of the three-dimensional groundwater-flow system.

A long-term record of water-level measurements should encompass the period between the natural and developed states of aquifer systems. Other approaches, such as gravity measurements to estimate subsurface-water storage changes, should be considered in conjunction with the water-level monitoring program. Establishing links between water-level and water-quality requires an understanding of groundwater-flow systems. Studies of natural mixing in aquifers suggest that existing damage to groundwater quality may be lasting and could gradually extend deeper into aquifer systems, thereby reducing further groundwater availability.[29]

Data on changes in groundwater levels provide essential information about changes in groundwater storage and provide the simplest way to convey the extent of groundwater depletion. However, changes in groundwater storage are only part of the story. As noted in previous examples, the status of groundwater reserves should be placed in the context of the complete water budget for that aquifer system. Thus, the monitoring of surface water and groundwater should be linked, particularly measurements of streamflow during low-flow periods when groundwater discharge is the primary component of streamflow. This might require an increase in the number of streamgaging stations in targeted basins to estimate the groundwater contribution to streamflow.

In addition to monitoring data on natural systems, estimation of water withdrawals and consumptive use is an essential part of computing a water budget for a developed aquifer system. Groundwater pumping is one component of the water budget that is physically possible to measure; yet it is commonly one of the most uncertain components of the water budget. Water-use information should be an integral part of evaluations of groundwater quantity and quality and other environmental conditions. Where multiple overlying aquifers are used, efforts should be made to estimate withdrawals from each.

Groundwater systems are dynamic and adjust over decades or more to pumping and other stresses. Many aquifer systems have undergone several decades of intensive development and may be far from equilibrium. Thus, it is challenging to place current conditions in the context of the dynamic but slow changes that may be taking place. A simple snapshot of current conditions may not indicate, for example, how future streamflow depletion will evolve from the pumping that has already occurred.

During the past several decades, computer models for simulating groundwater and surface-water systems have played an increasing role in the evaluation of groundwater development and management alternatives. Groundwater modeling serves as a quantitative means of evaluating the water balance of an aquifer,

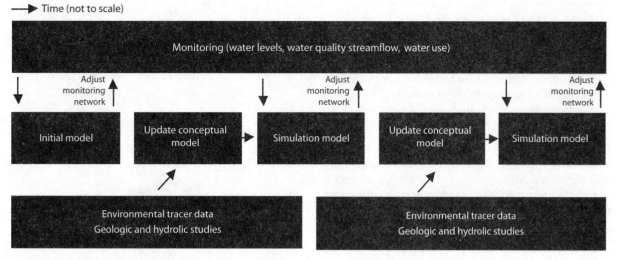

**Figure 5**  Tracking Groundwater Resources.

as it is affected by land use, climate, and groundwater withdrawals, and how these changes affect streamflow, lake levels, water quality, and other important variables. Generally, monitoring and computer modeling are treated as distinct activities, but to be most effective, the two should be linked. Such a framework is considered further below, and its essential elements are illustrated in Figure 5.

Monitoring groundwater reserves serves as primary information used in the development and calibration of computer models. Likewise, the process of model calibration and use provides insights into which components of the system are best known, which components are poorly known, and which components are more important than others. Thus, the experience gained from modeling should provide a basis for a periodic evaluation of the monitoring network.

As its basis, every simulation model has a conceptual model that represents the prevailing theory of how the groundwater system works. The appropriateness of this conceptual model is tested as a numerical model is built, and field observations are compared to the model simulations. Unfortunately, more often than not, data will fit more than one conceptual model, and good calibration of a model does not ensure a correct conceptual model. Thus, conceptual and numerical modeling should be viewed as an iterative process in which the conceptual model is continuously reformulated and updated as new information is acquired.[30] The importance of this approach and its link to monitoring data is recognized explicitly in Figure 5 as a key step prior to each new stage of groundwater modeling.

Additional scientific studies conducted at the time of modeling or during intervening periods can provide insights into the adequacy of the conceptual model that underlies the computer model as well as help in adjustment of model parameters. Such studies include use of environmental tracers, studies of the geologic framework, and geophysical studies. For example, an increasing number of chemical and isotopic substances are being measured in groundwater to identify water sources, trace directions of groundwater flow, and measure the age of the water (time since recharge). Comparison of the results from these environmental tracers with information from computer modeling can lead to either increased confidence in the conceptual model of a groundwater system or

recognition of the need for changes. The Middle Rio Grande Basin in central New Mexico (see box on page 170) provides an example of how long-term water-level monitoring combined with environmental and geologic studies has contributed to an evolving series of conceptual and simulation models used to help manage the groundwater resources of the basin.

Not all aquifer systems lend themselves to the exact same approach. For example, consolidated geological formations with fractures, joints, or solution cavities can be difficult to model, given the discontinuous nature of their permeability. These rocks commonly are highly vulnerable to contamination, and the wide range in water-level fluctuations can cause shallow domestic wells to go dry during extended droughts. Interpretation of water-level monitoring from individual wells is difficult in such terrain. It remains important, however, to have a conceptual model of the system as a driving force behind the monitoring network design with a goal to quantify that model as knowledge of the system improves.

One should not infer that the simulation model in Figure 5 is always the same. Indeed, a stepwise approach may be used in which simpler analytic codes are used in the initial phases before constructing three-dimensional numerical models. Also, it is likely that further groundwater research will develop multiple models addressing different roles and objectives. Each model provides a means to reevaluate the monitoring network from a different perspective and to advance understanding of how the water balance of the aquifer system responds to human development.

Generalized long-term monitoring will provide critical information for many uses but will not offset the need for very specific monitoring to address more localized issues, such as the effects of pumping on the ecology of a stream reach. Ideally, the broader scale monitoring programs provide a hydrologic context for the design of such studies.

## Conclusions

Groundwater monitoring data serve as a foundation that permits informed management decisions on many kinds of groundwater resource and sustainability issues. Unfortunately, data on groundwater conditions and trends are generally lacking worldwide: Groundwater is commonly undervalued, and there is a

# Middle Rio Grande Basin

The Middle Rio Grande Basin encompasses about 38 percent of the population of New Mexico and is the primary source of water supply to the City of Albuquerque and surrounding area. Some of the most productive parts of the aquifer system are in eastern Albuquerque, where, coincidentally, most of the initial groundwater development occurred. This led to the popular belief that the entire Middle Rio Grande Basin was underlain by a highly productive aquifer that was equivalent to one of the Great Lakes. During the 1980s and early 1990s, a combination of large water-level declines measured in monitoring wells (greater than 150 feet in some areas) and new insights into the geologic framework of the basin led to serious questions about this paradigm. In 1995, the New Mexico State Engineer declared the Middle Rio Grande Basin a "critical basin" faced with rapid economic and population growth for which there is less than adequate technical information about the available groundwater supply.

To fill some of the gaps in information, an intensive 6-year effort was undertaken to improve understanding of the hydrogeology of the basin.[1] Geological, geophysical, and environmental tracer studies provided new insights into the source areas for recharge to different parts of the aquifer; indicated that mountain-front recharge is less than previously estimated; showed that the hydraulic connection between the Rio Grande and the aquifer is less than previously thought in some areas; identified new faults that may affect groundwater flow; and suggested that the aquifer is less productive in some areas than previously thought (see the figure at right). The new information was incorporated into a revised groundwater model for the region. In conjunction with the study, new monitoring wells were established in the Albuquerque area, generally as nests of several wells completed at different depths in the aquifer and located to minimize short-term fluctuations caused by nearby high-capacity production

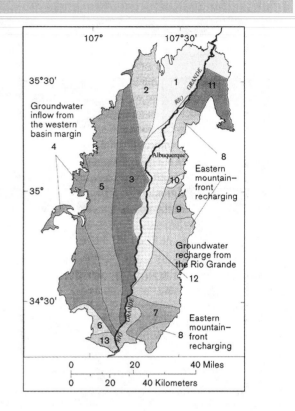

wells. The combined approach of monitoring, modeling, and scientific studies has been instrumental in helping the City of Albuquerque revise its water-use and future water-supply strategy.

1. J. R. Bartolino and J. C. Cole, *Ground-Water Resources of the Middle Rio Grande Basin, New Mexico,* U.S. Geological Survey Circular 1222 (Reston, VA, 2002), http://pubs.water. usgs.gov/circ1222.

deceptive time lag between withdrawals and the resultant impacts of those withdrawals. Long-term groundwater data from individual wells are useful primarily as part of a broader analysis of aquifer systems; thus, the value of data from individual wells is often as invisible as the resource they represent.

Long-term water-level monitoring needs to be integrated with analysis of other monitoring data and an underlying model of the water budget of the aquifer system (typically a simulation model) as a means for interpreting monitoring results and guiding the design of monitoring networks. Regular reassessment of monitoring objectives is necessary to ensure that monitoring programs provide the information needed by groundwater users and those who manage water resources. To enhance the value of groundwater data, managers and policymakers must also ensure the continuity of data-collection programs over time.

Within the United States, one might summarize the current situation as one in which we have some ability to track groundwater levels and water use for many aquifers, generally have limited ability to place these data in the context of groundwater sustainability for most aquifers, and often lack an integrated approach with

feedback among monitoring, simulation, scientific studies, and management approaches. Similar issues exist in managing groundwater resources throughout the world.[31] Fortunately, the ability to access groundwater data on the Internet and to portray them in a spatial context should continue to enhance their visibility and value in the coming years.

## Notes

1. Single geologic units may define aquifers. Alternately, multiple aquifers and surrounding lower permeability units may be collectively referred to as "aquifer systems." The two terms are used somewhat interchangeably in this article, with "aquifer systems" used when an emphasis is placed on aquifers as hydrologic systems. The *Ground Water Atlas of the United States* describes many of the important aquifers of the nation and can be found at http://capp.water.usgs.gov/gwa/.

2. W. M. Alley, R. W. Healy, J. W. LaBaugh, and T. E. Reilly, "Flow and Storage in Groundwater Systems," *Science* 296, 5575 (14 June 2002): 1985–90.

3. National Research Council, *Confronting the Nation's Water Problems: The Role of Research* (Washington, DC: National Academies

Press, 2004), 187; and L. F. Konikow and E. Kendy, "Groundwater Depletion: A Global Problem," *Hydrogeology Journal* 13, no. 1 (2005): 317–20.

4. G. M. Hornberger, "A Water Cycle Initiative," *Ground Water* 43, no. 6 (2005): 771.

5. Saturated thickness is the vertical thickness of the aquifer in which the pore spaces are filled (saturated) with water. Specific yield is the ratio of the volume of water that a saturated rock will yield by gravity drainage to the volume of the rock. Specific yield typically ranges from 0.05 to 0.3.

6. B. B. Wilson, D. P. Young, and R. W. Buddemeier, *Exploring Relationships Between Water Table Elevations, Reported Water Use, and Aquifer Subunit Delineations,* Kansas Geological Survey Open File Report 2002-25D (Lawrence, KS, 2002).

7. For example, in a well-known study, researchers found that most of the water pumped from the confined Dakota sandstone aquifer in South Dakota has come from confining beds. J. D. Bredehoeft, C. E. Neuzil, and P. C. D. Milly, *Regional Flow in the Dakota Aquifer: A Study of the Role of Confining Layers,* U.S. Geological Survey Water-Supply Paper 2237 (Washington, DC, 1983).

8. G. L. Bertoldi, R. H. Johnston, and K. D. Evenson, *Ground Water in the Central Valley, California—A Summary Report,* U.S. Geological Survey Professional Paper 1401-A (Washington, DC, 1991). Land subsidence is a gradual settling or sudden sinking of the Earth's surface. Several different processes can cause it. Most water-related subsidence occurs as a result of compaction of aquifer materials (as in the Central Valley), drainage and oxidation of organic soils, and the dissolution and collapse of limestone and other susceptible rocks forming sinkholes and similar features.

9. Ibid., page 27.

10. J. D. Bredehoeft, "Safe Yield and the Water Budget Myth," *Ground Water* 35, no. 6 (1997): 929.

11. R. J. Glennon, *Water Follies: Groundwater Pumping and the Fate of America's Fresh Waters* (Washington, DC: Island Press, 2004).

12. U.S. Department of the Interior, *Water Management of the Regional Aquifer in the Sierra Vista Subwatershed, Arizona—2004 Report to Congress,* prepared in consultation with the Secretaries of Agriculture and Defense and in cooperation with the Upper San Pedro Partnership in response to Public Law 108-136, Section 321, 30 March 2005.

13. Arizona Department of Water Resources, *Upper San Pedro Basin Active Management Area Review Report* (Phoenix, AZ, 2005) 3–25, http://www.azwater.gov/dwr/Content/Publications/files/UpperSanPedro/UpperSanPedroBasinAMAReviewReport.pdf.

14. M. A. Maupin and N. L. Barber, *Estimated Withdrawals from Principal Aquifers in the United States, 2000,* U.S. Geological Survey Circular 1279 (Reston, VA, 2005), http://pubs.water.usgs.gov/circ1279.

15. M. R. Llamas and P. Martinez-Santos, "Intensive Groundwater Use: A Silent Revolution that Cannot be Ignored," *Water Science and Technology* 51, no. 8 (2005): 167–74.

16. B. L. Morris et al., *Groundwater and its Susceptibility to Degradation: A Global Assessment of the Problem and Options for Management,* United Nations Environment Programme (UNEP) Early Warning and Assessment Report Series, RS 03-3 (Nairobi, Kenya, 2001).

17. National Research Council, *Estimating Water Use in the United States* (Washington, DC: National Academy Press, 2002).

18. W. M. Alley, T. E. Reilly, and O. L. Franke, *Sustainability of Groundwater Resources,* U.S. Geological Survey Circular 1186 (Denver, CO, 1999), http://pubs.water.usgs.gov/circ1186; and W. M. Alley and S. A. Leake, "The Journey from Safe Yield to Sustainability," *Ground Water* 42, no. 1 (2004): 12–16.

19. W. A. Abderrahman, "Should Intensive Use of Non-renewable Groundwater Resources Always Be Rejected?" in R. Llamas and E. Custodio, eds., *Intensive Use of Groundwater: Challenges and Opportunities* (Lisse, Netherlands: A. A. Balkema, 2002), 191–203.

20. D. L. Galloway, W. M. Alley, P. M. Barlow, T. E. Reilly, and P. Tucci, *Evolving Issues and Practices in Managing Ground-Water Resources: Case Studies on the Role of Science,* U.S. Geological Survey Circular 1247 (Reston, VA, 2003), http://pubs.water.usgs.gov/circ1247.

21. C. J. Taylor and W. M. Alley, *Ground-Water-Level Monitoring and the Importance of Long-Term Water-Level Data,* U.S. Geological Survey Circular 1217 (Denver, CO, 2001), http://pubs.water.usgs.gov/circ1217.

22. O. E. Meinzer, "Introduction" in R. M. Leggette, et al., *Report of the Committee on Observation Wells, United States Geological Survey,* (unpublished manuscript on file in Reston, VA, 1935), 3.

23. H. John Heinz III Center for Science, Economics and the Environment, *The State of the Nation's Ecosystems: Measuring the Lands, Waters, and Living Resources of the United States* (Cambridge, UK: Cambridge University Press, 2002), http://www.heinzctr.org/ecosystems/report.html.

24. U.S. General Accountability Office, *Watershed Management: Better Coordination of Data Collection Efforts Needed to Support Key Decisions,* GAO-04-382 (Washington, DC, 2004).

25. M. Rodell and J. S. Famiglietti, "Detectability of Variations in Continental Water Storage from Satellite Observations of the Time Dependent Gravity Field," *Water Resources Research* 35 (1999): 2705–23.

26. For examples of the use of InSAR to understand groundwater systems, see G. W. Bawden, M. Sneed, S. V. Stork, and D. L. Galloway, *Measuring Human-Induced Land Subsidence from Space,* U.S. Geological Survey Fact Sheet 069-03 (Sacramento, CA, 2003), http://pubs.water.usgs.gov/fs-069-03/.

27. National Ground Water Association, *Ground Water Level and Quality Monitoring* (2005) http://www.ngwa.org/pdf/monitoring7.pdf (accessed 8 February 2006).

28. P. M. Barlow et al., *Concepts for National Assessment of Water Availability and Use,* U.S. Geological Survey Circular 1223 (Reston, VA, 2002), http://pubs.water.usgs.gov/circ1223.

29. G. E. Fogg, "Groundwater Quality Sustainability, Creeping Normalcy, and a Research Agenda," *Geological Society of America Abstracts with Programs* 37, no. 7 (2005): 247.

30. J. D. Bredehoeft, "The Conceptualization Model Problem—Surprise," *Hydrogeology Journal* 13, no. 1 (2005): 37–46.

31. Recently, the International Groundwater Resources Assessment Center (IGRAC) has been established to share data and information on groundwater resources worldwide, http://www.igrac.nl/.

**William M. Alley** is chief of the Office of Ground Water at the U.S. Geological Survey. He is an active participant in groundwater conferences and has served on national and international committees for the American Geophysical Union, National Ground Water Association, UNESCO, and the National Research Council. He can be reached at walley@usgs.gov.

From *Environment,* Vol. 48, no. 3, April 2006, pp. 11–25. This article is in the public domain. Published by Heldref Publications, 1319 Eighteenth St., NW, Washington, DC 20036-1802.

# How Much Is Clean Water Worth?

## A lot, say researchers who are putting dollar values on wildlife and ecosystems—and proving that conservation pays

Jim Morrison

The water that quenches thirsts in Queens and bubbles into bathtubs in Brooklyn begins about 125 miles north in a forest in the Catskill Mountains. It flows down distant hills through pastures and farmlands and eventually into giant aqueducts serving 9 million people with 1.3 billion gallons daily. Because it flows directly from the ground through reservoirs to the tap, this water—long regarded as the champagne of city drinking supplies—comes from what's often called the largest "unfiltered" system in the nation.

But that's not strictly true. Water percolating through the Catskills is filtered naturally—for free. Beneath the forest, fine roots and microorganisms break down contaminants. In streams, plants absorb nutrients from fertilizer and manure. And in meadows, wetlands filter nutrients while breaking down heavy metals.

New York City discovered how valuable these services were 15 years ago when a combination of unbridled development and failing septic systems in the Catskills began degrading the quality of the water that served Queens, Brooklyn and the other boroughs. By 1992, the U.S. Environmental Protection Agency (EPA) warned that unless water quality improved, it would require the city to build a filtration plant, estimated to cost between $6 and $8 billion and between $350 and $400 million a year to operate.

Instead, the city rolled the dice with nature in a historic experiment. Rather than building a filtration plant, officials decided to restore the health of the Catskills watershed, so it would do the job naturally.

What's this ecosystem worth to the city of New York? So far, $1.3 billion. That's what the city has committed to build sewage treatment plants upstate and to protect the watershed through a variety of incentive programs and land purchases. It's a lot of money. But it's a fraction of the cost of the filtration plant—a plant, city officials note, that wouldn't work as tirelessly or efficiently as nature.

"It was a stunning thing for the New York City council to think maybe we should invest in natural capital," says Stanford University researcher Gretchen Daily.

Daily is one of a growing number of academics—some from economics, some from ecology—who are putting dollar figures on the services that ecosystems provide. She and other "ecological economists" look not only at nature's products—food, shelter, raw materials—but at benefits such as clean water, clean air, flood control and storm mitigation, irreplaceable services that have been taken for granted throughout history. "Much of Mother Nature's labor has enormous and obvious value, which has failed to win respect in the marketplace until recently," Daily writes in the book *The New Economy of Nature: The Quest to Make Conservation Profitable.*

Ecological economist Geoffrey Heal, a professor of public policy and business responsibility at Columbia University, became interested in the field as an economist who was concerned about the environment. "The idea of ecosystem services is an interesting framework for thinking why the environment matters," says Heal, author of *Nature and the Marketplace: Capturing the Value of Ecosystem Services.* "The traditional argument for environmental conservation had been essentially aesthetic or ethical. It was beautiful or a moral responsibility. But there are powerful economic reasons for keeping things intact as well."

Daily notes that beyond providing clean water, the Catskills ecosystem has value for its beauty, as wildlife habitat and for recreation, particularly trout fishing. Such values are not inconsequential. While no one has assessed the total worth of the watershed, even a partial look reveals that habitat and wildlife are powerful economic engines.

Restored habitat for trout and other game fish, for example, attracts fishermen, and angling is big business in this state. According to a report by the U.S. Fish and Wildlife Service (FWS), more than 1.5 million people fished in New York during 2001, yielding an economic benefit to the state of more than $2 billion and generating the equivalent of 17,468 full-time jobs and more than $164 million in state, federal, sales and motor fuel taxes. Though not as easily measured, individual Catskills species also have value. Beavers, for instance, create wetlands that are vital to filtering water and to biodiversity.

Ecological economists maintain that ecosystems are capital assets that, if managed well, provide a stream of benefits just as any investment does. The FWS report, for example, notes that 66 million Americans spent more than $38 billion in 2001 observing, feeding or photographing wildlife. Those expenditures resulted in more than a million jobs with total wages and salaries of $27.8 billion. The analysis found that birders alone spent an estimated $32 billion on wildlife watching that year, generating $85 billion of economic benefits. In Yellowstone National Park, the reintroduction of gray wolves that began in 1995 has already increased revenues in surrounding communities by $10 million a year, with total benefits projected to reach $23 million annually as more visitors come to catch a glimpse of these charismatic predators.

When it comes to water quality, EPA projects that the United States will have to spend $140 billion over the next 20 years to maintain minimum required standards for drinking water quality. No wonder, then, that 140 U.S. cities have studied using an approach similar to New York's. Under that agreement, finalized in 1997, the city promised to pay farmers, landowners and businesses that abided by restrictions designed to protect the watershed. (The city owns less than 8 percent of the land in the 2,000-square-mile watershed; the vast majority is in private hands.) "In the case of the Catskills, it was a matter of coming up with a way to reward the stewards of the natural asset for something they had been providing for free," Daily says. "As soon as they got paid even a little bit, they were much happier and inclined to go about their stewardship." There's no guarantee this experiment will work, of course; it may be another decade before the city finds out.

Elsewhere, other governing bodies are also recognizing the value of ecosystem services. The U.S. Army Corps of Engineers, for example, bought 8,500 acres of wetlands along Massachusetts' Charles River for flood control. The land cost $10 million, a tenth of the $100 million the Corps estimated it would take to build the dam and levee originally proposed.

## Nature's Services

### How the Experts Categorize Them

- Ecosystem goods, the traditional measure of nature's products such as seafood, timber and agriculture
- Basic life support functions such as water purification, flood control, soil renewal and pollination
- Life-fulfilling functions, the beauty and inspiration we get from nature, including activities such as hiking and wildlife watching
- Basic insurance, the idea that nature's diversity contains something—like a new drug—the value of which isn't known today, but may be large in the future

To fight floods in Napa, California, county officials spent $250 million to reconnect the Napa River to its historical floodplains, allowing the river to meander as it once did. The cost was a fraction of the estimated $1.6 billion that would have been needed to repair flood damage over the next century without the project. Within a year, notes Daily, flood insurance rates in the county dropped 20 percent and real estate prices rose 20 percent, thanks to the flood protection now promised by nature.

Even insects supply vital ecosystem services. More than 218,000 of the world's 250,000 flowering plants, including 70 percent of all species of food plants, rely on pollinators for reproduction—and more than 100,000 of these pollinators are invertebrates, including bees, moths, butterflies, beetles and flies. Another 1,000 or more vertebrate species, including birds, mammals and reptiles, also pollinate plants. According to University of Arizona entomologist Stephen Buchmann, author of *The Forgotten Pollinators,* one of every three bites of food we eat comes courtesy of a pollinator.

A Cornell University study estimated the value of pollination by honeybees in the United States alone at $14.6 billion in 2000. Yet honeybee populations are dropping everywhere, as much as 25 percent since 1990, according to one study. Now many farms and orchards are paying to have the bees shipped in.

Today's interest in assigning dollar values to pollination and other ecosystem services was spawned by publication of a controversial 1997 report in *Nature* that estimated the total global contribution of ecosystems to be $33 trillion or more each year—roughly double the combined gross national product of all countries in the world. The study became a lightning rod. Detractors scoffed at the idea that one could put a dollar value on something people weren't willing to purchase. One report by researchers at the University of Maryland, Bowden College and Duke University called the estimate "absurd," noting that if taken literally, the figure suggests that a family earning $30,000 annually would pay $40,000 annually for ecosystem protection.

Other researchers, including Daily and Heal, charged that the $33 trillion figure greatly underestimates nature's value. "If you believe, as I do, that ecosystem services are necessary for human survival, they're invaluable really," Heal says. "We would pay anything we could pay."

Daily doesn't believe the absolute value of an ecosystem can ever be measured. Heal agrees, yet both scientists say that pricing ecosystem services is an important tool for making decisions about nature—and for making the case for conservation. "Valuation is just one step in the broader politics of decision making," she says. "We need to be creative and innovative in changing social institutions so we are aligning economic forces with conservation."

Indeed, as dollar values for nature's services become available, environmentalists increasingly use them to bolster arguments for conservation. One high-profile example is the contentious dispute over whether to tear down four dams on the lower Snake River in southeastern Washington to restore salmon habitat, and thus the region's lucrative salmon fishery. Ed Whitelaw, a professor of economics at the University of Oregon, notes that estimates of the economic impact of breaching

# Natural Capital

## What's the Annual Dollar Value of . . .?

- Recreational saltwater fishing in the United States: $20 billion
- Wild bee pollinators to a single coffee farm in Costa Rica: $60,000
- Tourism to view bats in the city of Austin, Texas: $8 million
- Wildlife watching in the United States: $85 billion
- U.S. employment income generated by wildlife watching: $27.8 billion
- State and federal tax revenues from wildlife watching: $6.1 billion
- Natural pest control services by birds and other wildlife to U.S. farmers: $54 billion

the dams range from $300 million in net costs to $1.3 billion in net benefits, largely due to the wide range of projections about recreational spending.

A 2002 report by the respected, nonprofit think-tank RAND Corporation concluded the dams could be breached without hurting economic growth and employment. Energy lost as a result of the breaches could be replaced with more efficient sources, including natural gas, resulting in 15,000 new jobs. Further, the report noted that recreation, retail, restaurants and real estate would experience a marked growth. Recreational activities alone would increase by an estimated $230 million over 20 years.

There's no question that returning the salmon runs would have a major impact on the region. When favorable ocean conditions increased the runs in 2001, Idaho's Department of Fish and Game estimated the salmon season that year alone generated more than $90 million of revenue in the state, most of it in rural communities that badly needed the funds.

"Some people think it sounds crass to put a price tag on something that's invaluable, careening down the slippery slope of the market economy," says Daily. "In fact, the idea is to do something elegant but tricky: to finesse the economic system, the system that drives so much of our individual and collective behavior, so that without even thinking it makes natural sense to invest in and protect our natural assets, our ecosystem capital."

What Daily and other ecological economists want is to insinuate consideration of ecosystem services into daily decision making, whether it takes the form of financial incentive or penalty. "At a practical level, decisions are made at the margin, not at the 'should we sterilize the Earth' level," she says. "It's in all the little decisions—whether to farm here or leave a few trees, whether to build the shopping mall there or leave the wetland, whether to buy an SUV or a Prius—that ecosystem service values need to be incorporated."

Heal agrees. "Although ecosystem services have been with us for millennia," he says, "the scale of human activity is now sufficiently great that we can no longer take their continuation for granted."

Virginia journalist **JIM MORRISON** wrote about polar bears and global warming in the February/March 2004 issue.

# Searching for Sustainability

## *Forest Policies, Smallholders, and the Trans-Amazon Highway*

Eirivelthon Lima, et al.

It is a powerful and disturbing image: loggers driving roads deep into the forest to remove a few mahogany trees, with slash-and-burn settlers following closely on their heels. However, it no longer captures the whole picture of logging in the Brazilian Amazon. So, then, what is the role of logging in the impoverishment or potential conservation of the Amazon rainforest? The answer to this question is deceptively complex: To achieve a sustainable future in Amazon forestry, policymakers and stakeholders must understand the physical, economic and political dimensions of competing land use options and economic interests. They must provide effective governance for multiple agendas that require individual oversight.

For simplicity's sake, suppose that forest governance can be approached from two angles: a preservation approach in which the land is tucked away, never to be used again; and a "use-it-or-lose-it" approach in which a well-managed forest estate becomes part of a sustainable economic development scenario and competes successfully with other land use options. In fact, 28 percent of the Brazilian Amazon is already listed as some form of park, or as a protected or indigenous area.[1] But what of the forest without protection, found mainly on private lands or on as-yet undesignated government lands? For many, selective logging of these forests is a form of forest impoverishment that is only slightly less devastating than forest clear-cutting.[2] For others, the selective harvest of timber is the best way to make the long-term protection of standing forests economically and politically viable.[3]

Opponents base their argument on two points: The long-term selective logging of primary tropical forest is financially impracticable, and selective logging is the first step in a vicious cycle of degradation that includes settlement and land clearing.[4] Advocates say that selective logging, when done well (called "reduced impact logging"), is renewable, economically viable, and may provide an important stream of revenue for government and private landowners that would encourage the maintenance of forest cover.[5] These proponents contend that if tropical forestry is to compete successfully with other land use options and essentially push back against the encroaching line of deforestation, some conditions must be met: the removal of subsidies to other land use options; the breakdown of barriers to entry, such as complex forest management plans; the dissemination of information on forestry to all potential market participants;

and the elimination of perverse incentives for deforestation—in particular the establishment of land titles through clearing to demonstrate active use. Furthermore, if forestry is deemed the least cost-effective approach to maintaining forest cover outside parks and protected areas, then subsidies to forest management activities might also be appropriate.

There are vast forested areas in the Brazilian Amazon located outside parks and protected areas, and a multitude of landowners, including state and federal governments, are controlling that forest. As a result, policies to manage forest resources must necessarily be comprehensive, flexible, and appropriate to varying conditions and agents. The Brazilian government has recently identified, and is now beginning to implement, a strategy of timber concessions that should help to corral some part of the industry into a controlled region. This should make it easier to monitor and will hopefully reduce illegal logging. The policy, however, mostly ignores the sticky issue of forestry on private land, which, although complex, could provide the engine for sustainable economic development among the disenfranchised settlers of the Amazon frontiers. The settlers may straggle onto the frontier individually, but they eventually form communities, control large areas of land, and become an increasingly important component of the timber industry.

A major economic corridor in the Brazilian Amazon—the Trans-Amazon Highway—illustrates how logging can be transformed from a force driving forest impoverishment to one driving forest conservation, and how this transition, in turn, carries important potential benefits for the semi-subsistence farmers who live along this corridor. To fully describe this transformation, it is necessary to place it in light of the history and current context of the timber industry of the Amazon, and with the understanding that the complexity inherent in the largest and most diverse tropical forest in the world makes forest governance a mighty task.

## A Brief History of the Amazon Timber Industry

Understanding logging along the Trans-Amazon Highway depends upon the historical context of the timber industry in the Amazon, which can be roughly divided into three periods.[6]

The early production period lasted from the 1950s to the early 1970s and was followed by a transition or boom period, which lasted from the mid 1970s to the late 1980s. A third period, industry consolidation and migration to new frontiers, started in the early 1990s but is now coming to an end. The current timber industry is in such disarray from political mismanagement that in October 2005 the federal police temporarily suspended the transport of all logs from the Amazon.

## Early Days (1950s to Mid-1970s)

In the 1950s, the island region of the Amazon delta in the state of Pará was the center of the wood industry in the Amazon. Through the 1960s, there were three large plywood mills and six large sawmills that controlled production. With no connection to the large domestic markets of southeastern Brazil and the dependence on fluvial transport to access raw materials and deliver products, these mills produced only for the export market. Limited shipping capacity and irregular delivery schedules hindered sales to ports in northeastern Brazil, which could be reached by ship along the Atlantic coast. The primary source of raw material was smallscale landowners who sold logs along the banks of rivers. The environmental impact of logging was minimal, as timber extraction was an integrated part of diverse smallscale family farming systems on the Amazon River floodplain. The two popular tree species harvested were Virola (*Virola surinamensis*) for plywood and Andiroba (*Carapa guianensis*) for sawnwood (the first stage of the log processing sequence in which logs are cut into boards, but not planed).

In the early and mid-1970s, a number of smaller sawmills began to appear in the island region and farther up along the upper Amazon River. Into the mid-1970s, the Amazon remained disconnected from domestic markets but the export market flourished. Estimated log consumption was in the region of 2.5 million cubic meters per year—all harvested by axe. Early reports on timber production in the Brazilian Amazon suggest this was a period of poor market access, poor quality of laborers, obsolete equipment, insufficient knowledge of local tree species, and scarce information on prices and markets for products.[7]

## Transition Period (Late 1970s to Early 1990s)

A period of dramatic transition in the timber sector began in the late 1970s to early 1980s. Several highways were completed to link the Amazon to domestic southeastern and northeastern Brazilian markets. The states of Rondônia, Mato Grosso, and Pará became connected through the BR364, BR163, and BR010 highways. Large public investment programs for the construction of dams, hydropower plants, a railroad for the Carajás mining program, and the settlement of migrants from southern and northeastern Brazil changed the interfluvial forests of the Amazon, passively protected until that time by their inaccessibility.

Deforestation during this time was largely a response to government actions that either directly promoted or enabled land conversion from forests to other uses. The number and size of sawmills increased in response to the inexpensive primary resource and newly accessible markets, growing local demand, and the availability of cheap labor. Mechanization of harvesting, transport, and processing also contributed to the growth of sawnwood output.

By the early 1980s, Paragominas (a city in Pará) became the most important mill center in the Amazon, producing mostly for the domestic market. The state of Mato Grosso also produced lumber for the domestic market, with important logging centers appearing in the towns of Sinop and Alta Floresta, Meanwhile, the island region continued to produce for the export market. In all, the transition period during the 1970s and mid-to-late 1980s was a turning point in the timber industry of the Brazilian Amazon.

## Consolidation and Migration (Mid-1990s to 2000s)

After the transition period, another (less dramatic) period of consolidation and expansion ensued along old and new logging frontiers.[8] Old frontiers can now be found in eastern Pará (Paragominas and Tailandia) and in northern Mato Grosso (Sinop). In these areas, virgin forests have become increasingly scarce, and the logging industry became more diverse and efficient. The more inefficient logging firms exited the market, and those that remained became vertically integrated in an effort to capture value added in downstream processing.

Access to the old frontiers is generally good given the high density of paved roads. In contrast, new frontiers are characterized by a rapid inflow of mills and producers from the old frontier, poor government regulation, and high transport costs. The notable new logging frontier is in western Pará along the northern section of the Santarém-Cuiaba Highway, the BR163.

## The Industry Today

The current volume of wood produced in the Legal Amazon is between 20 and 30 million cubic meters, of which more than 50 percent is sold in the domestic Brazilian market.[9] Prior to 2003, legal timber harvest was possible through the preparation of a forest management plan submitted to the government agency and approved with a temporary land title. All that was required by IBAMA, the Brazilian government environmental agency, was proof that a firm or individual had initiated a land legalization process with Brazilian land titling institutions such as the Institute of Colonization and Agrarian Reform (INCRA).[10] Generally, land titling procedures took years, and they did not always result in legalization. By the time the land titling institution had made its decision, the harvest was already complete and the loggers had moved on to the next native forest stocks.

In 2003, the Brazilian government abruptly decided that management plans could no longer be approved on lands where property rights were not well established. That year, nearly all forest management plans were rejected.[11] The government, however, did not have an alternative readily available for the nearly 2,500 logging companies based in the Amazon, and an unintended side effect of the policy has been that more companies now simply operate illegally in such areas. Conflicts, protests, and widespread unregistered logging are now the norm.

To solve the problem of legalizing timber harvest and controlling the timber industry, the Brazilian government has proposed implementing forest concessions on public lands.[12] While this approach has some merit, and indeed has been debated extensively in the Brazilian public arena, large concessions controlled by a few companies may not be the best economic option in the regions where smallholders and other private landowners, including a large number of migrant settlers, are the predominant land users.[13]

# The Case of the Trans-Amazon Highway

For two weeks in August 2003, the Trans-Amazon Highway was impassable: Angry loggers had blocked the road, stopping traffic to protest a government-imposed timber shortage. A similar display occurred outside the town of Santarém, Pará, in January 2005 and recently on the BR163 Highway in western Pará. Tragically, access to timber was also one of the underlying reasons for the murder of Sister Dorothy Stang in the municipality of Anapú.[14] Timber scarcity is a startling concept for the Amazon. How is it that a resource so apparently abundant can be the root cause of violent conflicts and protests?[15] The answer lies partially in the sudden requirement by the Brazilian government that loggers provide proper legal documentation for land rights in areas where logs are extracted. But who owns the forests and logs along this frontier highway?

Built by General Emílio Garrastazu Médici (president of Brazil from 1969–1974), the main part of the Trans-Amazon Highway stretches approximately 1,000 kilometers from the town of Marabá to Itaituba on the banks of the Tapajós River.[16] The highway is largely unpaved and virtually impassable for four months of the year during the rainy season. Homesteaders are usually allocated demarcated lots of 100 hectares apiece (approximately 250 acres) and then often battle the elements and wealthy land speculators to continue occupying the land.[17] Still, migration to the region is relentless, as a constant stream of formal and informal land control followed early colonization projects in the late 1970s.[18] INCRA, the federal land settlement agency, has formally settled approximately 30,000 families and an unknown number of informal squatters.

While it is commonly accepted that smallholders control vast areas of land along the Trans-Amazon Highway, the exact quantity of land is debatable. This question is taken up under the auspices of the Green Highways Project, an international multi-institutional project, led by the Brazilian nongovernmental organization Instituto de Pesquisa Ambiental da Amazônia (IPAM, Amazon Institute of Environmental Research) with the support of the Massachusetts-based Woods Hole Research Center (WHRC).[19] An area 100 kilometers (km) on either side of the Trans-Amazon Highway from the municipality of Itupiranga to Placas was mapped using satellite imagery and secondary data from Brazilian government sources. Land distribution was

mapped and deforestation measured using 30-meter spatial resolution satellite images and secondary data from INCRA and the Brazilian Institute of Geography and Statistics (IBGE).[20] Images were classified into forest and non-forest classes by supervised classification and visual interpretation.[21] The objective was to identify where smallholders are located and where they will be located in the future.

Of the total 15.7 million hectares located within this buffer, 7.9 million are under the control of or are promised to smallholders. Of the total area within the 100-km study area, the land distribution is: 1.1 percent in demarcated settlements, 5.4 percent in current settlements, 11.4 percent as squatters (posseiros), 13.2 percent in old colonization projects, and 19.5 percent destined for future settlements by INCRA. Four percent of the land is in conservation areas, 7.6 percent in informal medium and large-scale land holdings, 15.4 percent in indigenous reserves, and a final 21.2 percent is unclaimed government land.[22] The number of smallholders currently residing in the 100-km zone was estimated by summing the area with active settlements, which includes current settlements, colonization, and squatters, and then dividing by an 82.6-hectare average lot size from survey results (see below), giving a total area of approximately 4.7 million hectares held by 57,000 smallholder families.

Given the observed distribution of smallholders from the spatial analysis, the next logical question for the Green Highways Project was whether these agents could potentially supply the timber industry with wood. Demand for timber in the area is strong; the demand for logs on the Trans-Amazon Highway more than doubled over 12 years, increasing from roughly 340,000 cubic meters in 1990 to approximately 840,000 cubic meters in 2002. To determine whether smallholders can provide this quantity it is important to first estimate the growing stock potential of the forest held by smallholders, assuming that smallholders will in fact sell wood (this assumption will be revisited below). Using conservative (high) deforestation assumptions (for example, a range of 60 percent deforested for old colonization areas to 15 percent deforested for INCRA land allocated to future settlement) and a conservative stand volume of ten cubic meters per hectare, forest stock in active settlement areas is estimated to be 25.8 million cubic meters.[23] Using a harvest cycle of 30 years, this would give a sustained harvest volume of approximately 860,000 cubic meters, which matches current demand. At an estimated stumpage price of 10 Reais (R$10) per cubic meter of standing trees (approximately US$3.33 per cubic meter), this volume would generate R$8.6 million per year.[24] To put this in perspective, if the smallholder forests within current settlements were used to their full potential right now, and the benefits distributed evenly to every family (recall there are an estimated 57,000), each smallholder household could receive R$150 per year—a large sum given the discussion below.

Assuming that smallholders will eventually settle in areas set aside by INCRA, there will be an estimated forest stock of 52.6 million cubic meters, which could render a sustainable harvest of approximately 1.7 million cubic meters per year, more than double the current regional demand. Thus, there appears to be sufficient potential forest stock to meet the demand, and a tremendous opportunity for a redistribution of wealth to the poor, should smallholders have an unhampered market to sell wood [see Table 1 below].[25]

However, one needs to ask if these estimates based upon government census data are consistent with data on the ground. To answer this question we make use of data generated from a recent comprehensive socioeconomic survey of smallholders along the Trans-Amazon Highway. Between June and December 2003, a total of nearly 3,000 families were interviewed, of which 2,441 lived within the 100-km zone.[26]

In the survey, smallholders were asked about their forest production, and socioeconomic data were collected. The results add to the discussion above, showing that 26 percent had sold wood, and those sales had occurred largely within the last 5 years. There had been only one sale per lot. Ninety-six percent of the smallholders sold standing trees, and the average number of trees sold was 20 per smallholder, which corresponds to a harvest rate of approximately 1 tree per 5 hectares and, assuming an average volume of 5 cubic meters of log per tree, an average sale volume of 100 cubic meters. The average total sale value was R$173, which corresponds to R$8.65 per tree or R$1.73 per cubic meter.[27]

Comparing these observations with the results from the geospatial analysis above, based on timber produced through legal deforestation and harvest of legal forest reserves (the area of smallholder land prohibited from clearing for crops), smallholders are selling approximately 1 cubic meter per hectare, and only 26 percent of them actually sell wood. At this harvest volume, it would take the harvest from 10,000 families per year—about 18 percent of the estimated total smallholder families—to sustain current demand from the area industry at current prices. This amounts to a harvest volume that is only 4 percent of current estimates for Amazon timber production from other studies. At a harvest intensity of 10 cubic meters per hectare, this participation requirement would be drastically reduced to only 1,000 families per year (which represents 1.8 percent of all families). This level of participation could be easily achieved without undue

change in the smallholder system by subcontracting the timber industry to do much of the technical work associated with logging. The production of logs is dramatically low on smallholder lots because smallholders have limited knowledge of the forest potential and limited access to the financial resources required to manage the forests. This barrier can be overcome with a partnership between smallholders and the timber industry. For the successful implementation of such a partnership, however, it will be important for smallholders to understand the logging process and have adequate access to production information so that they can maintain a check on their industrial partners.

## Holding Back the Tide of Smallholder Forestry

From the perspective of community foresters, the current ideal is that individuals within the communities must work collectively and must control the entire chain of production through to sales of the final product. Formal interaction with the timber industry is anathema. Also, there is still the idea that forest management must happen in large, undisturbed, contiguous tracts of forests. This closely held and restrictive view has undermined the potential of community forestry in the Amazon. The reality is that there are more than 500,000 settlement families in the Brazilian Amazon who work individually or in community associations and who specialize (though perhaps not yet efficiently) in the supply of standing timber by working closely with logging companies.

However, most of the community-based forestry operations have two key problems. First, when dealing with smallholders on an individual basis, the loggers hold all the cards. They have more information about the species and value of timber, and they exploit the immediate financial needs of cash-poor smallholders. Second, logging on smallholder lots is legal only under two premises: smallholders have deforestation licenses that allow the clearing of 3 hectares per year and the sale of 60

**Table 1**  Timber Potential from Smallholder Lots on the Trans-Amazon Highway

| Smallholders | Total Area (hectares (ha)) | Percent Land Area | Forest Cover (percent) | Total Forest Area (ha) | Timber Stock (m³) | Potential Timber Flow (m³/year) |
|---|---|---|---|---|---|---|
| Future settlement projects | 3,055,000 | 19.5 | 85 | 2,596,000 | 25,965,000 | 865,000 |
| Colonization projects | 2,063,000 | 13.2 | 40 | 825,000 | 8,252,000 | 275,000 |
| Informal settlement | 1,792,000 | 11.4 | 60 | 1,075,000 | 10,750,000 | 358,000 |
| INCRA settlements | 852,000 | 5.4 | 80 | 682,000 | 6,815,000 | 227,000 |
| Demarcated settlements | 169,000 | 1.1 | 50 | 85,000 | 847,000 | 28,000 |
| Total smallholders | 7,931,000 | 50.6 | – | 5,263,000 | 52,629,000 | 1,753,000 |

Note: INCRA is Brazil's National Institute of Colonization and Agrarian Reform. The entire buffer area is 15,643,000 ha. The area not occupied by smallholders is comprised of unclaimed government land (21.2 percent), indigenous land (15.4 percent), medium and large informal settlement (7.6 percent), and conservation units (4.2 percent).

Source: Instituto de Pesquisa Ambiental da Amazônia (the Amazon Institute of Environmental Research).

cubic meters per year (up to 20 percent of the land area owned), or they may have the option to develop a forest management plan that must be approved by IBAMA. Of the sales registered in the surveys, 26 percent came from deforestation permits, and a startling 79 percent came from the "legal reserve" on each plot.[28] Because no formal forest management plans have been developed for these smallholder systems, this would imply that nearly 80 percent of log sales from smallholders are currently illegal by government rules; in addition, few smallholders get legal deforestation permits. Why are there no formal plans? A forest management plan requires that the landowner hold legal title, and although 95 percent of smallholders surveyed claimed to be the landowner, we found only 26 percent held formal title; a statistic supported by previous research in the region.[29] This lack of coordination between agencies and resource users is a major barrier to overcoming illegal logging within smallholder systems and to the integration of smallholders into the formal timber market.

# Small-Farm Family Forestry in the Amazon

Coordination between ministries is not an impossible task, however. For example, IBAMA, INCRA, and the Ministry of Public Works of the town of Santarém (in Pará) operating with limited resources but in partnership with loggers and smallholders, found a creative solution to this problem in the form of an equitable partnership between industry and smallholders. In this case, the community associations subcontract the loggers to plan and implement harvesting, while the government ministries have the responsibility of expediting title and management approval. The land is owned individually, and management plans are done for each private 100-hectare lot, but the negotiations are between the logger and the community association. The community can demand higher prices by selling as a group, and the logger is assured of a long-term supply of timber. As a result, legal forest operations are taking place and smallholders are capturing a fair share of the benefits from the timber harvest on their land [see box on next page].[30]

However, changes in government personnel and extreme inefficiency (the project industry coordinator has had management plans under review at IBAMA for more than a year) has made even this promising partnership tenuous. These types of projects are in danger of failing because government oversight is inefficient, inadequate, corrupt, and contradictory.[31] There may be a partial solution to be found in timber concessions, but even with successful concessions, the large-scale problems of illegal logging will not disappear. Indeed, the problems of illegal logging will never be solved if IBAMA cannot control the industry or support it effectively, but there is no indication so far that IBAMA can do it alone. It is reasonable to assume, however, that the economic benefits of timber production on their private landholdings will stimulate smallholders to manage their forests and help control illegal activities.

# What Does the Future Hold?

What do the results of the Green Highways Project have to say about the issue of loggers and forest policies? As mentioned above, the main thrust of the new forest policy centers around timber concessions on public lands with some allowances given to communities. This is an effective program for a portion of the industry, but there are two problems with the idea. First, the evidence presented above indicates that this approach is inadequate for some major economic corridors where there are many smallholders, such as in the case of the region surrounding the Trans-Amazon Highway; of the 80 percent of land available for harvest (for example, excluding conservation units and indigenous areas), the Green Highways Project shows that 64 percent is under the control of, or is promised to, smallholders. Second, it also shows that forestry is highly underutilized in these smallholder systems. This and the fact that there are more than 500,000 families settled in the Amazon region mean these results imply a very large economic loss to Brazilian society from not capturing a potential timber supply that would almost do away with the need for timber concessions on public lands.

Further, by excluding smallholders from access to the timber industry through current management plan requirements, smallholders are denied what could amount to a substantial and vital source of economic development. In some settlements, research has shown that the value of a single harvest can equal more than 15 years of agricultural production.[32] And finally, even if only some portion of the demand for logs is met by concessions harvesting on public government lands, it may have a negative socioeconomic impact on the potential for small farm forestry by depressing overall prices.

To promote sustainable forestry, the evidence indicates that the government has to realistically deal with land titling, facilitate institutional coordination, and commit to stopping illegal logging through better enforcement. Invariably, the causes of policy failure and poor governance are related to corruption and political auction of important positions in government institutions. An intricate net of political obligation, to the detriment of technical decisions, is commonplace, and even those individuals fiercely committed to their tasks (and there are many) struggle to make quality strategies a reality. A lack of efficiency in government agencies, whether through poor coordination or delays, increases transaction costs and makes formal forest management difficult. Also, by neglecting secure property rights, or making these difficult and costly to obtain, the government inadvertently creates incentives for smallholders and loggers to engage in illegal logging.

Forest management projects on smallholder settlement lots in the Brazilian Amazon will, if widely adopted, help move the region toward equitable forest-based economic development and a peaceful resolution to the problems now facing migrant families. This is not the only solution for the Amazon, but it is a step forward and one well within the reach of the current administration. Without change, however, we can expect further illegal degradation of the forest and a continuing struggle for economic development and social justice on the Amazon frontier.

# André Da Silva Dias Reflects on the Forest Families Program

Forest management models that can contribute to the social, environmental, and economic development of smallholders and traditional populations have been the subject of many recent initiatives in the Amazon. The "Forest Families" program [in Santarém] works with a specific relationship that appears to be very common but little studied: smallholders and the timber industry. It is interesting to note some of the fundamental characteristics around which the program is built: the relationship between the smallholder and the industry already exists; its foundation is market-based; its actors are well-defined; and [it] is based on uncommonly strong legal and ethical rigor. The last characteristic alone makes one pay attention.

One can question whether this is community forest management or not, A pertinent doubt, but, in the end, there exists a forest and its resources and a people organized, or organizing, in communities. In fact, the smallholders are not directly managing their forests: they delegate this activity to a subcontractor and his team. And when they delegate they relinquish some personal control of the forest. However, they exercise their rights to the forest in a free manner, in a negotiation process that strengthens the local organization, generates collective responsibility, creates a commonly used infrastructure, provides income and, most importantly, gives value to the standing forest. All of which are the principles that underlie community forest management.

It is possible to imagine a scenario in which they should manage their own forests in accordance with their capacity, limitation, abilities, and interests. Perhaps this will happen one day. But for right now, the reality is different. No better and no worse, this is just different than many other community forest management initiatives where the local residents play the role of managers. The fact is that they, the owners, are who should say whether this is how it should be. And they seem to be making this [decision] in an informed way, understanding their limitations, and identifying opportunities. It is interesting to observe a community and its people (in this case Santo Antonio) started barely two years ago by families of different origin, who until this point never knew each other, but who already have solid development plans and a growing autonomy in the formulation of local projects, rather than just hope of better days.

I believe that one of the principal contributions that this program can lend to the discussion of local forest management is to define criteria and indicators of a healthy and egalitarian relationship between smallholders and the timber industry. To get there, some challenges that deserve more attention are:

- Improving local knowledge of good forest management practices.
- Identifying the impact of timber harvest on the supply of hunting and non-timber forest products.
- Analyzing the socioeconomic impact of the timber income on the smallholder systems.

Source: André da Silva Dias, Executive Manager, Fundação Floresta Tropical, December 2003. This box was translated from the Portuguese by Frank Merry and first published in a report by Instituto de Pesquisa Ambiental da Amazônia (IPAM) for the International Institute for Environment and Development (IIED) as part of the IIED Power Tools Initiative: Sharpening Policy Tools for Marginalized Managers of Natural Resources. F. Merry, E. Lima, G. Amacher, O. Almeida, A. Alves, and M. Guimares, *Overcoming Marginalization in the Brazilian Amazon Through Community Association: Case Studies of Forests and Fisheries,* (Edinburgh, UK, 2004). It is reprinted with permission.

# Notes

1. This approach has been plied very successfully by large conservation organizations, in particular Conservation International, which solicits funds to buy up biodiversity "hot-spots." For an analysis of the effects of parks and protected areas on fires in the Amazon, see D. Nepstad et al., "Inhibition of Amazon Deforestation and Fire by Parks and Indigenous Reserves," *Conservation Biology.* In press, expected publication February 2006.

2. I. Bowles, R. E. Rice, R. A. Mittermeier, and G. A. B. da Fonseca, "Logging On in the Rain Forests," *Science,* 4 September 1998, 1453–58; R. Rice, C. Sugal, and I. Bowles, *Sustainable Forest Management: A Review of the Current Conventional Wisdom.* (Washington, DC: Conservation International, 1998); R. Rice, R. Gullison, and J. Reid, "Can Sustainable Management Save Tropical Forests?" *Scientific American,* April 1997, 34–39.

3. D. Pearce, F. E. Putz, and J. Vanclay, "Sustainable Forestry in the Tropics: Panacea or Folly?" *Forest Ecology and Management* 172, no. 2 (2003): 229–247; M. Verissimo, A. Cochrane, and C. Sousa Jr., "National Forests in the Amazon," *Science,* 30 August 2002, 1478; F. E. Putz, K. H. Redford, J. G. Robinson, R. Fimbel, and G. Blate, *Biodiversity Conservation in the Context of Tropical Forest Management* (Washington, DC: Biodiversity Studies, The World Bank, 2000), http://world-bank.org/biodiversity.

4. G. Asner, et al., "Selective Logging in the Amazon," *Science,* 21 October 2005: 480–481. The Asner study claimed that the selective logging of the Amazon is far more widespread than previously thought. The authors suggest that the source of logs in the Amazon is not slash-and-burn deforestation—those logs are simply burned—but conventional poor-quality selective logging and that this is the first step in the economic and ecological degradation of the forest. According to the data, this is more widely practiced and perhaps more damaging than previously thought.

5. Forest management and reduced impact logging (FM-RIL) guidelines are available from many sources: the Suriname Agricultural Training Center (CELOS); the International Tropical Timber Organization (ITTO); the Food and Agricultural Organization of the United Nations (FAO); the Institute of Humans and the Environment of the Amazon (IMAZON); and the Fundação Floresta Tropical (FFT, Tropical Forest Foundation). In addition, field models in Brazil demonstrate the improvements of FM-RIL practices over conventional selective logging. See the FFT

website at http://www.fft.org.br and click "Research." There have been several studies on the economic benefits of reduced impact logging and comparisons with "conventional" selective logging. For a few examples see: S. Armstrong and C. J. Inglis, "RIL For Real: Introducing Reduced Impact Logging Techniques into a Commercial Forestry Operation in Guyana," *International Forestry Review* 2, (2000): 264–72; F. Boltz, D. R. Carter, T. P. Holmes, and R. Perreira Jr., "Financial Returns Under Uncertainty for Conventional and Reduced-Impact Logging in Permanent Production Forests of the Brazilian Amazon," *Ecological Economics* 39 (2001): 387–98; P. Barreto, P. Amaral, E. Vidal, and C. Uhl, "Costs and Benefits of Forest Management for Timber Production in Eastern Amazonia," *Forest Ecology and Management* 108, no. 1 (1998): 9–26; and T. P. Holmes et al., *Financial Costs and Benefits of Reduced Impact Logging Relative to Conventional Logging in the Eastern Amazon* (Washington, DC: Tropical Forest Foundation, 1999).

6. Thanks to Johan Zweede of the Instituto Florestal Tropical in Belém, Brazil, and Benno Pokorny of the University of Freiburg, Germany, for valuable comments on the history and context of the timber industry.

7. For an excellent review, see I. Sholtz, *Overexploitation or Sustainable Management: Action Patterns of the Tropical Timber Industry: The Case of Pará, Brazil, 1960–1997* (London: Frank Cass Publishers, 2001).

8. By definition the term "frontier," when applied to forests, implies the point at which new logging occurs. It is, however, common in the literature of logging in the Amazon to differentiate frontiers by age. This is done partially out of custom, but also because logging on all "frontiers" is relatively new; even old frontiers are less than 30 years old.

9. The "Legal Amazon" is a geo-political definition of the Amazon region in Brazil and comprises the states of Amapá, Amazonas, Acre, Maranhão, Mato Grosso, Pará, Rondônia, and Tocantins. The volume of sawnwood destined for export is different across frontiers. More than 60 percent of logs from new frontiers are destined for the export market, whereas on the intermediate and old frontiers, that level dips to 50 and 15 percent, respectively, according to F. Merry et al., "Industrial Development on Logging Frontiers in the Brazilian Amazon," *International Journal of Sustainable Development,* in review. For a recent discussion of production volumes, see G. Asner et al., note 4 above.

10. The Instituto Brasileiro do Meio Ambiente e dos Recursos Naturais Renováveis (http://www.ibama.gov.br) is the Brazilian government's environmental agency responsible for the forest sector and all issues of environmental control in the country. The federal land-titling agency is the Institute of Colonization and Agrarian Reform (INCRA). For more information, see http://www.incra.gov.br. Each state also has a local agency.

11. The forest management process includes a formal management plan that essentially states that the company intends to harvest in a given area (with accompanying maps and documentation) and subsequently an annual operating plan that delivers the details of each year's harvest operation. The term "forest management plan" includes both of these components of logging.

12. For more information on forest concessions, see A. Veríssimo, M. A. Cochrane, and C. Sousa Jr., National Forests in the Amazon," *Science,* 30 August 2002, 1478.

13. The forest concessions issue has long been debated in the scientific literature. Both sides of the argument for Brazil can be explored in F. D. Merry et al., "A Risky Forest Policy in the Amazon?" *Science,* 21 March 2003, 1843 and in F. D. Merry et al., "Some Doubts About Concessions in Brazil," *Tropical Forestry Update* 13, no. 3 (2003): 7–9 (see http://www.itto.or.jp/live/contents/download/tfu/TFU.2003.03.English.pdf). See also F. D. Merry and G. S. Amacher, "Forest Taxes, Timber Concessions, and Policy Choices in the Amazon," *Journal of Sustainable Forestry* 20, no. 2 (2005): 15–44; and Veríssimo, Cochrane, and Sousa, note 12 above. For earlier discussion on concessions see J.A. Gray, *Forestry Revenue Systems in Developing Countries,* FAO Forestry Paper 43 (Rome, 1983); R. Repetto and M. Gillis, eds., *Public Policies and the Misuse of Forest Resources,* (Cambridge, UK: Cambridge University Press, 1988); J. R. Vincent, "The Tropical Timber Trade and Sustainable Development," *Science,* 19 June, 1992, 1651–1655; and J. A. Gray, "Underpricing and Overexploitation of Tropical Forests: Forest Pricing in the Management, Conservation and Preservation of Tropical Forests," *Journal of Sustainable Forestry* 4, no. 1/2 (1997): 75–97. The Ministry of Environment has created a new law on public forest management (Law 4776/05), which was approved by Brazil's Chamber of Representatives in July, is still awaiting the vote of the Senate. This law would create the national forest service, the forest development fund, and would regulate timber harvest on public lands. Three kinds of harvest are sought for production forests: direct government management of conservation units (such as national forests); local community use (such as extractive reserves); and forest concessions.

14. Dorothy Stang, a 73-year-old nun from Dayton, Ohio, a practitioner of liberation theology, and an ardent supporter of local settlers, was assassinated in broad daylight in February 2005 in a remote farm community near her home of 25 years in Anapú on the Trans-Amazon Highway. Her battle for equal rights for the poor, including legal land and resource ownership, brought her in direct conflict with loggers and ranchers. Her death triggered an avalanche of government response. Two thousand soldiers were sent to the region to crack down on illegal loggers and land speculators, and five million hectares of forest (an area the size of Costa Rica) were designated as parks and reserves in what may be the world's single greatest act of tropical rainforest conservation.

15. The estimate of forest stock for the Amazon is approximately 60 billion cubic meters. There are varying estimates of the flow from the forest: The IBGE, which is the government institute of geography and statistics (http://ibge.gov.br), estimates log demand in the north of Brazil to be about 17 million cubic meters; IBAMA, the environmental regulation agency of Brazil, estimates it to be around 25 million; and IMAZON, a local nongovernmental research organization, estimates it at about 24 million—down from 28 million in 1999.

16. The entire Trans-Amazon Highway runs approximately 3,300 kilometers, connecting the state of Tocantins to the state of Acre near the Peruvian border. Continuing westward from Itaituba to the town of Humaitá (a stretch which lies to the west of the Tapajós River) is virtually uninhabited, but may be the future frontier on which this story is replayed some years hence.

17. For an excellent discussion on property rights, violence and settlement on the Trans-Amazon Highway see L. J. Alston,

G. D. Libecap, and B. Mueller, *Titles, Conflict, and Land Use: The Development of Property Rights and Land Reform on the Brazilian Amazon Frontier* (Ann Arbor: The University of Michigan Press, 1999); and L. G. Alston, G. D. Libecap, and B. Mueller, "Land Reform Policies, the Sources of Violent Conflict in the Brazilian Amazon," *Journal of Environmental Economics and Management* 39, no. 2 (2000): 162–188.

18. For more discussion of smallholder settlement in new and old settlements and the roles of community associations in economic development in migrant settlements see F. Merry and D. J. Macqueen, *Collective Market Engagement* (Edinburgh, International Institute for Environment and Development, 2004), http://www.iied.org/docs/flu/PT7_collective_market_engagement.pdf.

19. Other institutions working on the Trans-Amazon Highway within the Green Highways Project are the Fundação Viver, Produzir e Preservar (FVPP) and the Instituto Floresta Tropical (IFT).

20. The principal source of government statistics for Brazil is the Brazilian Institute of Geography and Statistics (Instituto Brasiliero de Geografia e Estatistica, IBGE). Their website can be accessed at http://www.ibge.gov.br.

21. A supervised classification is a procedure for identifying spectrally similar areas on an image by pinpointing training sites of known targets and then extrapolating those spectral signatures to other areas of unknown targets. The signatures are quantitative measures of the spectral properties at one or several wavelength intervals. These measures include class maximum, minimum, mean and covariance matrix values. Training areas, usually small and discrete compared to the full image, are identified through visual interpretation and used to "train" the classification algorithm to recognize land cover classes based on their spectral signatures, as found in the image. The training areas for any one land cover class need to fully represent the variability of that class within the image.

22. The total area for squatters was 19 percent of the buffer zone, of which local extension agents estimated 60 percent to be smallholders. The remaining 40 percent were said to be medium- and large-size holdings.

23. The evidence also indicated that only one percent of the buffer area is currently deforested, so these estimates could be considered very conservative for deforestation.

24. The price of R$10 is based on a conservative estimate of a formal logging contract between smallholders and the industry near the town of Santarém and the example of a forest concession (3-year cutting contract) in the Tapajós national forest—an ITTO project run by IBAMA—where the average stumpage fee for three price categories in 2003 was R$11.73. The exchange rate for the period of the survey was approximately R$3 per US$1, but is now at R$2.2 per US$1. For further commentary on the timber markets of Brazil, see A. Veríssimo and R. Smeraldi, *Acertando O Alvo: Consumo da Madeira no Mercado Interno Brasileiro a Promocao da Certificacao Florestal* (Finding the Target: Consumption of Wood in the Brazilian Domestic Market and the Promotion of Forest Certification) and M. Lentini, A. Verissimo, and L. Sobral, *Fatos Florestais da Amazônia* (Forest Facts of the Amazon) (Belém, Brazil: Imazon, 2003); E. Lima, and F. Merry, "Views of Brazilian

Producers—Increasing and Sustaining Exports," in D. Macqueen, ed., *Growing Timber Exports: The Brazilian Tropical Timber Industry and International Markets* (London: IIED, 2003), 82–102.

25. For an economic model of smallholder decision-making, production, and labor allocation, see F. D. Merry and G. S. Amacher, "Emerging Smallholder Forest Management Contracts in the Brazilian Amazon: Labor Supply and Productivity Effects," Environment and Development Economics. Invited to revise and resubmit, expected publication 2006.

26. The preliminary results of the survey were presented in seminars to the smallholders in June 2004. Further details of this survey are available from the authors.

27. In comparison, the estimated price for logs at the mill gate in 2002 on the Trans-Amazon was R$58 per cubic meter, and an unadjusted five-year average price for logs from 1998 to 2002 was R$39 per cubic meter, but this is before accounting for harvest costs—which for intermediate frontiers such as the Trans-Amazon can run between 30 and 40 Reais per cubic meter and transportation costs; transport distances can run as far as 80 or 90 kilometers from log deck to mill.

28. The legal reserve (Reserva Legal) of a smallholder lot, or for that matter any private land holding in the Brazilian Amazon, is 80 percent of the total land area. This "reserve" area can only be used for forestry with approved forest management plans or the collection of non-timber forest products.

29. Alston, Libecap, and Mueller, note 17 above. Only 11 percent of land owners hold formal title. In our survey, individuals were asked whether they held "definitive title," not formal records.

30. This example is well documented. See D. Nepstad et al., "Managing the Amazon Timber Industry," *Conservation Biology* 18, no. 2 (2004): 575–577; D. Nepstad et al., "Governing the Amazon Timber Industry," in D. Zarin, J. R. R. Alavalapati, F. E. Putz, and M. Schmink, eds., *Working Forests in the American Tropics: Conservation through Sustainable Management?* (New York: Columbia University Press, 2004), 388–414.

31. Another example is the project Safra Legal (Legal Harvest) on the Trans-Amazon Highway. The objective of this project was to make use of the legal deforestation options available to smallholders. The idea of this project came from the forest management projects near Santarém and presented a wonderful alternative to smallholders who would have simply burned the trees where they planned to conduit agricultural activities. The project, however, has recently become embroiled in scandal as a conduit of illegal logging, see L. Coutinho, "More Petista Mud in the Ibama," *VEJA*, 15 June 2005, 70. The problems behind the Safra Legal program were also described in L. Rohter, "Loggers, Scorning the Law, Ravage the Amazon Jungle," *The New York Times,* 16 October 2005. These articles illustrate the far-reaching negative effects of corrupt government on the sustainable management of natural resources.

32. F. Merry et al., "Collective Action Without Collective Ownership: the Role of Formal Logging Contracts in Community Associations on the Brazilian Amazon Frontier," *International Forestry Review,* in review. Drafts available from the authors.

**EIRIVELTHON LIMA** is an associate researcher at the Instituto de Pesquisa Ambiental da Amazônia (IPAM, the Amazon Institute of Environmental Research), headquartered in Belém, Pará, Brazil, and doctoral student in Forest Economics at the Virginia Polytechnic Institute and State University (Virginia Tech). **FRANK MERRY** is an associate researcher at IPAM, research fellow in environmental studies at Dartmouth College, and visiting assistant scientist at the Woods Hole Research Center (WHRC). **DANIEL NEPSTAD** is a senior researcher at IPAM and senior scientist at WHRC. **GREGORY AMACHER** is an associate professor of forest economics at Virginia Tech and associate researcher at IPAM. **CLÁUDIA AZEVEDO-RAMOS** is a senior researcher at IPAM. **PAUL LEFEBVRE** is a senior research associate at WHRC. **FELIPE RESQUE JR.** is a GIS technician at IPAM. We gratefully acknowledge funding from (in alphabetical order) the European Union; the Gordon and Betty Moore Foundation; the William and Flora Hewlett Foundation; NASA Large Scale Biosphere and Atmosphere Project; the National Science Foundation; and the United States Agency for International Development–Brazil program.

From *Environment,* January/February 2006, pp. 26, 28–36. Reprinted by permission of the Helen Dwight Reid Educational Foundation. Published by Heldref Publications, 1319 Eighteenth St., NW, Washington, DC 20036-1802. Copyright © 2006. www.heldref.org

# UNIT 6

# The Hazards of Growth: Pollution and Climate Change

## Unit Selections

## Key Points to Consider

- Why are agricultural pesticides that have been banned in the United States and other developed countries still used in the lesser-developed regions of the world? Do the benefits of increased food production outweigh the costs of using agricultural pesticides?

- Why is ozone, a chemical that we view as necessary at high elevations of the atmosphere, so dangerous when concentrated near the ground? Why, despite the recent reductions in emissions of other air pollutants such as volatile organic compounds and nitrous oxides, does ozone pollution persist in being a problem?

- What are some of the potential relationships between the beginnings of agriculture and the first human alterations of global climates? Is there a connection between the possible climate-altering practices of farmers thousands of years ago and the potential global climate change of the present and future?

- Explain the concept of "thresholds" or "tipping points" in environmental systems. Why do you think the rate of change in global climate systems is more rapid than scientists predicted a decade ago? If scientists were wrong about when global warming would begin to be a problem, is it possible that they were wrong about what has caused abrupt climate change?

- How has one of the atmospheric scientific community's most respected voices been systematically discredited? What is the current nature of the scientific consensus over global warming and to what extent does it represent political opinion or scientific examination of atmospheric phenomena?

## Student Web Site
www.mhcls.com/online

## Internet References
Further information regarding these Web sites may be found in this book's preface or online.

**Persistent Organic Pollutants (POP)**
*http://www.chem.unep.ch/pops*
**School of Labor and Industrial Relations (SLIR): Hot Links**
*http://www.lir.msu.edu/hotlinks*
**Space Research Institute**
*http://arc.iki.rssi.ru/eng/index.htm*
**Worldwatch Institute**
*http://www.worldwatch.org*

Of all the massive technological changes that have combined to create our modern industrial society, perhaps none have been as significant for the environment as the chemical revolution. The largest single threat to environmental stability is the proliferation of chemical compounds for a nearly infinite variety of purposes, including the universal use of organic chemicals (fossil fuels) as the prime source of the world's energy systems. The problem is not just that thousands of new chemical compounds are being discovered or created each year, but that their long-term environmental effects are often not known until an environmental disaster involving humans or other living organisms occurs. The problem is exacerbated by the time lag that exists between the recognition of potentially harmful chemical contamination and the cleanup activities that are ultimately required.

A critical part of the process of dealing with chemical pollutants is the identification of toxic and hazardous materials, a problem that is intensified by the myriad ways in which a vast number of such materials—natural and man-made—can enter environmental systems. Governmental legislation and controls are important in correcting the damage produced by toxic and hazardous materials such as DDT, PCBs, or CFCs; in limiting fossil fuel burning; or in preventing the spread of living organic hazards such as pests and disease-causing agents. Unfortunately, as evidenced by most of the articles in this unit, we are losing the battle against harmful substances regardless of legislation, and chemical pollution of the environment is probably getting worse rather than better.

The first article in this unit deals with one of the most serious of all international pollution problems: that of soil and water pollution from the widespread application of agricultural pesticides. In "Agricultural Pesticides in Developing Countries" Sylvia Karlsson of Yale University's Center for Environmental Law and Policy notes that it was the use of agricultural pesticides in North America and Europe, along with their undesired and unanticipated side effects, that produced the first alerts (largely through Rachel Carson's *Silent Spring*) that modern society can have major impacts on Earth's ecological systems. While many of the most dangerous chemicals have been banned in countries like the United States, their benefits in terms of increased food production outweighs their perceived dangers in countries in the lesser-developed world. As a consequence, increases rather than decreases in biological pollution from pesticides have been noted in Africa and Asia, posing massive challenges for governments that now, whether they like it or not, operate in a large, interconnected system. Another form of pollution on a more local scale is dealt with in the second article in this selection. In "Smog Alert: The Challenges of Battling Ozone Pollution," environmental policy analyst Mark Bernstein and *U.S. News & World Report* contributing editor David Whitman describe the success (or lack thereof) of controlling ozone pollution in the United States. Ozone, a primary component of more chemically complex "smog", has been a major public health issue in the United States for decades. Since the passage of the Clean Air Act in 1970 the nation has been battling ground-level ozone pollution. Once thought of as an exclusively urban problem, air analysis of many of the "crown gems" of the National Park system suggest that ozone pollution exceeding safe levels is also a problem in rural areas. What scientists term "the ozone paradox" is that, although legislation has substantially reduced levels of nitrogen oxides and volatile organic compounds (both components of smog) in the atmosphere, no similar reduction has been noted in levels of ozone depletion. The paradox suggests that there is no "one-size-fits-all" answer to air pollution and that policies directed specifically at ground-level ozone pollution need to be developed.

The section's next three articles also deal with the unwanted injection of harmful substances into the environment—but at the other end of the scale. Where the previous articles dealt with local contamination, these final selections deal with global contamination and what it might mean at all scales from local to worldwide. In "How Did Humans First Alter Global Climate?," climate historian William Ruddiman ponders whether the recent acknowledgment that human activities over the last century or two have had a warming effect on the global climate goes far enough. Ruddiman suggests that farming practices begun as early as 8000 years ago began increasing global carbon dioxide and contributing to a global warming trend that began slowly and—as industrialization increased the use of fossil fuels and as human populations increased—gathered momentum to the point where it became noticeable over the last few decades. Ruddiman suggests anomalies in atmospheric carbon dioxide that are large enough to have prevented the world from entering a new glacial state several thousand years ago. However, the warming trend of the past—which has proven generally beneficial for humankind—now runs the risk of accelerating into a course of increasing global temperatures that will lead us into completely unknown territory. The next article in this section also deals with the issue of global warming produced

by pollution. In "The Tipping Point" science writer and reporter Jeffrey Kluger notes that while the scientific community has long acknowledged the problem of global warming, it was anticipated that global climate change would be gradual enough to allow human societies to adapt and to develop strategies for mitigating and even reversing the process. Events of the last few years, however, have demonstrated that the process of climate change has not only been more rapid than assumed but more dramatic as cool, wet forests have dried, warmed, and burned, as more severe atmospheric events such as hurricanes have occurred, and as Arctic ice floes, continental glaciers, and the ice sheets covering Greenland and Antarctica have begun to melt at unanticipated rates. Concentrations of the chief greenhouse gas, carbon dioxide, will continue to increase for the next few decades because of inaction in the past decades. But curbing global warming is possible. It may, Kluger notes, be an order of magnitude more difficult than putting a man on the moon or eliminating smallpox. But it is immoral not to try if we are to save humankind's only home.

In the final article of the section, NASA's James Hansen, the atmospheric scientist who was one of the first to blow the whistle on global warming two decades ago, discusses the approach of the current federal administration to stifle scientific discussion and public action on the global warming problem. In "Swift Boating, Stealth Budgeting, and Unitary Executives," Hansen accuses "contrarian" scientists who continue to deny global warning as behaving like lawyers defending a client rather than like scientists seeking the truth. He also takes on those (like novelist Michael Crichton) who have taken quotes out of context or purposely distorted Hansen's research to discredit one of the greatest of all "whistle-blowers." Hansen also decries the administration's attempt to muzzle NASA scientists by cutting back on NASA's budget and even removing the first line from NASA's mission statement: "To understand and protect our home planet." Like nearly all serious and dedicated scientists who study the earth's systems, Hansen understands the dangers of inaction.

We now possess the knowledge and the tools to ensure that environmental cleanup is carried through. It will not be an easy task, and it will be terribly expensive. It will also demand a new way of thinking about humankind's role in the environmental systems upon which all life forms depend. If we do not complete the task, however, the support capacity of the environment may be damaged or diminished beyond our capacities to repair it. The consequences would be fatal for all who inhabit this planet.

# Agricultural Pesticides in Developing Countries

## A Multilevel Governance Challenge

Sylvia I. Karlsson

The use of agricultural pesticides in North America and Europe and their non-desired side effects were among practices and subsequent effects that set alarm bells ringing—bells that, in the 1960s and 1970s, awakened the public and, eventually, government agencies to the impact our modern human life can have on the Earth's life-sustaining ecological systems. For the past 30–40 years in developing countries, agricultural pesticide use has set off a continuously ringing alarm, alerting us to a heavy toll on human health and the environment. It is an issue that clearly demonstrates links—between the local and the global and between developed and developing countries—often associated with environmental and societal problems. This creates a complex picture with many challenges to address in terms of governance.[1]

Following the trajectory of modern agriculture in developed countries, developing countries have, over the past half century, increasingly adopted a pest management approach that centers on the use of chemical pesticides. The pesticide world market's total value surpasses US$26 billion. Developing countries' share of this is approximately one-third.[2] It is reasonable to expect that developing countries will continue to experience similar—if not identical—patterns of negative human-health and environmental side effects that motivated industrialized countries to develop the pesticide regulatory apparatus in the 1970s. Looking at potential effects on human health, a variety of factors is likely to make local populations in developing countries—in particular farmers and agricultural workers but in general all food consumers—more vulnerable to the toxicological effects of pesticides.[3] Examples of such factors include

- low literacy and education levels;
- weak or absent legislative frameworks;
- climatic factors (which make the use of protective clothing while spraying pesticides uncomfortable);
- inappropriate or faulty spraying technology; and
- lower nutritional status (less physiological defense to deal with toxic substances).

The first four of these factors increase the likelihood of higher exposure; the last factor increases the toxic effects from that exposure on the human body. In addition, it has been found that when most organochlorine pesticides are banned or restricted, developing countries have often turned to substances that exert higher toxicity.[4]

According to a World Health Organization/United Nations Environment Programme (WHO/UNEP) working group report from 1990, unintentional acute poisonings with severe symptoms exceed 1 million cases each year, out of which 20,000 are fatal.[5] Additionally, there are estimated to be 2 million intentional poisonings (mainly suicide attempts) resulting in 200,000 deaths per year.[6] The vast majority of both unintentional and intentional intoxications occur in developing countries. However, good intoxication data are sparse in developing countries and these global estimates are often contested: Some stakeholders argue that they are too high, others insist that they are too low.[7] The data for estimating the toll of less acute effects, including longterm effects such as cancer, are even more sparse.[8]

Looking at potential environmental impacts of pesticide use in developing countries, many of the same factors that increase the vulnerability to health impacts are likely to exacerbate the release of the chemicals into the environment. Furthermore, in these regions there are unique ecosystems and species of ecological and economic importance and a range of managed systems (agricultural, silvicultural, and aquacultural) that are only marginally present in developed countries.[9] Because most species in developing countries are not subject to any tests before an industrialized country approves a particular pesticide, they may exhibit previously unknown sensitivities to pesticide exposure. Studies of pesticide impact on local environments in developing countries are few and far between: Data collection and research in these regions is extremely limited.[10] However, observed effects include contaminated ground, river, and coastal waters; fish kills; and impacts on cattle.[11] Environmental effects from organochlorine pesticides, a few of which are still in use, may also extend well beyond the region of use.[12] In the last decade, a hypothesis regarding the transport of such substances has received stronger support: It is believed that, due to their chemical characteristics, organochlorine pesticides are partially transported via the atmosphere from warmer tropical regions toward colder regions (the Arctic and Antarctica). According to this theory, the polar regions are thought to function as sinks for these substances as they condense and accumulate in ecosystems and food chains—ultimately affecting human health as well.[13]

# Governance Challenges, from Local to Global

Pesticide use in developing countries is seen to produce benefits to society. At the same time, however, it has the potential to produce negative consequences on different scales, ranging from immediate health effects for those who spray them to environmental effects in remote regions. The reasons, or driving forces, behind these negative effects can be found in policies and actions covering levels of governance from the local to the global. The responses by stakeholders to reduce the negative effects are also found at each governance level. It is thus an issue that well illustrates an interconnected, globalized world and a multilevel governance challenge.[14]

To search for appropriate responses to such a globalized problem, it is necessary to examine the human activity—the use of pesticides in agriculture—that acts as the direct driving force for various undesired effects. In addition, it is helpful to analyze governance at the local, national, and global levels. It is equally important to determine how the diversity of stakeholders at these levels understand and structure the "pesticide problem" and address the problem through strategies of risk reduction.

> **Compared with other nations, Costa Rica produces the highest amount of coffee per hectare. Along with a number of other factors (many having to do with the banana trade), this has contributed to relatively high levels of pesticide use.**

Such research—examining many aspects of pesticide use and governance in developing countries—was carried out in association with Linköping University, Sweden, in 1997–1999.

A close look at the stakeholders included in the study reveals three major categories: organizations (government and intergovernmental organizations (IGOs)), civil society (nongovernmental organizations (NGOs), private companies, and academia), and individuals (farmers and workers).[15]

At the global level of governance, some key stakeholders involved in pesticide use in developing countries (and which were examined in the 1997–1999 study) include IGOs such as the Food and Agriculture Organization of the United Nations (FAO) in Rome and the United Nations Environment Programme (UNEP) in Nairobi and Geneva (a more complete list appears below in the discussion of the activities of such organizations).

On the national level, two countries, Kenya and Costa Rica, served as important case studies of the issue. Both countries

have enacted relatively ambitious pesticide legislation and other governance measures compared with other countries on their respective continents. Taken together, the countries' heavy dependence on—and the involvement of large businesses in—export agriculture, the type of export crops grown, and a number of other factors have made Kenya and Costa Rica substantial users of agricultural pesticides. The trend in this regard has been increasing for both nations in the past two decades.[16] In terms of product category, fungicides dominate import figures by volume in Costa Rica, followed by insecticides and fumigants.[17] In Kenya, fungicides account for half the market, insecticides 20 percent, and herbicides 18 percent.[18]

The high and increasing use of pesticides in Costa Rica has been attributed to such factors as the switch to horticultural crops; the expansion of land under banana production, a particularly difficult problem in banana plantations with the leaf spot disease *Mycosphaerella fijiensis*; the highest production of coffee per hectare (ha) in the world; and the fact that some transactions with pesticides have been exempted from taxes.[19] The volume of pesticide use is usually closely linked to the type of crop. In Costa Rica the average amount of active ingredient applied per ha was, according to one study, 0.25 kilograms (kg) for pasture, 3.5 kg for sugarcane, 6.5 kg for coffee, 10 kg for rice, 20 kg for vegetables/fruits, and 45 kg for bananas.[20]

Kenyan agricultural exports have been traditionally dominated by coffee and tea, but nontraditional crops such as pineapples, vegetables, and ornamentals have expanded quickly in recent years.[21] This expansion has contributed to increasing pesticide use. Historically, pesticide use per ha on the large agricultural farms has been much higher than on small farms. However, a substantial percentage of smallholder farmers in Kenya use pesticides, mainly on their export crops.[22]

When comparing approaches to pesticide use at the local level in different countries, it is helpful to look at the same crop. In the case of Kenya and Costa Rica, coffee is a good choice, and two prominent coffee growing regions are the Meru District in Kenya and the Naranjo *cantón* (district) in Costa Rica. Kenya's Meru District, which is located about 300 kilometers northeast of Nairobi (in the Eastern Province, covering the slopes of Mount Kenya), is one of the most fertile areas of the country with a range of subsistence and cash crops. Naranjo *cantón*, in Costa Rica's Alajuela province, is located in the hills northwest of San José. Coffee is the dominant crop, but there is also sugarcane and livestock production.[23] The land distribution is similar in both districts, with many small-scale landholders and a few medium- or large-scale landowners. However, the differences are substantial: Naranjo has significantly larger-sized farms, a higher level of modernization, and much higher literacy rates and general education levels than does Meru. Although the same variety of coffee, *Coffee arabica*, is grown in Meru and Naranjo, the answers farmers in the two districts gave to inquiries about pest problems share only some similarities. (Farmers in Meru were interviewed in early 1999; Naranjo farmers were interviewed later that year.) In both areas, farmers said that the dominating pest problems had been fungi. However, Meru was also beset by a number of insect pests. In Naranjo, farmers said that insect pests were largely absent, but they mentioned that a number of nematodes (commonly known as roundworms) and different fungi have affected coffee production. In Meru, at least a quarter of the farmers responding to interviews were not spraying anything on their coffee. This was a new situation that had occurred for a few years prior to the interviews and were due primarily to falling coffee prices. Farmers in Meru have been spraying their coffee trees since the crop was introduced in the area in the 1950s. The interviews in Naranjo found that virtually all of the farmers there sprayed pesticides. Coffee was introduced in Costa Rica in the early 1900s—long before pesticides were available—but since the arrival of these chemicals in the 1950s, they have been applied ubiquitously.[24]

# Which Problems and Whose Problems?

The stakeholders mentioned above raised five problem categories from pesticide use in developing countries: economic, production, human health, environment, and trade.[25] At the global level, pesticides are primarily seen as a health issue, especially for the poor, uneducated farmers and workers in developing countries who apply them. The pesticide-related trade problems for developing countries emerge in some agencies involved with regulations for pesticide residues in food crops. Persistent categories of pesticides—organochlorines—have emerged as a transboundary environmental issue in multilateral discussions and agreements, but there has been little mention of potential local and regional environmental effects in developing countries themselves. While the lack of data and research makes it impossible to give solid numbers and figures on health and environmental impacts of pesticides on a large scale in developing countries, the discussion above showed that there is enough isolated data coupled with higher risk factors in developing countries for IGOs to be taking the problem seriously.

At the national level in Kenya and Costa Rica, stakeholders involved in the study tended to either associate the entire assortment of pesticides with health and environmental effects or claimed there was no evidence for substantive negative effects. The government in Kenya tilted toward the latter position; various environmental NGOs in the country referred to substantial problems.[26] Costa Rican stakeholders made more references to significant health effects, and although environmental effects were usually discussed in general terms, the situation in banana plantations was often lifted out as potentially more serious. The trade problem has definitely remained high on the agenda for both countries, but stakeholders largely referred to it as a problem of the past, before appropriate policies had been put in place.[27]

There is little data on pesticide-related health effects in Kenya to substantiate opinions expressed regarding their magnitude. There was (at the time of the study) no working system to report intoxications to the authorities. In fact, one of the main areas of concern regarding chemical use generally was the absence of documentation on the risks in the country.[28] Figures surfaced in a national debate in the early 1990s claiming that 7 percent of the people in the agricultural sector—about 350,000 people—suffered pesticide poisoning each year.[29] Other sources reported

that, in 1985, three major hospitals treated an average of two cases of pesticide poisoning every week.[30] Kenya's Ministry of Health estimated in 1996 that 700 deaths per year were caused by pesticide-related poisonings.[31] However, there were virtually no data at all on subacute poisonings from pesticides.[32] A few studies investigated residue levels of organochlorine pesticides in food and human tissue, but none of these looked at possible symptoms from such chronic, low-level exposure.[33]

---

**Farmers and workers interviewed in Kenya's Meru District said the biggest problem with pesticides was their cost: None mentioned the possibility of long-term health effects.**

---

Costa Rica had better data on pesticide-related health effects. In 1991, a new law there made it obligatory to report intoxications from pesticides to the Ministerio de Salud (Ministry of Health).[34] In the period 1980–1986, 3,347 cases of intoxications were treated in hospitals.[35] The figures for the years 1990–1996 indicate a significant increase of intoxications until 1995, followed by a slight decrease in 1996—when the reported number was 792.[36] Out of all the reported intoxications in 1996, farmers and workers associated with banana cultivation suffered the highest number (64 percent), followed by those working with ornamental plants (3.7 percent).[37] In terms of chemical classes, insecticides and nematicides were attributed to more than one-half of the intoxication cases, followed by herbicides.[38] One study identified young workers (under 30) and women as groups particularly affected by occupational poisonings.[39] The same study concluded that on a yearly basis, 1.5 percent of the agricultural workforce is medically treated for occupational poisonings.[40] Intoxications from cholinesterase-inhibiting pesticides (organophosphates and carbamates) and the herbicide paraquat together accounted for a majority of poisonings identified in hospitalization and fatality records in the 1980s in Costa Rica.[41] Unique studies—in a developing country context—have been done in Costa Rica on subacute health effects such as cancer. The organochlorines that were used in agriculture until the 1980s have shown a rather strong association with breast cancer in areas where rice—a crop that has been intensely sprayed with such products—is grown.[42] Other associations between pesticides and cancer include paraquat and lead arsenate, which have been linked to skin-related cancers; formaldehyde and leukemia; and dibromochloropropane (DBCP), which has been associated with lung cancer and melanoma.[43] Banana workers in particular showed increased incidence of some cancer types, specifically melanoma in men and cervical cancer in women.[44] Table 1 shows pesticide import figures for Kenya and Costa Rica in the early 1990s, classified in categories according to acute toxicity. (Such classification gives an indication for assessing potential health risks.) Nearly one-quarter of the pesticides used are moderately hazardous, and more than 10 percent are highly hazardous.

Unfortunately, the lack of data on environmental effects is the rule and not the exception. The few studies that were found

**Table 1**  Toxicity Classification of Imported Pesticides, Kenya and Costa Rica, 1993

| World Health Organization (WHO) Toxicity class | Percent Total Kenya | Percent Total Costa Rica |
| --- | --- | --- |
| 1a-Extremely hazardous | — | 10 |
| 1b-Highly hazardous | 3 | 8 |
| Highly hazardous volatile fumigants | 11 | — |
| II-Moderately hazardous | 22 | 24 |
| III-Slightly hazardous | 24 | 8 |
| Unlikely to present acute hazard | 25 | 40 |
| Not classified by WHO | 3 | — |
| Unidentified | 10 | 10 |

Note: It is very difficult to compare and interpret these kinds of import figures from different countries: Some of the products are imported as technical-grade material (concentrated) and others as ready-to-use formulations. For example, after formulation in Kenya the proportion of highly hazardous pesticides will be significantly higher, because a substantial portion of those substances are imported as technical-grade material. The figures show that nearly one-quarter of the pesticides used are moderately hazardous. More than 10 percent are highly hazardous.

Source: H. Partow, *Pesticide Use and Management in Kenya* (Geneva: Institut Universitaire D'Etudes du Développement (Graduate Institute of Dévelopement Studies), 1995); and F. Chaverri, and J. Blanco, *Importación, Formulación y Use de Plaguicidas en Costa Rica*. Periodo 1992–1993 (Importation, Formulation, and Uso of Pesticides in Costa Rica between 1992–1993) (San José, Costa Rica: Programa de Plaguicidas: Desarollo, Salud y Ambiente, Escuela de Ciencias Ambientales Universidad Nacional (Pesticide Program: Development, Health, and Environment, School of Environmental Sciences, National University, 1995).

in Kenya only look at pesticide levels in the environment and do not indicate significant problems.[45] In Costa Rica, the Ministry of Environment and Energy had commissioned a report on the environmental effects of pesticide use on banana plantations in three regions in the Atlantic zone. The report, however, could also mainly refer to pesticide residues detected in the environment that have unknown effects: The only concrete direct negative effects reported were some incidences of fish killed in rivers and declining fish populations.[46]

At the local level, the chief complaint farmers in Meru expressed was an economic problem: They were unable to purchase pesticides due to low coffee prices. Interview revealed that farmers in neither Meru nor Naranjo saw production problems from pesticides. Respondents said they had not seen pests develop resistance, and with few exceptions, all pests could be adequately controlled. Health effects that workers associated with using pesticides varied substantially. Some said they suffered problems such as dizziness, nausea, headache, skin problems, eye problems, or fever; others said they never had any problems.[47] There was hardly any mention in Meru of the possibility of more long-term health effects. However, it was not uncommon for respondents in Naranjo to mention such effects, including cancer. The general environment as a possible victim of negative effects was practically absent in the interviews in Meru, while respondents in Naranjo were slightly more aware of its vulnerability. Table 2 shows some of the most commonly used pesticides in the districts. Many are moderately hazardous; some have notorious records as intoxicants in developing countries.

The upshot of all this is that there were, at the time of the study, substantial divergences in the understanding of the pesticide-linked problems among different stakeholder groups and at varying governance levels. This lack of common understanding is not surprising considering the diversity of priorities among stakeholders and prevalent lack of knowledge on so many aspects of the potential impacts.

# Which Risk-Reduction Efforts, Where?

Three basic risk-reduction strategies emerge for addressing the risks with pesticides: reducing their use, using less-toxic types of substances, and using them in a more precautionary fashion or mode?[48]

These approaches can be linked to three factors of pesticides that contribute to their health and environmental risks—the volume of use, the type of pesticide used, and the mode in which they are used. Table 3 on page 191 shows which kind of risk-reduction efforts existed at each governance level and to which strategy they belong. At the global level, the predominant risk-reduction efforts focused on the type of pesticides, applying a chemical-by-chemical approach in many activities. Several IGOs have assisted developing countries to regulate and control the use of pesticides. This guidance has come in the form of facilitating the development of national legislation, helping to build capacity in chemical management, and providing scientific data on individual pesticides. Several international agreements have been developed to address problems associated with pesticides and other chemicals by targeting individual substances—by phasing them out, for example.[49] Some IGOs have also made significant efforts to support a better mode of use. For example, FAO developed the International Code of Conduct on the Distribution and Use of Pesticides, and there have been several efforts promoting its universal observance.[50] In addition, a number of technical guidelines have been made on how pesticides should be used.[51] Finally, there have been some efforts to reduce the overall use of pesticides. For instance, some IGO programs have encouraged the implementation of integrated pest management (IPM) techniques rather than supporting increased use of pesticides in agricultural projects.[52] However, there has been very low support for organic farming.[53] Table 4 on pages 192 and 193 gives more details on

**Table 2**  Pesticides Most Frequently Mentioned by Local-area Coffee Farmers

| Product Name | Pesticide Category | Active Ingredient | WHO Hazard Classification[1] |
|---|---|---|---|
| **Meru** | | | |
| Copper (various) | Fungicide | Copper | III |
| Lebaycid® | Insecticide | Fenthion | II |
| Sumithion® | Insecticide | Fenitrothion | II |
| Gramoxone® | Herbicide | Paraquat | II |
| Ambush® | Insecticide | Permethrin | II |
| Karate® | Insecticide | Lamba-cyhalothrin | II |
| **Naranjo** | | | |
| Atemi® | Fungicide | Cyproconazole | III |
| Silvacur® | Fungicide | Triadimenol | III |
| Gramoxone® | Herbicide | Paraquat | II |
| Roundup® | Herbicide | Glyphosphate | -[2] |
| Counter® | Nematicide | Terbufos | 1a |
| Furudan® | Nematicide | Carbofuran | 1b |

[1]The World Health Organization (WHO) classification places products' active ingredients into the following categories: 1a-extremely hazardous; 1b-highly hazardous; highly hazardous volatile fumigants; II-moderately hazardous; III-slightly hazardous; and unlikely to present acute hazard.

[2]Unlikely to present hazard under normal use.

Note: Coffee farmers in Meru, Kenya, and Naranjo, Costa Rica, responded to questions about which pesticides they apply with a few product names: The table above lists the most commonly mentioned. In total, farmers in Meru identified more than 20 different products; in Naranjo more than 30 were mentioned. Fungicides, the products sprayed most frequently, belong to the less acutely toxic categories (II and III). The nematicides—used only in Costa Rica—stand out as having highly or even extremely toxic ingredients. The three pesticides most commonly involved with intoxications in Costa Rica during the years 1994–1996 were paraquat (bipyridylium), carbofuran (carbamate), and terbufos (organophosphate).

Source: S. Karlsson, *Multilayered Governance. Pesticide Use in the South: Environmental Concerns in a Globalised World* (Linköping, Sweden: Linköping University, 2000); and R. Castro Córdoba, N. Morera González, and C. Jarquín Núñez, *Sistema de Vigilancia Epidemiologica de Intoxicaciones con Plaguicidas, la Experiencia de Costa Rica, 1994–1996* (Epidemiological Monitoring System of Pesticide Intoxications, Costa Rica's Experience, 1994–1996) San José, Costa Rica: Departamento de Registro y Control de Sustancias y Medicina del Trabajo del Ministerio de Salud (Ministry of Health Department of Registration and Control of Occupational Substances and Medicine, 1998).

**Table 3**  Strategies to Reduce Health and Environmental Risks, by Governance Level

| Level | Country/District | Use | Type | Mode |
|---|---|---|---|---|
| Global | | Integrated pest management Organic farming[a] | Phase-out Information exchange Risk assessment Toxicity classification | Codes of conduct Guidelines |
| National | Kenya | Integrated pest management Organic farming[a] | Registration Banning | Regulation and training |
| | Costa Rica | Integrated pest management Organic farming | Registration Banning | Regulation and training |
| Local | Meru, Kenya | Resistant coffee variety Getting others to spray | —— | Safe-use training |
| | Naranjo, Costa Rica | Integrated pest management | —— | Safe-use training |

(a) Organic farming exists as a strategy on this level but is not strongly stressed.
Note: Strategies to reduce health and environmental risks are categorized under "use," "type," or "mode;" that is, they reduce pesticide use, address certain types of pesticides use related to their individual characteristics, or they address modes of the chemicals' application.

Source: S. Karlsson, *Multilayered Governance. Pesticides in the South: Environmental Concerns in a Globalised World* (Linköping, Sweden: Linköping University, 2000).

some of the relevant organizations and their activities as well as agreements on the international level.[54]

Most risk-reduction efforts by governments and industry at the national level in Kenya and Costa Rica belong to the type and mode categories, although IPM efforts have been increasingly encouraged. To address risks arising from the types of pesticides used, both countries have developed extensive laws and regulations on pesticides. The core of these have been registration processes.[55] Within these processes, pesticide companies submit each product for approval before it can be used in the country.[56]

**Table 4**  Global-level Actors, Activities, and Agreements on Developing Country Pesticide Risks

| Actor/Organization | Activities and International Agreements |
| --- | --- |
| **Intergovernmental Forum on Chemical Safety (IFCS)**<br>A noninstitutional arrangement where governments meet with intergovernmental and nongovernmental organizations every three years. Smaller meetings are held every year.<br>Geneva, Switzerland (small secretariat)<br>Established in 1994<br>www.who.int/ifcs | • Provides advice to governments, international organizations, intergovernmental bodies, and nongovernmental organizations on chemical risk assessment and environmentally sound management of chemicals.<br>• Sets priorities and promotes coordination mechanisms at the national and international level. |
| **Food and Agricultural Organization of the United Nations (FAO)**<br>Specialized agency<br>Rome, Italy<br>Established in 1945<br>www.fao.org | • Supports integrated pest management (IPM) projects with Farmer Field Schools in Asia and Africa (for example) at national and regional levels for various crop systems.<br>• Assists countries with development of national pesticide legislation.<br>• Produces technical guidelines on various aspects of pesticide risks.<br>• Runs projects to clean up obsolete stocks of pesticides (wastes) in developing countries.<br>• Facilitated the negotiation of the International Code of Conduct on the Distribution and Use of Pesticides (the FAO Code of Conduct) and monitors its observance.<br>www.fao.org/ag/agp/agpp/pesticid/ |
| **Global IPM Facility**<br>Cosponsored by FAO, UNEP, United Nations Development Programme (UNDP), and the World Bank<br>FAO Headquarters, Rome, Italy<br>Established in 1995<br>www.fao.org/globalipmfacility/home.htm | • Assists governments and nongovernmental organizations to initiate, develop, and expand IPM.<br>• Strengthens IPM programs through, for example, initiation of pilot projects around the world. Programs include policy development and capacity building.<br>• Applies a farmer-led, participatory approach to IPM. |
| **United Nations Environment Programme (UNEP)**<br>Nairobi, Kenya<br>Established in 1972 | • Hosts the Interim Secretariat for the Stockholm Convention on Persistent Organic Pollutants (POPs), which initially targets 12 chemicals (out of which 9 are pesticides) for reduction and eventual elimination. It also sets up a system for identifying further chemicals for action. Signed by 151 countries, it will enter into force 17 May 2004.<br>www.pops.int |
| **UNEP Chemicals**<br>Geneva, Switzerland<br>www.unep.org<br>www.chem.unep.ch | • Hosts (with FAO) the Interim Secretariat for the Rotterdam Convention on Prior Informed Consent for Certain Hazardous Chemicals and Pesticides in International Trade, which prevents export of harmful pesticides (and industrial chemicals) unless the importing country agrees to accept them. Signed by 73 countries, it was entered into force 24 February 2004.<br>www.pic.int<br>• Produces the Legal File, which contains information on regulatory actions on hazardous chemicals in 13 countries and 5 international organizations.<br>www.chem.unep.ch/irptc/legint.html |

*(continued)*

**Table 4** *(Continued)*

| Actor/Organization | Activities and International Agreements |
|---|---|
| **International Programme on Chemical Safety (IPCS)**<br>A joint program of the World Health Organization (WHO), FAO, and the International Labour Organization (ILO)<br>WHO Headquarters, Geneva, Switzerland<br>Established in 1980<br>www.who.int/pcs/ | • Produces and disseminates evaluations of the risk to human health and the environment from exposure to chemicals (including pesticides) and produces guideline values for exposure.<br>• Carries out projects with governments to support their capacity in chemical safety.<br>• Since the early 1990s, has carried out activities to develop a project for collecting data on pesticide poisoning. Several countries are testing the harmonized approach of data collection. |
| **Pesticide Action Network (PAN)**<br>A network of more than 600 participating nongovernmental organizations, institutions, and individuals in more than 60 countries.<br>Five regional centers (San Francisco, California; Santiago, Chile; London, United Kingdom; Dakar, Senegal; and Penang, Malaysia)<br>www.pan-international.org/ | • Works to replace the use of hazardous pesticides with ecologically sound alternatives.<br>• Working from five autonomous regional centers, the network's programs include research, policy development, and media and advocacy campaigns. |
| **CropLife International**<br>(formerly (pre-2000) Global Crop Protection Federation (GCPF))<br>Represents the crop-protection product manufacturers and their regional associations<br>Brussels, Belgium<br>www.gcpf.org | • Supports the FAO Code of Conduct.<br>• Initiated three pilot projects in 1991 to promote the safe use of pesticides (in Guatemala, Kenya, and Thailand). Now continuing to support safe-use programs through its National Associations.<br>• Endorses and supports IPM.<br>• Has carried out several projects addressing obsolete pesticide stocks. Supports a nongovernmental organization-initiated program to eliminate stocks of POPs in Africa. |

Source: Sylvia I. Karlsson, 2004.

Both countries have banned a number of primarily organochlorine pesticides over the last two decades.[57] In both countries, training programs for the safe use of pesticides has involved government agencies as well as the pesticide industry.[58] The pesticide laws in both countries cover, for example, conditions for storage of pesticides by retailers, the training of pesticide retailer staff, and the application of pesticides by farmers (prescribing that they must be used safely).[59] In Kenya, there have been small efforts to establish IPM as a national policy, but there has been very little implementation in this regard. However, it appears that IPM has received more official support in Costa Rica.[60] On the margin, some NGOs in both countries—and even the Costa Rican government—have encouraged organic farming.[61]

On the local level in the two coffee-growing districts, safe-use training has been the only explicit approach to reduce the risks of pesticides. More than 20,000 farmers in Meru were trained in a project sponsored by the pesticide industry in the first half of the 1990s, but this number is still a small part of Meru's agricultural population: The project neglected groups such as women and casual and permanent farm workers, and it is likely that such farm workers are the group most heavily exposed to pesticides. In Naranjo, training on protective measures has been present for a number of years, although not in the form of an explicit safe-use project: It was incorporated in the general training from the national agricultural extension system. Despite these training efforts, the adoption of the safe-use message among interview respondents was low in both districts. In Meru, a few said they used improvised or partial protective clothing, but most respondents neither owned nor had access to these. Farmers and workers interviewed in Meru said they were too expensive. In Naranjo, access was generally not a problem, but farmers said they were uncomfortable: Few used the protective wear, or if they did they wore it only in the early morning before it got too hot. When asked, most farmers said they believed there were no alternatives to pesticides. The few references to nonchemical alternatives to combat pests were primarily made by farmers in Naranjo and included pruning, hand weeding, and general soil conservation measures. In Meru, the extension system had provided a coffee variety that was resistant to the two most prominent fungal pests, but only some farmers at the time of the interviews had planted them. Interviewees had not applied any of these measures to address either health-risk or environmental problems. Accumulated experience and research in Costa Rica had, at the end of the 1990s, begun to show the negative consequences of herbicide use both on the productivity of the coffee plants and erosion levels. Consequently, the extension service had begun to discourage farmers from using herbicides. Many coffee farms were de facto organic: They could afford

## Table 5a Matching Institutions

| | Global | | National | | Local | |
|---|---|---|---|---|---|---|
| | **Negative Effects** | **Driving Forces** | **Driving Forces** | **Capacity to Act** | **Negative Effects** | **Driving Forces** |
| **Example** | Global transboundary pollution of pesticides | Institutions giving incentives for high-input agriculture, resulting in increased pesticide use | Regulations prescribing how toxic pesticides are allowed to be used in countries | Governmental authority to regulate pesticide storage, sales, and marketing | Local health and environmental effects | Influence that farmers and workers who handle pesticides have on how safely they are used |
| **Degree example is matched in pesticide risk reduction** | High: The Stockholm Convention addresses many persistent organic pollutant pesticides. | Low: Regulation is absent and there are few incentives for alternative agricultural systems. | Moderate: Many developing countries allow many highly toxic pesticides. | Moderate: These institutions are weak or very weakly enforced in many developing countries. | Low: Many places have poor healthcare facilities and no institutions that address potential environmental effects. | Low: Most countries have limited and/or ineffective safe-use education and training. |

neither pesticides nor chemical fertilizers. But because there was no market infrastructure for organic coffee, they could not receive a higher price for their crop. In Naranjo, some farmers were aware that it was possible to receive a significant premium price for growing certified organic coffee. However, even the promoters of organic coffee farming did not encourage existing coffee farms to switch to organic farming: Organic certification requires a farm to be without agrochemical inputs for three years, and the harvest slumps in the meantime. (Generally, long-abandoned coffee farms are preferred instead; they can most successfully be developed into an organic farming system.) There were no efforts at the local level to reduce risks by avoiding certain types of pesticides. Farmers bought and applied the products that were recommended by companies, cooperatives, or the extension system.

# Which Institutions, at What Level?

Despite large uncertainties in the precise nature and levels of risks, the description above of the scope of the health and environmental problems, globally and on a smaller scale in Kenya and Costa Rica, shows that there is by no means an effective governance system for pesticides in developing countries. As evidenced by the risk-reduction measures described above, considerable efforts of governance at different levels are not only insufficient but are also characterized by fragmentation and incoherence. One way to identify the reasons for this situation as well as potential options for change is to focus on institutions—here defined as formal and informal rules of human interactions—and their role in governance at various levels.[62] Institutions are at the core of governance: They influence who has access to what information, shape the incentives for various courses of action, and affect who has the capacity to act. In a multilevel governance context, the following question arises: Is it possible to identify some criteria to determine the types of institutions that should preferably be established, enforced, or changed at particular levels? Some suggestions for elements of such criteria can be found in theories of the management of collectively owned natural resources. To identify current gaps in the governance system, it is useful to explore two sets of criteria. The first relate to the potential effectiveness of institutions if they were to match up the level of effects, driving forces and capacity; the second set relates to how the possibility to change institutions may vary across levels. Tables 5a and b illustrate these criteria.

## Matching Institutions

It is particularly instructive to determine criteria that can address how well institutions and governance "match" the level where most negative effects occur, the driving forces behind the problems originate, and where there is capacity to take action.

- *Matching effects.* If institutions are not in place at the level to correspond with, or "fit" the geographical scope of the negative effects, their effectiveness can be limited.[63] The potential for persistent organic pollutants (POPs)—which include a number of organochlorine pesticides—to act as transboundary pollutants made

# Table 5b Changing Institutions

| Types of Institution | Example | Proneness to Change |
|---|---|---|
| Operational | Guidelines for safe use | • Relatively easy to change with few resources and within a short time span.<br>• Difficult to implement/enforce, particularly at global scale. |
| Collective-choice | Regulations prescribing which pesticides are banned at national and global levels | • Moderately difficult to change with large variances in time and resources required. |
| Constitutional-choice | Favored type of agricultural system (including system of pest management) | • Very difficult to change, requiring significant political will and involving multiple sectors. |

Source: S. I. Karlsson, 2004.

countries like Canada and Sweden push for an international agreement to ban them. The development of the Stockholm Convention can thus be seen as an effort to match the global scope of the problem with global governance. The efforts to establish a harmonized registration system for pesticides in Central America can be seen in a similar light. Banning a pesticide in Costa Rica while it is allowed in a neighboring country invites smuggling, black market sales, and potential for cross-border pollution via rivers.[64] However, pesticides also exert very local and context-dependent effects on health and environment—which by aggregation can be seen as global problems—and these are not at all well matched with institutions. Consider, for example, the common use of paraquat in both Meru and Naranjo despite its well-known (on the international level) intoxication record, and the use of category la and lb (extremely and highly hazardous, respectively) pesticides by small farmers in Naranjo (see Table 2).

**The banana trade is a huge market in Costa Rica for pesticides. Unfortunately, data show that banana workers are more likely to suffer pesticide intoxications than other workers.**

• *Matching driving forces.* If institutions are not sufficiently well established and implemented at the level where the driving forces for the problems originate, the governance measures at other levels will merely target symptoms. Moreover, from an ethical perspective, it would be preferable if those who are explicitly responsible for the problems would be more often targeted in governance.[65] There are layers of direct and indirect driving forces for pesticide problems and they differ depending on what risk factor is in focus. The strongest driving forces for using pesticides emerge at the global level, where the agrochemical industry, along with governments, promote incentives for modern high-input agriculture, and at the national level, where agricultural policy, research, and extension advice and marketing strategies of the agrochemical companies create similar

incentives. However, these are the drivers least addressed in pesticide governance.

The strongest driving forces that determine the types of pesticides used are also located at the global and national levels. At the global level, multinational corporations develop and choose which of their products to market in developing countries. The Stockholm and Rotterdam Conventions target specific pesticides to regulate or ban; if implemented, these rules can reduce risks in developing countries. However, the number of substances included is very small, making this process a weak match of institutions and driving forces. At the national level, the governments in Kenya and Costa Rica control which pesticides may be used in their respective countries—although this also depends on which products the pesticide industry chooses to market. Overall, developing countries have no influence on pesticide development: Most research in this regard centers around pests and crops in the temperate regions.[66] And while Kenya and Costa Rica have the authority to ban pesticides, the overwhelming majority are approved.[67] In terms of how pesticides are actually used, there are a number of driving forces, including images presented in marketing campaigns. Ultimately, however, individual farmers or workers are the ones who determine how they are used. This is where the major mismatch between institutions and driving forces lies. Even if many efforts are made at national and higher levels to make farmers use pesticides more safely, such measures have a long way to go before they actually reach individuals who spray—or to a sufficient degree effect a change in their behavior.

• *Matching capacity.* If there is no capacity to act at the level where effects or driving forces originate, then there is not much one can expect in terms of institution building.[68] Stakeholders at a governance level where there is capacity to act may not necessarily have contributed to the problem but could take on responsibility for governance because of a sense of concern and moral obligation.[69] The whole focus at the global level to assist developing countries with scientific and technical information on pesticide risk is largely due to the fact that IGOs have the capacity to assess those risks, while many developing countries have neither the expertise nor resources needed. The governments in Kenya and Costa Rica focus on deciding which products are allowed for use because they have the capacity to make those decisions, while local actors are considered not to have such capacity. Moreover, for the most part, farmers do not have the capacity to adopt risk-reduction strategies—either because they are not aware of risks or because they believe it is impossible to farm without pesticides and have no alternative pest-control strategies to adopt. Those stakeholders who do have the capacity to influence this—such as the pesticide industry and the international community—have not fully utilized their resources to this end.[70]

## Changing Institutions

Another key set of criteria relates to the cost (in monetary or human-resource terms, for example) required to create a new institutional arrangement or to enforce or change an existing one. Three types of institutions are considered here: operational, collective-choice, and constitutional-choice institutions.[71]

- *Operational institutions.* These institutions provide structures for making day-to-day decisions in a wide diversity of operational situations, which means a large number of individual actors are involved.[72] In a local context, it is assumed that these are the institutions that can be most quickly changed. However, this may not always be the case on the global scale, where the sheer number and diversity of multiple localities make the picture more complex. When pesticides continue to be the primary pest-management tool, and risk reduction follows the mode strategy of ensuring safe use, an effective approach to effect change must include a strong focus on operational institutions at the local level. A global code of conduct is not enough nor are national laws that make unsafe use illegal. Any institutions established at higher levels have to be implemented by a large number of stakeholders: farmers and workers—female and male alike; their families; and all others who handle pesticides throughout their life cycle. Even if changes can be initiated quickly, they are costly and time consuming to implement on a large scale.

---

**Government officials in Costa Rica recognized herbicides had negative impacts, including erosion. Many local officials later discouraged their use.**

---

- *Collective-choice institutions.* This category describes those institutions that indirectly affect the options for operational rules.[73] The number of stakeholders involved in designing these institutions are fewer but, depending on the governance level, can range from a single local

NGO to all the member states of the United Nations. The cost of changing such institutions, and the time it takes to do so, will vary accordingly. When risk-reduction efforts target the types of pesticides used, the focus is on collective-choice institutions at the national level. Here, decisions on which pesticides farmers will have access to involve only very few individuals—and in Costa Rica, for example, the pesticide registration process takes 6–12 months.[74] Implementation of these governmental decisions involves a smaller number of stakeholders, such as customs officers and pesticide retailers, who are charged with ensuring that the unwanted pesticides do not reach the farmers. This strategy also involves measures at the global level, through, for example, sharing toxicity information about certain chemicals or banning specific substances. Such global processes can take a very long time: For instance, the process to establish the Stockholm Convention began in 1997 and will not have entered into force until mid-May 2004. Change to collective-choice institutions can thus in some cases be made at relatively low cost within a short time frame, but in other cases the process is lengthy and cumbersome.

- *Constitutional-choice institutions.* These determine the specific institutions that create the collective-choice institutions.[75] These types of institutions may involve the smallest number of the most powerful and knowledgeable stakeholders when crafting formal institutions. Or, when institutions are part of deeply rooted structures in society, the number of stakeholders involved may be innumerable and not easy to pinpoint. These types of institutions are usually the slowest to change. The risk-reduction strategy to reduce or eliminate the use of pesticides requires changes in constitutional-choice institutions. These are institutions that favor one type of agricultural system: one dependent on pesticides, one less dependent on pesticides, or one completely independent of pesticides. Such institutions consist of consumer demands, national and international market and trade structures, government policies, and farmer attitudes (for example). The inter-linkages between sectors, the resistance from prevailing power structures, and the number of decisionmakers involved all present a considerable and time-consuming challenge for anyone attempting to change these institutions.

## Correcting Mismatches and Facilitating Change

In very general terms, it appears that institutions must be assigned primarily to those levels where driving forces originate and where stakeholders have the capacity to establish and enforce institutions. Furthermore, if rapid institutional change is desired, the focus should be on enforcing changes in operational institutions at local levels—although this may not be the most cost-effective or long-lasting governance strategy. If slower—but likely more enduring—change is to be achieved it is the constitutional-choice institutions that need to be targeted

—not only on global levels but across national and local levels as well. The more pragmatic and manageable approach in time and resources is to target collective-choice institutions.

## Many agricultural workers, unaware of the risks or lacking adequate resources, do not always wear protective gear. Effective education programs could change this.

In more specific terms, to address the mismatch between institutions and effects, institutions need to be established that enable developing countries and the international community to incorporate the concerns for local health and environmental effects from pesticides in their specific climatic, ecological, economic, and social context. Currently, developing countries rely on global institutions and knowledge when establishing their own institutions.[76] There is a need to create institutions that facilitate the collection of data locally and nationally—for instance, those that prescribe the monitoring of health and environmental effects after registration is approved.

To address the mismatch between driving forces and institutions, there are specific needs for each risk-reduction strategy. Institutions at national and global levels are needed that discourage the use of pesticides and provide alternative, economically viable farming strategies. Some national and international NGOs are currently involved in small-scale initiatives promoting organic farming, but these are very limited in scope and hard to upscale. The situation is better for IPM initiatives. Such higher-level institutions could include changing or developing new agronomist education programs, reducing hidden pesticide subsidies, increasing the market channels for organic products, or even influencing consumer demand for these products through education campaigns that raise the general public's awareness. Institutions are also needed at the global level to support those countries that can neither create institutions nor enforce them—a situation that results in a substantial black market of smuggled and substandard pesticides—to prevent the most toxic products from entering their countries.[77] Only global-level phase-outs can, in such cases, keep the most toxic types of pesticides out of the hands of smallholder farmers. This implies that the criteria for the type of products to be banned globally would need to be expanded to include those that are known to produce locally occurring severe health and environmental effects. Finally, to strengthen safer modes of pesticide handling, existing higher-level institutions need to radically upscale or fundamentally change their implementation efforts. A local culture of safe pesticide use can only emerge with long-term, more effective education efforts that reach all groups who come in contact with pesticides. To address the mismatch between capacity and institutions, one step is to look at the pesticide industry—a stakeholder that is a strong driving force and that holds considerable capacity to effect change. They could be charged with greater responsibility to become a stronger player in risk reduction. An example of how this could be implemented is illustrated with the cases in Kenya and Costa Rica: Pesticide companies now

pay for field tests of the efficacy of their products on the crops in both countries—the governments require this for registration.[78] National institutions could hold these companies accountable to contribute resources toward national data-gathering programs examining the impacts of their products in each respective country. The international community—the only entity that can regulate global production and trade—could also take on a larger share of responsibility by creating more encompassing institutions, for example by phasing out more substances globally (see above) and making such institutions more effective with stronger mechanisms of monitoring and enforcement. Combining the two sets of criteria puts in focus the collective-choice institutions targeted at eliminating the pesticides that pose the highest risks. Changing these institutions would take considerable time and effort, particularly in building up a better knowledge base and banning substances on a global scale. Nevertheless, it would be the most accessible "fast track" to reduce risks pending a global-scale mustering of resources for widespread implementation of safe use and changes in agricultural systems.

## Conclusions

The final choice of strategies for risk reduction, institution building, and change will be heavily dependent on answers to such questions as

- What are the major contributing factors to risks?
- What are the inherent toxic properties of pesticides?
- What are the exposure patterns under conditions of use?
- What is the acceptable level of risk? and
- Who should be responsible to address the risks?

Some of these questions could be resolved with more monitoring and research, others are more value laden, and these kinds of issues divide stakeholder groups substantially. Stakeholder views range from NGO movements that consider all pesticides inherently toxic to many in the pesticide industry who assert that all pesticides—as long as used as prescribed—are safe. The former are convinced that the conditions in developing countries—especially considering the human and financial input necessary to effect change—make it impossible to change the operational institutions and ensure safe use. The latter fear that changing the constitutional-choice institutions—striving to establish IPM or organic farming on a global scale, for example—would endanger food and economic security. There is no easy way to resolve the debate between the widely diverging knowledge bases and value judgments that underlie these different views. However, some of the recommendations above can help clarify and structure the available options for governance and the institution building and change that they would require.

In addition, the pesticide case and the policy-relevant conclusions drawn from it have much to teach us in other policy areas. Pesticides are one of the first groups of toxic chemicals introduced on a large scale in developing countries, and the complexities involved with their use illustrate many of the challenges and possibilities for the management of other groups of chemicals in these regions. As one of the environmental issues that exhibits a number of local-global linkages, the impacts of pesticide use illuminate directions that future research and policy discussions need to take. Governance needs to be analyzed and addressed with a much more holistic approach, viewing the efforts at all levels—local, national, regional, and global—as elements of one system of governance. Only then can we evaluate how individual policies operate in the context of a large, interconnected system. Only then can research start identifying the most important elements of establishing multi-layered governance, with a nested hierarchy of mutually supportive policies and institutions initiated at all governance levels.[79]

# Notes

1. The term "governance" has emerged as one of the most-used concepts when discussing measures taken in society to address a particular issue—specifically when stressing that there are many more actors than just governments involved. It is particularly useful for the global level where there is no world government but still a lot of governance. The Commission on Global Governance defined governance as "the sum of the many ways individuals and institutions, public and private, manage their common affairs." See Commission on Global Governance, *Our Global Neighbourhood* (Oxford, UK: Oxford University Press), 2.

2. This figure is for 2001, a year in which the market suffered a 7.4 percent decline. See Phillipps McDougall (2002) quoted in CropLife International, *Facts and Figures,* accessed via http://www.gcpf.org on 21 January 2004. In 1999, the market was more than US$30 billion. See "World Agrochemical Market Held Back by Currency Factors," *Agrow,* 11 June 1999, 19–20.

3. See, for example, World Health Organization (WHO), *Public Health Impact of Pesticides Used in Agriculture* (Geneva: WHO, 1990); P. N. Viswanathan and V. Misra, "Occupational and Environmental Toxicological Problems of Developing Countries," *Journal of Environmental Management* 28 (1989): 381–86; L. A. Thrupp, "Exporting Risk Analysis to Developing Countries," *Global Pesticide Campaigner* 4, no. 1 (1994): 3–5; Health Council of the Netherlands, *Risks of Dangerous Substances Exported to Developing Countries* (Den Haag: Health Council of the Netherlands, 1992); and C. Wesseling, R. McConnell, T. Partanen, and C. Hogstedt, "Agricultural Pesticide Use in Developing Countries: Health Effects and Research Needs," *International Journal of Health Services* 27, no. 2 (1997): 273–308. For a special analysis of the impact on women, see, M. Jacobs and B. Dinham, eds., *Silent Invaders: Pesticides, Livelihoods and Women's Health* (New York: Zed Books, 2003).

4. Food and Agriculture Organization of the United Nations (FAO), *Analysis of Government Responses to the Second Questionnaire on the State of Implementation of the International Code of Conduct on the Distribution and Use of Pesticides* (Rome: FAO, 1996). For example, highly toxic insecticides is the main pesticide category in use in many less developed countries. Wesseling, McConnell, Partanen, and Hogstedt, note 3 above, page 276.

5. WHO, note 3 above, pages 85–86. This estimate is calculated by using a 6:1 ratio between nonhospitalized (unreported) and hospitalized (reported) cases.

6. WHO, note 3 above, page 86. The background for this is that pesticides in rural areas are among the most accessible types of toxic substances. Because these data are based on hospital registers, they probably overestimate the proportion of suicides. Wesseling, McConnell, Partanen, and Hogstedt note 3 above, page 283.

7. The figures from the WHO 1990 report have been strongly challenged by the pesticide industry. Anonymous official, International Programme on Chemical Safety (IPCS), interview by author, Geneva, 25 June 1998. IPCS has a project to support developing countries to gather data on pesticide intoxications more systematically. IPCS, *Pesticide Project: Collection of Human Case Data on Exposure to Pesticides* (Geneva), accessed via http://www.intox.org/pagesource/intox%20area/other/pesticid.htm on 22 January 2004. However, the project has not yet resulted in new global estimates. In the first stage, studies were carried out in India, Indonesia, Myanmar, Nepal, and Thailand, based on hospital records (some results of these are available at http://www.nihsgo.jp/GINC/meeting/7th/profile.html), but the results were not satisfactory. In a second phase, they will use community-based studies in pilot countries. Dr. Nida Besbelli, IPCS, e-mail message to author, 10 February 2004. While there are some developing countries where intoxications have to be reported to the authorities, overall there is limited data on pesticide health impacts in developing countries. Wesseling, McConnell, Partanen, and Hogstedt, note 3 above, page 284.

8. The lack of data provides a significant obstacle for global estimates of the number of people suffering from chronic effects. WHO, note 3, page 87. The few studies that have been done have demonstrated neurotoxic, reproductive, and dermatologic effects. Wesseling, McConnell, Partanen, and Hogstedt, note 3, page 273.

9. See, for example, P. Bourdeau, J. A. Haines, W. Klein and C. R. K. Murti, eds., *Ecotoxicology and Climate With Special Reference to Hot and Cold Climates* (Chichester, UK: John Wiley and Sons Ltd., 1989); and T. E. Lacher and M. I. Goldstein, "Tropical Ecotoxicology: Status and Needs," *Environmental Toxicology and Chemistry* 16, no. 1 (1997): 100–11.

10. For example, 89 percent of the 60 developing countries who responded to an FAO questionnaire in 1993 reported that they are not studying the effects of pesticides on the environment. FAO, note 4, page 61. Research in disciplines that are essential for detecting and understanding environmental degradation—such as biology, ecology, and ecotoxicology—is very limited in sub-tropical and tropical regions compared to that in nontropical latitudes. Bourdeau, Haines, Klein and Murti, note 9 above; and Lacher and Goldstein, note 9 above. This situation reflects a general knowledge divide in the environmental field between developed and developing countries, which in turn reflects the generic divide in resources (human and financial) available for monitoring and research. S. Karlsson, "The North-South Knowledge Divide: Consequences for Global Environmental Governance," in D. C. Esty and M. H. Ivanova, eds., *Global Environmental Governance: Options & Opportunities* (New Haven, CT: Yale School of Forestry & Environmental Studies, 2002), 53–76. It is estimated that about 5 percent of the world's scientific production comes from developing countries. International Development Research Centre, "The Global Research Agenda: A South-North Perspective," *Interdisciplinary Science Reviews* 16, no. 4 (1991): 337–4. The number of scientists/engineers per million inhabitants in developed countries is 2,800 on average; in developing countries it is 200. T. H. I. Serageldin, "The Social-Natural Science Gap in Educating for Sustainable Development," in T. H. I. Serageldin, J. Martin-Brown, G. López Ospina, and J. Dalmatian, eds., *Organizing Knowledge for Environmentally Sustainable Development* (Washington, DC: The World Bank, 1998).

11. See B. Dinham, *The Pesticide Trail: The Impact of Trade Controls on Reducing Pesticide Hazards in Developing Countries* (London: The Pesticide Trust, 1995), which summarizes case

studies from a number of countries. One of the few larger studies on environmental impact is the Locustox project studying the impact from large-scale locust sprayings in Africa. See J. W. Everts, D. Mbaye, and O. Barry, *Environmental Side-Effects of Locust and Grasshopper Control, Volume I and II,* (Senegal: FAO and the Plant Protection Directorate, Ministry of Agriculture 1997, 1998). As part of that study ponds were treated with deltamethrin (a synthetic pyrethroid) and bendiocarb (a carbamate). Delthamethrin had considerable acute effects on most macroinvertebrates, and bendiocarb affected a number of zooplankton.

12. This does not mean that organochlorines have not caused environmental effects in the tropics. For some examples, see F. Bro-Rasmussen, "Contamination by Persistent Chemicals in Food Chain and Human Health," *The Science of the Total Environment* 188 Suppl. 1 (1996): S45–60.

13. The proposed process of long-range transport in the atmosphere of certain organic compounds is called "global distillation," or "global fractionation." F. Wania and D. Mackay, "Global Fractionation and Cold Condensation of Low Volatility Organochlorine Compounds in Polar Regions" *Ambio 22,* no.1 (1993): 10–18. The theory has received support by modeling and monitoring data. H. W. Vallack et al, "Controlling Persistent Organic Pollutants—What Next?" *Environmental Toxicology and Pharmacology* 6 (1998): 143–75; and S. N. Meijer, W. A. Ockenden, E. Steinnes, H. P. Corrigan, and K. C. Jones, "Spatial and Temporal Trends of POPs in Norwegian and UK Background Air: Implications for Global Cycling;" *Environmental Science and Technology* 37, no. 3 (2003): 454–61. Atmospheric deposition is considered to be the major source of POPs in the Arctic. A. Godduhn and L. K. Duffy, "Multi-generation Health Risks of Persistent Organic Pollution in the Far North: Use of the Precautionary Approach in the Stockholm Convention," *Environmental Science & Policy* 6 (2003): 341–53. There is thus a growing scientific consensus for the global distillation/fractionation hypothesis. Vallack et al, this note. This has also been reflected in policy where the Stockholm Convention includes in its screening criteria for adding further substances to the convention the potential for long-range transport through air, for example (one of the criteria for such substances is that their half-life in air must be greater than two days). Stockholm Convention on Persistent Organic Pollutants (POPs): Texts and Annexes (Geneva: Interim Secretariat for the Stockholm Convention on Persistent Organic Pollutants, 2001), 46–47.

14. An alternative concept for what has been referred to here as governance levels is levels of social organization. O. Young, *Institutional Dimensions of Global Environmental Change (IDGEC) Science Plan* (Bonn: International Human Dimensions Programme on Global Environmental Change, 1999).

15. Several methods were combined to solicit the perspectives and approaches of these stakeholders, including semistructured interviews and policy document analyses. A total number of 204 interviews were carried out during 8.5 months of fieldwork in the years 1997–1999. For further details on methodology, theoretical framework, and results of the study that are associated with this article, see S. Karlsson, *Multilayered Governance. Pesticides in the South: Environmental Concerns in a Globalised World* (Linköping, Sweden: Linköping University, 2000).

16. In Costa Rica, the average quantity of imported formulated pesticides rose from 8,100 tons to 15,300 tons between the second half of the 1980s and the first half of the 1990s. A. C. Rodríguez, R. van der Haar, D. Antich, and C. Jarquín, *Desarollo e Implementacion de un Sistema de Vigilancia de Intoxicaciones*

*con Plaguicidas, Experenica en Costa Rica, Informe Tecnico Proyecto Plagsalud Costa Rica, Fase 1* (Development and Implementation of a Pesticide Intoxication Monitoring System, Costa Rica's Experience, Technical Report, Plagsalud Costa Rica Project, Phase I), (San José, Costa Rica: Ministerio de Salud Departamento de Sustancias Toxicas y Medicina del Trabajo (Ministry of Health, Department of Toxic Substances and Occupational Medicine), 1997). The nominal value of pesticide imports rose approximately 50 percent between 1990 and 1994. In 1994, the value of pesticide imports reached US$84.2 million and in the same year on average more than US$170 were spent on pesticides per ha agricultural land. S. Agne, *Economic Analysis of Crop Protection Policy in Costa Rica, Publication Series No. 4* (Hannover, Germany: Pesticide Policy Project, 1996): 6, 12. Because of the substantive formulating industry, the trade figures on the value of pesticide purchases are gross underestimates of the amount spent on pesticides in Costa Rican agriculture. Agne, this note, page 12. Statistics are limited on the use of pesticides in the African continent. J. J. Ondieki, "The Current State of Pesticide Management in Sub-Saharan Africa" *The Science of the Total Environment* 188, Suppl. no. I (1996): S30–34; and S. Williamson, *Pesticide Provision in Liberalised Africa: Out of Control? Network Paper No. 126* (London: Overseas Development Institute Agricultural Research & Extension Network, 2003). As many as 47 percent of African countries responding to an FAO questionnaire do not collect any statistics on pesticide import and use. FAO, note 4, page 71. One study reported that in Kenya the average annual import of pesticides for 1989–1993 was just over 5,000 tons at a value of US$28 million. H. Partow, *Pesticide Use and Management in Kenya* (Geneva: Institut Universitaire D'Etudes du Développement (Graduate Institute of Development Studies), 1995): 205–6. The data reported for Kenya and Costa Rica only goes to the mid-1990s. (See note 2 on falling pesticide sales in 2001.) The increasing trend is not taking place in all developing countries. Furthermore, studies in Africa have shown significant variation in impacts of, for example, liberalization on pesticide prices, access, and use. See A. W. Shepherd and S. Farfoli, "Export Crop Liberalisation in Africa: A Review," *FAO Agricultural Services Bulletin No. 135* (Rome: FAO, 1999); and Williamson, this note.

17. Agne, ibid., page 8. One source listed the main groups of pesticides used in the Central American countries: the insecticides organophosphates, carbamates, and pyrethroids; fungicides, mainly dithiocarbamics; and the herbicides phenoxyacids, dipyridyls, and more recently, triazines. L. E. Castillo, E. de la Cruz, and C. Rupert, "Ecotoxicology and Pesticides in Tropical Aquatic Ecosystems of Central America," *Environmental Toxicology and Chemistry* 16, no. 1 (1997): 41–51.

18. Partow, note 16 above, page vii. Inorganic pesticides accounted for 21 percent of imported pesticides, organophosphates accounted for 15 percent, organochlorines accounted for 11 percent, thiocarbamates accounted for 7 percent, and phtalimides accounted for 7 percent. These figures do not distinguish between products imported as technical grade or formulated product. For example, around 25 percent of the organophosphates are imported as technical grade material. Partow, note 16 above, pages vii, 39.

19. Agne, note 16 above and C. Conejo, R. Díaz, E. Furst, E. Gitli, and L. Vargas, *Comercio y Medio Ambiente: El Caso de Costa Rica* (Trade and the Environment: The Case of Costa Pica) (San José, Costa Rica: Centro Internacional en Política Económica Para el Desarollo Sostenible (International Center of Political Economy for Sustainable Development), 1996). The banana

sector uses 45 percent of all pesticides and at the end of the 1970s the disease commonly known in Spanish as *sigatoka negra* (the scientific name is *Mycosphaerella fijiensis*), arrived in the country, which affected production severely and led to significant pesticide use.

20. Castillo, de la Cruz, and Rupert note 17. Permanent crops like coffee, oilpalm, and cacao are sprayed less often (1–5 times/year) compared to annual crops such as tobacco, potatoes, or vegetables and products like banana, melon, watermelon, or flowers—extreme cases of which can be sprayed up to 39 times per cycle. J. E. García, *Introducción a los Plaguicidas* (Introduction to Pesticides) (San José, Costa Rica: Editorial Universidad Estatal a Distancia (Publishing Trust of the State University for Distance Education), 1997).

21. For a long time, coffee was the main export earner in agriculture, but tea has taken over the lead position. Since 1985, horticulture has grown substantially. In 1990 it came in third place as an export earner. General Agreement on Tariffs and Trade (GATT, *Trade Policy Review Kenya Volume 1* (Geneva: GATT, 1994). Between 1991 and 1994 the value of horticultural exports increased by 113 percent, the principle horticultural crops being cut flowers, French beans, mangoes, avocados, pineapples, and Asian vegetables. Republic of Kenya, *National Development Plan 1997–2001* (Nairobi: Government Printer, 1996).

22. While most pesticides are used on export crops, studies have shown that in some areas where the coffee economy has introduced a modernized agricultural system—for example, by using pesticides, their use on food crops has increased. A. Goldman, "Tradition and Change in Postharvest Pest Management in Kenya" *Agriculture and Human Values* 8, no. 1–2 (1991): 91–113. This is a phenomenon found in other countries as well. See, for example, Williamson, note 16.

23. Karlsson, note 15, pages 209–13 and 238–40.

24. In Naranjo, coffee farmers apply fungicides on average three times a year, nematicides a maximum of once per year and herbicides once or twice a year. Karlsson, note 15, page 262. About 7 percent of all pesticide purchases in Costa Rica are used on the 20 percent of agricultural land that is under coffee production. See Agne, note 16, page 13. In Meru, on the other hand, those who still spray their coffee apply fungicides either 2–4 times or 8–12 times a year. Herbicides are sprayed occasionally there. Karlsson, note 15, page 262. It is important to note that the comparisons here are only made from the self-reported numbers of sprayings per year. The dose in each application is not addressed.

25. This result is based on careful analysis of a large number of interviews with stakeholders and study of policy documents. Thus, the summary conclusions cannot be attributed to a single source. Karlsson, note 15. The problem categorization is not clear-cut since several of the areas are closely interrelated. The environment serves as a medium for transport of pesticides and their metabolites, which may expose humans to these substances via air and water (for example) and potentially affect human health. Pesticides, as a trade issue, emerge in the regulations established to address the concern of long-term, low-level exposure of pesticide residues in food. Pesticides that exert effects on non-target organisms—those on the farm and surrounding environment—can disrupt populations of natural enemies of the original pest, leading to increased and different pest attacks and thus production problems.

26. In addition to evidence from the interviews (see Karlsson, note 15, pages 144–45), references to environmental considerations were absent in pesticide management plans. Partow, note 16, page xiii. A study from the early 1990s on the legislation and institutional framework for environmental protection and natural resource management in Kenya concluded that there was no stress on environmental problems in the agricultural areas. S. H. Bragdon, *Kenya's Legal and Institutional Structure for Environmental Protection and Natural Resource Management—An Analysis and Agenda for the Future* (Washington, DC: Economic Development Institute of the World Bank, 1992). However, the Kenya National Environment Action Plan urges the adoption of as many nonchemical measures as possible and it urges the use of the least toxic chemicals as a last resort. Ministry of Environment and Natural Resources, *The Kenya National Environment Action Plan* (NEAP) (Nairobi, 1994).

27. For example, in Kenya, stricter European Union pesticide residue limits on agricultural products had caused significant concern both in government and in the pesticide industry. Standing Committee on the Use of Pesticides, *Interim Report of the Standing Committee on the Use of Pesticides* (Nairobi, 1996); and "Editorial," *Newsletter for the Pesticide Chemicals Association of Kenya*, February 1995. This was also one reason the existence of 100 tons of pesticide wastes (including many from banned substances) raised concern that these may find their way on to horticultural produce. Standing Committee on the Use of Pesticides, this note. It should be emphasized that the trade issue is very sensitive for national governments because of the high economic stakes involved, and they are likely to be very reluctant to discuss possible problems openly.

28. Republic of Kenya, "Chemical Safety Aspects in Kenya" (A country paper presented by the Kenyan delegation during IPCS Intensive Briefing Session on Toxic Chemicals, Environment and Health for Developing Countries, held in Arusha, United Republic of Tanzania, 1997).

29. These figures are quoted in M. A. Mwanthi and V. N. Kimani, "Patterns of Agrochemical Handling and Community Response in Central Kenya," *Journal of Environmental Health* 55, no. 7 (1993): 11–16. This figure—7 percent of the agricultural population suffering poisonings annually—lies within the range between 2 and 9 percent reported in various studies in developing countries. Wesseling, McConnell, Partanen, and Hogstedt, note 3, page 283.

30. V. W. Kimani, "Studies of Exposure to Pesticides in Kibirigwi Irrigation Scheme, Kirinyaga District" (Submitted Ph.D. thesis, Department of Crop Science, University of Nairobi, 1996).

31. Ondieki, note 16, page S32.

32. Kimani, note 30 above, page 23.

33. A study in the 1980s of organochlorine residues in domestic fowl eggs in Central Kenya showed levels of dichlorodiphenyltrichloroethane (DDT) and dieldrin especially high, exceeding the acceptable daily intake (ADI) for children. (This ADI standard was developed by a panel of experts linked to WHO as part of the FAO/WHO Joint Meeting on Pesticide Residues (JMPR). Residues of dieldrin exceeded ADI for adults. J. M. Mugambi, L. Kanja, T. E. Maitho, J. U. Skaare, and P. Lökken, "Organochlorine Pesticide Residues in Domestic Fowl (*Gallus domesticus*) Eggs from Central Kenya," *Journal of the Science of Food and Agriculture* 48, no. 2 (1989): 165–76. Another study found organochlorines in mothers' milk exceeding the ADI for infants; except for lindane the exposure occurred long ago. Kimani note 30 above, pages 224–25. Because these substances have been banned, the levels will decline.

34. Regulation No. 20345-S in E. Wo-Ching Sancho and R. Castro Córdoba, *Compendio de Legislacion Sobre Plaguicidas* (Compendium of Pesticide Legislation) (San José, Costa Rica: Organizacion Panamericana de la Salud Proyecto Plag-Salud, Centro de Derecho Ambiental y de los Recursos Naturales (CEDARENA) (Pan American Health Organization Plag-Salud Project, Center for Environmental Rights and Natural Resources), 1996). A project was started in 1993 to improve reporting and help to prevent intoxications. R. Castro Córdoba, N. Morera González, and C. Jarquín Nuñez, *Sistema de Vigilancia Epidemiologica de Intoxicaciones con Plaguicidas, la Experiencia de Costa Rica, 1994–1996* (Epidemiological Monitoring System of Pesticide Intoxications, Costa Rica's Experience, 1994–1996) (San José, Costa Rica: Departamento de Registro y Control de Sustancias y Medicina del Trabajo del Ministerio de Salud (Ministry of Health Department of Registration and Control of Occupational Substances and Medicine), 1998).

35. Rodriguez, van der Haar, Antich, and Jarquín, note 16.

36. Ministerio de Salud, División de Saneamiento Ambiental, Departamento de Registro y Control de Sustancias Tóxicas y Medicina del Trabajo (Ministry of Health, Division of Environmental Health, Department of Registration and Control of Toxic Substances and Occupational Medicine), *Reporte Oficial intoxicaciones con Plaguicidas 1996* (Official Report of Pesticide Intoxications 1996) (Costa Rica, 1997).

37. Castro Córdoba, Morera González, and Jarquín Núñez, note 34 above. A study of the percentage of underreporting of intoxications in the system of monitoring gave an underreporting of 43 percent of symptoms that should have been linked to pesticides. Most of these were either dermal lesions or intoxications occurring outside the working environment, Rodríguez, van der Haar, Antich, and Jarquín, note 16, page 19.

38. Ministerio de Salud, note 36 above, page 13.

39. In the 1980s, even workers younger than 15 had very high occupational incidence rates. Today such instances are less frequent due to better enforcement of legislation that prohibits children under 18 to work with pesticides. Catharina Wesseling, Instituto Regional de Estudios en Sustancias Tóxicas (Central American Institute for Studies on Toxic Substances), e-mail message to author, 22 March 2004.

40. C. Wesseling, *Health Effects From Pesticide Use in Costa Rica—An Epidemiological Approach* (Stockholm: Institute of Environmental Medicine, Karolinska Institute, 1997).

41. C. Wesseling, L. Castillo, and C. F. Elinder, "Pesticide Poisoning in Costa Rica," *Scandinavian Journal of Work and Environmental Health* 19 (1993): 227–35. The data on hospitalizations and occupational accidents included a significant number of cases where the pesticide substance had not been identified.

42. Wesseling, note 40 above, page 42

43. Wesseling, note 40 above, page 50.

44. Wesseling, note 40 above, page 50

45. For example, a study of organochlorine pesticides along the Kenyan coast only revealed very low levels compared to other areas, including tropical areas, J. M. Everaarts, E. M. van Weerlee, C. V. Fischerm, and T. J. Hillebrand, "Polychlorinated Byphenyls and Cyclic Pesticides in Sediments and Macroinvertebrates from the Coastal Zone and Continental Slope of Kenya," *Marine Pollution Bulletin* 36, no. 6 (1998): 492–500.

In a study of the concentration in water of organochlorine pesticides in coffee- and tea-growing areas that was made in 1994–1995, the mean pesticide levels did not exceed WHO or U.S. Environmental Protection Agency limits for drinking water. M. A. Mwanthi, "Occurrence of Three Pesticides in Community Water Supplies, Kenya" *Bulletin of Environmental Contamination and Toxicology* 60, no. 4 (1998): 601–8. However, the Development Plan of 1989–1993 reported that agrochemicals had led to severe pollution effects in the Tana and Athi Rivers regimes. B. D. Ogolla, "Environmental Management Policy and Law" *Environmental Policy and Law* 22, no. 3 (1992): 164–75. There had also been a case of paraquat contamination in the water supply for one town (Anonymous official, Ministry of Land Reclamation, Regional Development and Water, interview by author, Nairobi, Kenya, 15 October 1997).

46. L. Corrales and A. Salas, *Diagnóstico Ambiental de la Actividad Bananera en Sarapiquí Tortuguero y Talamanca, Costa Rica 1990–1992 (Con Actualizaciones Parciales a 1996)* (Environmental Assessment of Banana Cultivation in Sarapiquí, Tortuguero, and Talamanca, Costa Rica 1990–1992 (with Partial Updates until 1996) (San José, Costa Rica: Oficina Regional para Mesoamérica, Unión Mundial para la Naturaleza (Regional Office for Mesoamerica, International Union for the Conservation of Nature (IUCN)), 1997). The environmental pollution reported included: high concentrations of heavy metals in coral reef along the coast that could partly be due to pesticides; detection of DDT residues in some fish species in the late 1980s; detection of hexachlorobenzene (HCB), dieldrin, DDT, 1, 1-dichloro-2,2-bis(p-chlorophenyl)ethylene (DDE), paraquat and lindane in soil; detection of organochlorines and organophosphates in rivers and along the coast of the Atlantic; residues of various pesticides in sediments primarily chlorothalonil; the detection of chlorothalonil in ground water around banana plantations.

47. Karlsson, note 15, page 264. The study was not a quantitative survey with the rigidity that is implied in random farmer selection. The more than 30 farmers who were interviewed in each district were selected with purposeful sampling, with the goal of seeking the broadest range of views rather than averages. It is thus not possible to give figures on the number of experienced intoxications and pesticide products used (for example), but qualitative judgments on differences between districts can be made.

48. When looking at how individuals and organizations address pesticide-associated problems, the study focuses on those measures that aim to reduce primarily the health and environmental risks. Evidently, such measures can be of relevance for reducing trade, production, and economic problems, and conversely measures taken to address these may have positive effects on health and environment as well. Karlsson, note 15.

49. In addition to the Rotterdam and Stockholm Conventions described in Table 4, there is, for example, the Basel Convention on the Control of Transboundary Movements of Hazardous Waste and Their Disposal (www.basel.int) and the ILO Convention on Safety in the Use of Chemicals (www.ilo.org/public/english/protection/safework/cis/products/safetytm/c170.htm). Furthermore, the Codex Alimentarius (linked to WHO and FAO) establishes recommended maximum residue limits (MRLs) of pesticides in traded food products, but with the Agreement of Sanitary and Phytosanitary Measures these standards have indirectly become legally binding on the member countries of the WTO. Karlsson, note 15, page 96.

50. The FAO Code was first adopted by the FAO General Assembly in 1985 and has since been amended twice, in 1989 and 2002. FAO, International Code of Conduct on the Distribution and Use of Pesticides, Revised Version (Rome: FAO, 2003), accessed via http://www.fao.org/ag/agp/agpp/pesticid/ on 19 January 2004.

51. Examples of guidelines include FAO, *Guidelines for Legislation on the Control of Pesticides* (Rome: FAO, 1989); and FAO, *Revised Guidelines on Environmental Criteria for the Registration of Pesticides* (Rome: FAO, 1989).

52. Support for agricultural pesticide use has been a commonplace element in development projects, supported by various multi- and bilateral donors, but data on these are often missing from government import data. Williamson, note 16, page 3. While this practice is now less common, it is still taking place. Williamson, note 16, page 11. The IPM concept had been present in IGO discussions for many decades. Originally it was focused on controlling pest populations through a combination of all suitable techniques below thresholds that would cause economic damage. FAO, *International Code of Conduct on the Distribution and Use of Pesticides, Amended Version,* (Rome: FAO, 1990). In the 1990s, however, the IPM concept came to be understood as a means to minimize the use of pesticides and increase reliance on alternative pest management technologies. See "Agenda 21," Rio de Janeiro, 2002, in Report of the United Nations Conference on Environment and Development, Rio de Janeiro, 3–14 June 1992, Volume I Resolutions Adopted by the Conference, A/CONF.151/26.Rev.1.

53. For example, it took until 1998 for FAO to have a first meeting with the International Federation of Organic Agriculture Movements (IFOAM), whose member organizations around the world are involved in research on alternatives to pesticides and in the design and manufacture of technology for controlling weeds, Karlsson, note 15, pages 107–8.

54. Organizations and IGOs left out of Table 4, but which were part of the study, include the International Labour Organization (ILO), United Nations Institute for Training and Research (UNITAR), World Health Organization (WHO), and the World Trade Organization (WTO).

55. Kenya's Pest Control Products Act came into force in 1984 and is often referred to as one of the most comprehensive laws in Africa. Republic of Kenya, The Pest Control Products Act Chapter 346 (Nairobi: Government Printer, 1985). In many respects it conforms to the guidelines for pesticide legislation from FAO. In 1997, 241 pesticide products had been registered and the rest were being screened. Republic of Kenya, note 28. The first effort to regulate pesticides in Costa Rica was made in 1954 and a modern pesticide registration process was established in 1976. R. Castro Córdoba, *Estudio Diagnostico Sobre la Legislación de Plaguicidas en Costa Rica* (Diagnostic Study of Pesticide Legislation in Costa Rica) (San José, Costa Rica: CEDARENA, 1995). The most recent revision was made in 1995 when law No. 24337MAG-S was published. Wo-Ching Sancho and Castro Córdoba, note 34 above. In 1993, there were 1,213 pesticides registered in Costa Rica, 347 generics, and 38 mixtures. Garcia, note 20, page 235.

56. As a basis for the government agency's decision, the companies need to submit data that supports the efficacy of the product to control pests on the crops it was intended to be used as well as a long list of physical, chemical, toxicological and ecotoxicological data. Pest Control Products Board (PCPB), *Data Requirements for Registration of Pest Control Products, Legal Notice N46* (Nairobi, 1994); and Ministerio de Salud (Ministry of Health), *Pesticide Registration in Costa Rica* (Costa Rica, n.d.).

57. Kenya has banned the following pesticides: in 1986, dibromochloropropane, ethylene dibromide, 2,4,5 Trichlorophenoxyacetic acid (2,4,5-T), chlordimeform, hexachlorocychlohexane (HCH), chlordane, heptachlor, endrin, toxaphene; in 1988, parathion; and in 1989, captafol. In addition, in 1986, lindane was restricted use for seed dressing only; aldrin and diledrin were restricted for termite control in the building industry and DDT was restricted for use in public health. PCPB, Banned/Restricted Pesticides in Kenya (Nairobi, n.d.). Costa Rica has banned the following pesticides: in 1987, 2,4,5-T; in 1988, alchin, captafol, chlordecone, chlordimeform, DDT, dibromochloropropane (DBCP), dinoseb, ethylendibromide (EDB), nitrofen, toxafen; in 1990, lead arsenate, cyhexatin, endrin, pentachlorophenol; in 1991, chlordane, heptachlor; and in 1995, lindane. Ministerio de Salud, *Lista de Plaguicidas Prohibidos y Restringidos en Costa Rica* (List of Banned and Restricted Pesticides in Costa Rica) (Costa Rica, n.d).

58. In Kenya, the pesticide industry, through the Global Crop Protection Federation (GCPF), cooperated with the government's extension system in a safe-use project, and between 1991 and 1993, about 280,000 people were trained, including 2,800 retailers. Anonymous official, GCPF-Kenya, interview by author, Nairobi, Kenya, 23 September 1997. In Costa Rica, cooperation between the government and the pesticide industry association has taken place throughout the 1990s, reaching 110,000 people. Anonymous official, Cámara Insumos Agropecuarios (Chamber of Agricultural and Livestock Inputs), interview by author, San Jose, Costa Rica, 4 February 1998. In 1991, the industry association and two ministries initiated the project "Teach," which was geared at training teachers in the rural areas so that they can teach children about pesticide issues. Conejo, Díaz, Furst, Gitli, and Vargas, note 19. Despite these efforts, there were a number of stakeholders in both countries who were concerned about the low effectiveness of the training measures. Karlsson, note 15. In Kenya, for example, results in the form of increased understanding of the toxic effects of pesticides were noted, but less than 30 percent of the farmers trained adopted the safety measures prescribed. Kimani, note 30, page 49. The Kenya Safe Use Project was also criticized for neglecting to train pesticide workers in the plantation sector. Partow, note 16, page xiv.

59. The Pest Control Products (Labeling, Advertising and Packaging) Regulation 3 (2) (n) requires each pesticide label to state that it is against the law to "use or store pest control products under unsafe conditions." Republic of Kenya, note 55 above. In Costa Rica, the banana sector is regulated by a special law (No. 7147), which obliges employers to train workers in appropriate use of pesticides and their associated risks. Castro Córdoba, note 55 above, page 21.

60. The Kenyan government has been involved in research on IPM and there were some pilot projects (often supported by donors). P. C. Matteson and M. I. Meltzer, *Environmental and Economic Implications of Agricultural Trade and Promotion Policies in Kenya: Pest and Pesticide Management* (Arlington, VA: Winrock International Environmental Alliance, 1995); O. Zethner, "Practice of Integrated Pest Management in Sub-Tropical Africa: An

Overview of Two Decades (1970–1990)," in A. N. Mengech, K. N. Saxena and H. N. B. Gopalan, eds., *Integrated Pest Management in the Tropics, Current Status and Future Prospects* (New York: John Wiley & Sons, 1995); and Organisation for Economic Co-operation and Development (OECD), *Report of the OECD/FAO Workshop on Integrated Pest Management and Pesticide Risk Reduction, Neuchâtel, Switzerland, 28 June—2 July 1998. ENV/JM/MONO (99)* 7 (Paris: OECD, 1999). In Costa Rica, the official extension service promotes IPM. Ague, note 16, page 27. Several programs have been introduced in the country to reduce the volume of pesticide use. A. Faber, *Study on Investigations on Pesticides and the Search for Alternatives in Costa Rica* (Guápiles, Costa Rica: Wageningen Agricultural University, 1997). However, these initiatives remained isolated in a dominating agricultural system where pesticide use has become considered essential to increase productivity in most crops Organización Panamericana de la Salud, Programa Medío Ambiente y Salud en el Istmo CentroAmericano (Pan American Health Organization, Program of Environment and Health in the Central American Isthmus), *Aspectos Ocupacionales y Ambientales de la Exposición a Plaguicidas en el Istmo Centroamericano PLAGSALUD-Fase II* (Environmental and Occupational Aspects of Pesticide Exposure in the Central American Isthmus PLAGSALUD-Phase II), PLG97ESP.POR (San José, Costa Rica: 1997).

61. In Kenya, the Kenya Institute of Organic Farming (KIOF) began in 1986, and in the first years they met resistance from the government. Anonymous official, KIOF, interview by author, Nairobi, Kenya, 26 September 1997. KIOF trains farmers' groups in the field and arrange exchange visits among groups. KIOF, *Organic Farming, A Sustainable Method of Agriculture* (Nairobi, n.d.). Organic farming has been through a period of significant growth in Costa Rica with estimates of about 3,000 ha being under organic agriculture. Faber, ibid., page 14. There was high demand for organic products in the export market but virtually no demand in the domestic market. Karlsson, note 15, page 180. The Costa Rican government supported organic farming, for instance, it created a law for organic agriculture included in the environmental law in 1995, and it established a special office in the Ministry of Agriculture. The Asociación Nacional de Agricultura Organica (ANAO) (National Association for Organic Agriculture) had received funds to establish a nationally based organic certification system in the second half of the 1990s. Karlsson, note 15, page 181.

62. This definition of institutions is often used in social science and economics and differs somewhat from the common-language use that defines institutions as organizations. For a discussion on the role of institutions in environmental governance, see, for example, Young, note 14.

63. The issue of fit is extensively explored in relation to common property resource management. See, for example, R. J. Oakerson, "Analyzing the Commons: A Framework," in D. W. Bromley, ed., *Making the Commons Work: Theory, Practice and Policy* (San Francisco: ICS Press, 1992); E. Ostrom, "Designing Complexity to Govern Complexity," in S. Hanna and M. Munasinghe, eds., *Property Rights and the Environment, Social and Ecological Issues* (Washington, DC: Beijer International Institute of Ecological Economics and the World Bank, 1995); M. McGinnis and E. Ostrom "Design Principles for Local and Global Commons," in O. R. Young, ed., *The International Political Economy and International Institutions Volume II* (Cheltenham, UK: Edward Elgar Publishing Ltd., 1996). For a general discussion of the concept of fit, see C. Folke, L. Jr. Pritchard, F. Berkes, J. Coiling and

U. Svedin, *The Problem of Fit between Ecosystems and Institutions* (Bonn, Germany: IHDP, 1998) accessed via www.uni-bonn.de/ihdp/wp02main.htm, on 27 June 2000.

64. Anonymous official, CEDARENA, interview by author, San José, Costa Rica, 10 March 1998. There have been efforts to harmonize pesticide registration in Central America under the umbrella of Organismo Internacional Regional de Sanidad Agropecuaria (OIRSA) (International Regional Organization for Plant and Animal Health) with some support from FAO, but at the time of the study (1998/1999) efforts seemed to have halted. See Karlsson, note 15, page 199.

65. This strategy is not straightforward, however, as there are usually layers of driving forces and responsible stakeholders, often at different levels. It can be difficult to identify the original causes (or ultimate drivers), and this confuses the allocation of responsibility between levels, J. Saurin, "Global Environmental Degradations, Modernity and Environmental Knowledge," in C. Thomas, ed., *Rio Unravelling the Consequences* (Essex, UK: Frank Cass, 1994).

66. Karlsson, note 15, page 87.

67. Karlsson, note 15, pages 156–60 and 184–89.

68. It has been argued that it is difficult to be held responsible for a problem if one does not have the means to respond to it. T. Princen, "From Property Regime to International Regime: An Ecosystems Perspective," *Global Governance* 4, no. 4 (1998): 395–413.

69. For a discussion on altruistic motivations for behavior, see, for example, J. J. Mansbridge, ed. *Beyond Self-Interest* (London: University of Chicago Press, 1990).

70. In an FAO questionnaire, 46 percent of the responding developing countries felt that the pesticide industry acted only partly responsibly or not responsibly in adhering to the provisions of the FAO Code of Conduct as a standard for the manufacture, distribution, and advertising of pesticides. FAO, note 4, page 9.

71. L. L. Kiser and E. Ostrom, "The Three Worlds of Action: A Metatheoretical Synthesis of Institutional Approaches," in E. Ostrom, ed., *Strategies of Political Inquiry* (London: Sage Publications, 1982); and E. Ostrom, *Governing the Commons, The Evolution of Institutions for Collective Action* (New York: Cambridge University Press, 1990). While these authors refer to them as rules, they fall within the definition of institutions and that term is used here for clarity.

72. C. C. Gibson, E. Ostrom, and T. K. Ahn "The Concept of Scale and the Human Dimensions of Global Change: A Survey," *Ecological Economics* 32, no. 2 (2000): 217–39.

73. Ostrom, note 71, page 52.

74. Karlsson, note 15, page 185.

75. Ostrom, note 71 above, page 52.

76. Karlsson, note 15 above; and S. Karlsson, "Institutionalized Knowledge Challenges in Pesticide Governance—The End of Knowledge and Beginning of Values in Governing Globalized Environmental Issues," *International Environmental Agreements* (forthcoming in 2004).

77. Williamsson, note 16.

78. In Kenya, the efficacy tests were made under some kind of cost-sharing arrangement with the government, while in Costa Rica the companies had to cover the full costs. Karlsson, note 15, pages 157, 186.

79. The term "multilayered governance" was constructed and defined as a system of coordinated and collective governance across levels in Karlsson, note 15, page 40. See also P. Hirst and G. Thompson, *Globalization in Question* (Cambridge, UK: Polity Press, 1996), 184, who argue for more 'sutured' governance across levels when discussing economic aspects of globalization, and Young, note 14, page 34, who argues for the need to influence behavior at all levels of governance.

SYLVIA I. KARLSSON is a postdoctoral fellow at Yale University's Center for Environmental Law and Policy and is a research fellow with the Institutional Dimensions of Global Environmental Change (IDGEC) project. Her current research focuses on cross-level aspects of global sustainable development governance. In 2001–2003 she worked as International Science Project Coordinator at the International Human Dimensions Programme on Global Environmental Change (IHDP) in Bonn, Germany. Sylvia I. Karlsson has worked for a short time at UNEP Chemicals in Geneva and the Economic Development Institute of the World Bank and as program officer for an action research project in Eastern Africa. Parallel to her studies and research, she has been actively engaged in the NGO processes of the Rio Conference in 1992, the World Summit for Social Development in 1995, and most recently, the World Summit on Sustainable Development, where she headed the delegation of the International Environment Forum, a scientific NGO accredited to the summit. The author wishes to thank Dr. Arthur L. Dahl and Ms. Agneta SundénByléhn and several anonymous reviewers for their helpful comments on earlier drafts. She further wishes to express her gratitude to all the people around the world who patiently answered questions during interviews—from the UN halls of Geneva to the soft grass of a Kenyan *shamba* (farm). The research was made possible through a grant from the Swedish International Development Agency (Sida). Karlsson's work has appeared in the peer-reviewed journals *International Environmental Agreements* and *The Common Property Resource Digest* and in a book edited by D. C. Esty and M. Ivanova, *Strengthening Global Environmental Governance: Options and Opportunities* (New Haven, CT: Yale School of Forestry & Environmental Studies, 2002). Karlsson can be contacted via e-mail at sylvia.karlsson@yale.edu.

From *Environment,* May 2004, pp. 24–41. Reprinted by permission of the Helen Dwight Reid Educational Foundation. Published by Heldref Publications, 1319 Eighteenth St., NW, Washington, DC 20036-1802. Copyright © 2004. www.heldref.org

# Smog Alert

## *The Challenges of Battling Ozone Pollution*

MARK BERNSTEIN AND DAVID WHITMAN

At 5 A.M. on 6 November 1958, Nathan Louis Gordon died of a heart attack at his home in Los Angeles. The death of the 73-year old former sales clerk would never have made the news—except that Gordon's physician, Dr. Peter Veger, was convinced there was more to his patient's sudden death than just a bum heart. Five years earlier, Gordon had developed atherosclerotic heart disease after living in Los Angeles for three decades. Dr. Veger believed that air pollution had inflamed Gordon's heart condition and contributed to his fatal heart attack. In what may be a first in the history of American epidemiology, Veger cited "Los Angeles Smog" as a contributing cause of death on Gordon's death certificate.[1]

At the time, the idea that air pollution kills was not new. For example, several thousand people had died in the notorious "Black Fog" of London in December 1952, and the noxious smog that beset Donora, Pennsylvania, in October 1948 killed 20 people and sickened thousands more.[2] However, these and other fatal air pollution episodes had been associated with cold-weather inversions that visibly trapped sulfurous soot particles from coal combustion. By contrast, Los Angeles' ozone-based smog, also known as photochemical smog, was a new and different breed of air pollution.

In 1952, biochemist Arie J. Haagen-Smit discovered that the primary ingredient in Los Angeles smog was ground-level ozone.[3] Unlike soot, ozone is not emitted directly from foundry smokestacks, tailpipes, or fireplaces. Instead, ozone is created in the atmosphere by the ultraviolet rays of the sun, which trigger a photochemical reaction between nitrogen oxides ($NO_x$), a combustion byproduct, and volatile organic compounds (VOCs), a series of highly reactive hydrocarbons that readily evaporate from solvents and other carbon-containing products. Primary sources of $NO_x$ are stationary point sources like power plants and industrial facilities and mobile sources such as cars, trucks, and construction equipment. Primary sources of VOCs include such mobile sources, as well as "non-point" stationary sources such as paints.

In the decades that followed Dr. Veger's controversial post-mortem diagnosis, the evidence has grown that summer spells of elevated ozone pollution do in fact contribute to the premature death of several thousand Americans each year.[4] At the same time, the threat of ground-level ozone pollution had increased the public awareness of pollution problems.[5] Ozone pollution is so ubiquitous that the terms "ozone" and "smog" are now often used interchangeably—although smog technically is more chemically complex than ground-level ozone alone.[6]

## The Ozone Paradox

How has the nation fared in its decades-long battle against smog? The good news is that ozone concentrations have declined substantially since the passage of the 1970 Clean Air Act, particularly in Los Angeles, the country's smog capital.[7] But the little-appreciated bad news is that progress in reducing ozone pollution has virtually halted in the last decade—and in some respects has shown the beginnings of a U-turn, with ozone pollution on the rise. In fact, from 1993 to 2002, 8-hour ozone levels edged up 4 percent nationwide (see Figure 1 below).[8]

The picture is slightly more favorable if an earlier reference year, 1990, is used as a basis of comparison: Between 1990

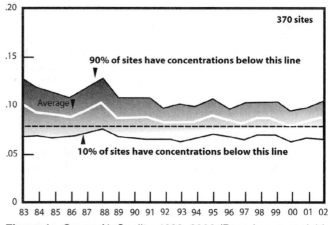

**Figure 1** Ozone Air Quality, 1983–2002 (Based on annual 4th maximum 8-hour average).

Note: 1983–2002: 14 percent decrease; 1993–2002: 4 percent increase.

Source: U.S. Environmental Protection Agency, *National Air Quality and Emissions Trends Report,* 2003 Special Studies Edition, EPA 454/R-03-005 (Washington, DC, 2003), http://www.epa.gov/airtrends/aqtrnd03/images/fig-2-23.gif.

and 2003, 8-hour ozone levels dipped 9 percent nationwide. Still, since 1990, ground-level ozone pollution has worsened in parts of the East and Midwest after adjusting for meteorological changes,[9] and it is not clear whether the modest improvements evident in other regions are due partly to changes in local weather conditions, such as surface temperature and wind speed. Across the country, numerous fast-growing cities are struggling to halt rising ozone levels. Even in rural areas, once-pristine national parks are befouled with ozone pollution. The federal government's air monitoring of 32 national parks and federal lands outside of urban areas shows that not a single one of the rural locations has had a significant reduction in ozone pollution since 1990. Six parks, moreover, suffered statistically significant increases in ozone pollution.[10] Areas that now exceed the 8-hour ozone air quality standard include crown gems of the park system such as the Great Smoky Mountains, Yosemite, Acadia, Rocky Mountain, Sequoia/Kings Canyon, and Shenandoah national parks.

The persistence of ground-level ozone pollution is particularly perplexing because emissions of the two ozone precursors—

# Ozone Creation

While chemists still do not fully understand the complicated process of ozone formation in many applied settings, they do know that ozone formation involves two important classes of chemical ingredients—nitrogen oxides ($NO_x$) and volatile organic compounds (VOCs)—and that ozone production increases under some weather and geographic conditions. In the troposphere, or lower atmosphere, and in the presence of sunlight, nitrogen dioxide ($NO_2$) separates or dissociates into nitrous oxide (NO) and an oxygen atom (O). The oxygen atom then combines with oxygen gas ($O_2$) to form ozone ($O_3$). What can happen next is that the NO can react again with the ozone molecule ($O_3$) to reform $NO_2$ and $O_2$. This relationship of creating and dissociating ozone would continue with little consequence except where the reaction that breaks up the ozone is blocked. VOCs serve to block the dissociation of the ozone molecules.

Thus, in tandem, $NO_x$ and VOCs appear to be responsible for the ground-level ozone problem. But it is not that simple—the concentration of ozone in the troposphere is not related linearly to emissions of $NO_x$ and VOCs alone but is rather a complex function of additional factors including warm temperatures, the presence of sunlight, and geography that traps these pollutants in such a way to promote mixing in an area, sometimes over several days. It is even possible under certain conditions to reduce one component ($NO_x$ or VOCs) and increase ozone.

The California Air Resources Board (CARB) reports that in smog chamber experiments where the VOC/$NO_x$ ratio is less than 8 to 10, reducing $NO_x$ tends to increase ozone formation. This is called a VOC-limited regime. Conversely, when the VOC/$NO_x$ ratio is higher than 8 to 10, reducing $NO_x$ tends to decrease ozone formation and is called $NO_x$-limited).[1] This nonlinear relationship is illustrated in the top figure at right, which shows combinations of $NO_x$ and VOCs and the different levels of ozone produced by different combinations. Each of the curved lines (isopleths) represents a different concentration of ozone, which increases as the lines move up and to the right in the graph. As noted previously, many different factors affect the interactions of $NO_x$ and VOCs and the formation of ozone. For example, moving in the figure from point A, which is VOC-limited, to point B represents reduced $NO_x$ but increased ozone (all other factors remaining the same). On the other hand, at point C, in a $NO_x$-limited region, small changes in VOCs will have no impact on ozone.

The bottom figure below shows average positions for more than a dozen cities in which data is publicly available, providing a notional picture of which urban areas tend toward $NO_x$-limited regimes and which tend toward VOC-limited regimes. While the data used to generate the chart are from 1992, the relative positions of the cities have not likely changed much. The VOC/$NO_x$ ratio is a convenient shorthand to characterize whether cities tend to be VOC- or $NO_x$-limited; however the ratios can vary across space and time. For example, the ratio can change as one moves from west to east in the Los Angeles Basin so that $NO_x$ reductions could both increase and decrease ozone across the region.

[1]California Air Resources Board (CARB), *ARB Report on the Ozone Weekend Effect in California: Executive Summary* (Sacramento, CA, 2003), 8, http://www.arb.ca.gov/aqd/weekendeffect/arb-final/web-executive-summary.pdf.

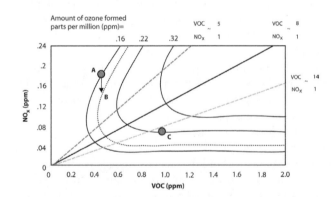

Notional relationship between $NO_x$ and VOCs for a few cities

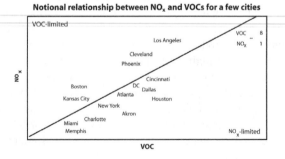

**Representation of the relationship between $NO_x$ and VOCs.**

Source: National Research Council, *Rethinking the Ozone Problem in Urban and Regional Air Pollution* (Washington, DC: National Academy Press, 1991).

$NO_X$ and VOCs—have been dropping in the United States and are tightly regulated by federal and state governments. Today, federal and state policymakers face what might be called "the ozone paradox": Emissions of the ingredients that lead to the formation of ground-level ozone are diminishing, yet the decline is not consistently producing comparable reductions in ozone pollution. This paradoxical pattern is unique to ozone: All other "criteria" air pollutants regulated by the U.S. Environmental Protection Agency (EPA)—particulates, lead, sulfur dioxide, nitrogen dioxide, and carbon monoxide—continued to decline throughout the 1990s, and nearly all have fallen by half or more since 1970.[11]

There are several explanations for the ozone paradox. In part, ozone regulation has been a victim of its success. Federal and state regulation has greatly reduced $NO_X$ and VOC emissions from the leading emission source, motor vehicles. At the same time, however, emissions have increased from other largely unregulated sources, such as off-road engines and equipment, as has transported pollution from other nations, including China and Mexico.

The distinctive and complex chemistry of ozone formation also hinders efforts to reduce smog. While anthropogenic emissions of VOCs are declining in the United States, natural or biogenic emissions of VOCs from trees and vegetative matter appear to be on the rise in a number of regions,[12] undermining the impact of the regulatory reductions. The mixing of $NO_X$ and VOC to form ozone ($O_3$) is also highly nonlinear. At certain $NO_X$-to-VOC ratios, lowering $NO_X$ concentrations can actually increase ozone pollution rather than reduce it (see box).

For policymakers, reversing the ozone U-turn may prove harder than imagined, particularly during the next decade. Since 1999, EPA has issued several major rules that will drastically reduce $NO_X$ emissions and, to a lesser extent, VOC emissions. The rules cover new diesel and new off-road engines (for instance, those used in tractor engines, backhoes, locomotives, and ships) and new off-road equipment (such as lawnmowers, chainsaws, and leaf blowers). Yet the existing inventory of off-road vehicles and equipment is enormous and durable and is expected to turn over slowly in coming decades. As a result, even aggressive regulation of new off-road equipment may reduce emissions of ozone precursors at a snail's pace. To take one example, EPA's 2004 far-reaching rule to cut emissions from off-road vehicles is not projected to start accruing its full benefits until 2030.[13]

International transport of ozone pollution could similarly complicate efforts to reduce smog. Not only has the impact of pollution from Asia and Mexico grown in recent decades, it is likely to worsen in the years ahead. China and Mexico could add more than a hundred million $NO_X$- and VOC-emitting cars and SUVs by 2030, and China is currently in the midst of vastly expanding its already abundant use of coal, the combustion of which produces $NO_X$.[14] And there is always the possibility of a weather wild card: Because photochemical smog flourishes in sunlight and dry, warm weather, there is a distinct possibility that in certain areas, climate change could exacerbate the threat of ozone pollution. At the very least, the history of battling ozone suggests that in the near term, even successful government

efforts to reduce $NO_X$ emissions may ironically aggravate ozone pollution in a number of cities.

## Public Health and Ozone Regulation

Ozone in the stratosphere, suspended 6 to 30 miles above the planet, beneficially shields the Earth from the harmful ultraviolet rays of the sun. By contrast, ground-level or tropospheric ozone is a respiratory tract irritant that can exacerbate asthma, decrease lung function, and inflame a variety of respiratory conditions. Sub-populations vulnerable to ozone pollution include children, the elderly, and individuals with respiratory conditions (see box).

While the morbidity (incidence of disease) of ground-level ozone is well established, the impact of ozone pollution on mortality was convincingly demonstrated only recently. For example, in a study published just last year, Michelle Bell, assistant

---

### Ozone and Children

Ground-level ozone is generally linked with respiratory problems in adults. Children, however, can be especially susceptible to the respiratory ailments associated with breathing ground-level ozone. In addition to inflaming asthma, bronchitis, and other breathing difficulties, unhealthy levels of ozone can impair lung function as well as lung development in children. In one study cited by the National Research Council, researchers estimated that three to five times more children had reduced lung function with short-term changes in ozone than when there were not changes in ground-level ozone.[1]

Large numbers of children are exposed to ozone pollution with significant health consequences. The American Lung Association estimates that 27 million children younger than 13 years of age are exposed to unhealthy levels of ozone and of those, almost 2 million have asthma.[2] In addition, the organization found that the highest percent of those children are minorities. Because minorities tend to have less access to quality healthcare, their ozone-related health problems often worsen, or they must seek care in emergency rooms—which can be expensive and time consuming. In Atlanta, for example, emergency department visits for asthma increase almost 40 percent when ozone levels are high.[3] Respiratory problems can also cause children to be absent from school. A study in Southern California showed that school absences for breathing difficulties increased over 60 percent following relatively modest increases in tropospheric ozone.[4]

[1]National Research Council, *Rethinking the Ozone Problem in Urban and Regional Air Pollution* (Washington, DC: National Academy Press, 1991).
[2]American Lung Association, "Children and Ozone Air Pollution Fact Sheet," September 2000.
[3]American Academy of Pediatrics, "Air Pollution: Children More at Risk than Adults," *Pediatrics* 114, no. 6 (2004).
[4]Ibid.

professor at the Yale School of Forestry and Environmental Studies, and her colleagues provided strong evidence of an association between mortality and short-term ozone exposure by tracking fatalities in 95 large U.S. cities from 1987 to 2000. Their analysis suggests that if summertime ozone spikes had raised ozone concentrations 10 parts per billion (ppb) daily, they would have contributed to nearly 3,800 premature deaths in 2000 in the 95 cities in the National Morbidity and Mortality Air Pollution Study (NMMAPS) Database.[15] Some 320 individuals in New York City alone would have died prematurely from the added ozone pollution. (For comparison, 287 individuals died in homicides in Manhattan and the Bronx in 2000.)

In the face of this and other evidence, public health authorities have come to recognize in the last decade that ground-level ozone has more serious impacts on mortality and morbidity than was once thought—and at lower exposure levels. In 1997, EPA changed the ozone air quality standard from the old 1-hour standard of 120 ppb to an 8-hour standard of 80 ppb—reflecting scientific evidence that ozone can cause health effects at lower ambient concentrations and is more dangerous over longer periods of exposure. Tightening the ozone standard has had the practical effect of increasing the official number of Americans exposed to unhealthy air, although peak ozone exposures are well below the levels of the 1970s and 1980s. Nearly 450 counties—home to 159 million people—fail the new 8-hour ozone standard.[16] In an effort to bring such counties into compliance with the new ozone standard, EPA is requiring most "nonattainment" areas to submit a state implementation plan (SIP) by April 2007 for reducing emissions of ozone precursors for approval. In previous decades, EPA and states often agreed on unrealistically optimistic SIP schedules for reducing ozone. The new ozone air quality standard provides longer deadlines for compliance, particularly in areas with the worst ozone pollution. "Severe" nonattainment areas (the Los Angeles air basin) will have 17 years, or until 2021, to attain the 8-hour ozone standard; "serious" violators, such as California's San Joaquin Valley, have until 2013 to clean up.

In addition to the 8-hour standard, EPA has also announced a far-reaching series of rules to reduce emissions of $NO_x$ and VOCs from diesel trucks and buses, off-road vehicles and equipment, and coal-fired power plants. But most of the rules will not achieve their full emission reductions for a decade or more. All told, EPA projects that by 2015 the new rules will cut $NO_x$ emissions by 7 million tons annually from 2001 levels (21.55 million tons)—a drop of about a third. The new air regulations are expected to cut VOC emissions by 3 million tons annually (from 17.1 million tons in 2001).[17] There will continue to be considerable uncertainty about how much reduction in ground-level ozone these regulations will achieve.

If EPA's new rules do in fact generate large reductions in ozone pollution, they could prove to be invaluable public health measures. A recent study indicates that attaining the 8-hour ozone standard would prevent about 840 deaths a year and almost one million school absences. The rules can also save about $5.7 billion a year in healthcare costs and premature deaths avoided.[18] Still, policymakers will not realize the full benefits of reducing

air pollution if they fail to navigate the potential pitfalls of the ozone regulatory paradox. The first of these stumbling blocks is achieving the sweeping reductions in $NO_x$ emissions called for by new EPA regulations.

## The Off-Road Riddle: Reducing $NO_x$ Emissions

One of the eventual triumphs of the 1970 Clean Air Act was the imposition of tailpipe controls on motor vehicles, the principal source of $NO_x$ emissions. Despite an enormous increase in cars and SUVs on the road since 1970, $NO_x$ emissions from highway vehicles have fallen 42 percent, from 12.62 million tons in 1970 to 7.38 million tons in 2003.[19] Yet at the time of the Clean Air Act, and in the two decades that followed, government officials essentially ignored the problem of $NO_x$ emissions from off-road vehicles and their potential contribution to ozone pollution. When the Clean Air Act was amended in 1990, Congress directed EPA to assess the impact of off-road engines on air quality and to regulate the equipment if it created violations. The subsequent EPA study showed that off-road vehicles and equipment were a surprisingly large source of $NO_x$ and particulate emissions. As it turned out, $NO_x$ emissions from off-road equipment had been rising rapidly at the same time that emissions from on-road vehicles were dropping: From 1970 to 2003, emissions from off-road sources increased almost 55 percent, from 2.65 million tons to 4.1 million tons. Collectively, off-road equipment now emits almost as much $NO_x$ as the nation's power plants.[20]

As part of its series of new regulations to reduce $NO_x$ emissions, in 2004 EPA issued its off-road diesel rule, which is projected to cut $NO_x$ emissions by 738,000 tons annually by 2030—the equivalent of cutting 700,000 diesel trucks from the fleet.[21] The groundbreaking regulation applies to the 650,000-plus pieces of off-road diesel equipment (such as tractors, backhoes, forklifts, and bulldozers) that are sold each year in the United States.

For all the promise of $NO_x$ reduction regulation, though, federal and state officials may find it considerably more difficult to ensure compliance for off-road equipment and diesel engines than for motor vehicles. As part of its rulemaking process, EPA conducts a regulatory impact analysis (RIA). The RIA for the off-road diesel rule anticipates that several cost increases will ultimately be passed on to consumers: EPA projects price bumps of approximately 20 percent, 7 percent, and 3 percent, respectively, for engines, fuel, and equipment through 2036.[22] But the RIA does not assess in detail how the price increases will affect manufacturer and consumer behavior. In the government's 30 years of regulating on-road engines, it has not achieved emission reductions nearly as smoothly or steadily as the RIA model suggests will be the case for off-road engines.

In several respects, motor vehicle regulation is dissimilar to off-road equipment control. Cars comprise a relatively homogenous vehicle fleet, making it easier to introduce technological breakthroughs like catalytic converters and electronic fuel injection. Enforcement of antipollution requirements in cars is also much simpler, because car owners are required to have valid registrations, which are tied to passing emission tests and smog-check inspections. Finally, EPA regulations apply only to new off-road equipment and diesel engines. However, consumers hold on to off-road vehicles and equipment for many more years than they typically keep a car. The American fascination with new car models and styles is not matched by an equally avid desire to acquire a new tractor or lawnmower. On average, car owners keep their vehicles for five to seven years, and often replace them while they are still in working order. Off-road vehicles and equipment are typically kept for longer periods of time.[23] This slow turnover in the inventory of off-road equipment—where old vehicles far outnumber the new—explains why the new EPA regulations only achieve their full cutbacks in $NO_x$ emissions in 2030.

Moreover, regulation of motor vehicles has increased the costs of new cars, yet this was an added expense that most consumers were ready to bear. The same cannot necessarily be said for off-road equipment, where the added purchase cost and maintenance expense for pollution-reduction technology could provide a perverse incentive for owners to hold onto their old "dirty" snowmobiles, motorbikes, and forklifts. After all, a lawnmower that dies in the front yard does not pose the same kind of danger as a family car that breaks down by the side of a busy highway.

## The Economy of Emissions Enforcement

Air quality enforcement that depends on emissions testing and vehicle registration is vastly more complicated with off-road equipment than with passenger vehicles. Consider the case of the Los Angeles air basin and the San Joaquin Valley, the two worst locales for ozone pollution in the nation. It is well known that a large number of off-road recreational vehicles, including all-terrain vehicles (ATVs), motorbikes, and snowmobiles, go unregistered in California.[24] But monitoring and enforcing off-road vehicle registration would likely entail a large expansion of the staffs of the California parks department, its department of fish and game, the U.S. Forest Service, and others. Harbor patrol would have to increase efforts to monitor emissions from boats. If California did register off-road vehicles, and assuming registration information was tracked by a division of the California Air Resources Board (CARB), what would regular emissions testing for a disparate range of off-road equipment look like? Emissions testing for a snowmobile or a stationary power generator would require different equipment than the dynamometer systems currently installed at motor vehicle smog-check stations. In much the same vein, a system of fines, penalties, and appeals for users of off-road equipment would prove administratively complex. What would happen, say, to a commercial gardener whose leaf blower failed an emissions test?

Even existing state and local $NO_x$ control initiatives are at best loosely enforced with the existing diesel fleet. CARB officials estimate that 60,000 heavy-duty trucks registered in the state and another 300,000 to 400,000 trucks that drive intermittently in California have "defeat devices"—computer software that allows trucks to save on fuel by bypassing smog controls at cruising speeds.[25] In the 1990s, California and EPA sued truck manufacturers over the installation of defeat devices, and in 1998, a group of manufacturers agreed to spend $1 billion as part of a retrofit program.

Since then, manufacturers have dawdled in installing the retrofits. In a briefing to state and regional air quality officials, EPA reported that only 7 percent of 1.3 million trucks had been retrofit.[26] In March 2004, CARB recognized the slow progress and negotiated a voluntary agreement to upgrade 35 percent of the state's registered trucks by November. After only 18 percent underwent the upgrade, an exasperated Air Resources Board ordered mandatory engine retrofits by the end of 2005. Diesel engine manufacturers are fighting the rule, but parts of it have been upheld in the State Superior Court. In other states, enforcement of anti-idling regulations for diesel trucks has been equally tepid.

To be fair, the sheer variety of $NO_x$-emitting off-road sources and off-road equipment will inevitably complicate enforcement of EPA's suite of off-road/diesel rules (see Figure 2). According to some estimates, there are more than 100 million pieces of nonfarm, non-construction equipment such as lawnmowers and chain saws.[27] Most off-road sources now in the beginning stages of government regulation—locomotives, marine diesel engines, airplanes, and snowmobiles—individually generate just a small portion of all $NO_x$ emissions. But collectively, off-road emissions can add up. Marine diesel engines generate about 8 percent of mobile $NO_x$ emissions nationwide, locomotives contribute 7 percent, and aircraft contribute about 1 percent.[28]

## VOCs: Seeing the Trees through the Forest

The complexity of VOC reduction is the second element of the ozone regulatory paradox. In 1980, Ronald Reagan set off an uproar when he claimed on the campaign trail that "[a]pproximately 80 percent of our air pollution stems from hydrocarbons released by vegetation. So let's not go overboard in setting and enforcing tough emission standards for manmade sources."[29] Reagan's insistence that trees cause pollution quickly became a punch line on late night talk shows and a shorthand confirmation that Reagan was an eco-nincompoop. Reagan's 80 percent figure was indeed wrong. But he was correct that trees and plants exacerbate air pollution by emitting VOCs—which interact with human-generated $NO_x$ emissions to form ozone.

The first scientific work connecting hydrocarbon emissions from trees and plants to smog was conducted in the 1960s by Oregon Graduate Institute of Science and Technology professor Reinhold Rasmussen, who wanted to explain the origins of the blue haze that lingered over Virginia's Blue Ridge Mountains.[30] Since then, research studies have confirmed that natural or biogenic emissions of VOCs are not only substantial but often exceed anthropogenic emissions.[31] This increase in biogenic VOC emissions mainly stems from a fortuitous ecological trend, reforestation, which is attributable mostly to natural conversion of agricultural lands back to woodlands. Contrary

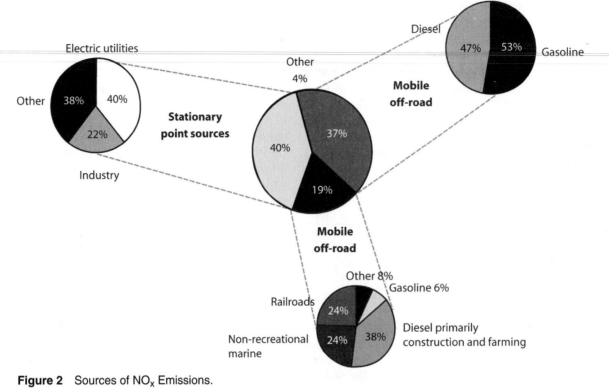

**Figure 2** Sources of NOₓ Emissions.

Source: U.S. Environmental Protection Agency National Emissions Inventory Database, http://www.epa.gov/air/data/neidb.html.

to conventional wisdom, the United States has more forest cover today than a decade ago, and the nation's forests now span about 750 million acres, 17 million more than in 1920.[32] The development of plantation forests has also played a part in fostering biogenic VOC emissions. In particular, southern plantation forests have a disproportionate number of high VOC-emitting trees, particularly loblolly pine and sweetgum.[33]

The policy implications of natural VOC emissions are sobering. In some regions of the country, reductions in anthropogenic VOC emissions can be cancelled out by biogenic increases—producing a rise in ozone pollution.[34]

Apart from the issue of biogenic emissions, VOC regulation is also complicated by the fact that VOC emissions stem from somewhat different sources than NOₓ emissions (see Figure 3 on next page). As is the case with NOₓ emissions, VOC emissions from motor vehicles plummeted 74 percent between 1970 and 2003, while off-road VOC emissions rose 37 percent. But unlike the case of NOₓ, nearly half of anthropogenic VOC emissions come from stationary non-point sources, chiefly solvents such as paints and paint thinners—which release more VOCs nationwide than even highway vehicles.[35]

Compared to cars, stationary non-point sources are more difficult to quantify and control. VOCs from solvents have only fallen about one-third nationwide since 1970, much less than the decline in VOC emissions from cars. In the San Joaquin Valley, California's agricultural heartland, VOC emissions from area non-point sources like livestock waste, belching cows, pesticides, and fertilizers may be about equal to VOC emissions from motor vehicles.[36]

## NOₓ versus VOC: The Weekend Effect

The third stumbling block to reducing ozone pollution is the conundrum of the weekend effect. This phenomenon runs directly counter to common sense: On weekends in a number of American cities, when less NOₓ and fewer VOCs are emitted from cars, diesel trucks, and industrial plants, ozone pollution nonetheless rises. In the Los Angeles area, for example, NOₓ concentrations are typically 40 percent lower on Sundays and VOC concentrations drop about 25 percent—yet ozone pollution jumps on average by 22 ppb, a 42 percent rise from Friday ozone levels.[37]

The weekend effect was first noted in Los Angeles in the mid-1970s.[38] But it is now being recognized in San Francisco, San Diego, Chicago, Philadelphia, Baltimore, Washington, and New York as well. There is no scientific consensus yet as to what causes the weekend effect or why it is evident in some large cities but not in others. In Philadelphia, for example, weekend ozone concentrations may rise due to transported pollution that blows in from other areas.[39] One theory postulates that diesel truck traffic diminishes on the weekend, reducing NOₓ emissions, which can increase ozone formation in VOC-limited regimes (see the box). Another idea is that on the weekend, peak driving hours are shifted, from morning and afternoon commutes to midday, when more ozone is created.[40] It might be that a combination of these two explanations is correct, making management of the weekend effect even more difficult.

Most EPA regulations slated to take effect over the next two decades concentrate on reducing NOₓ rather than VOCs. These

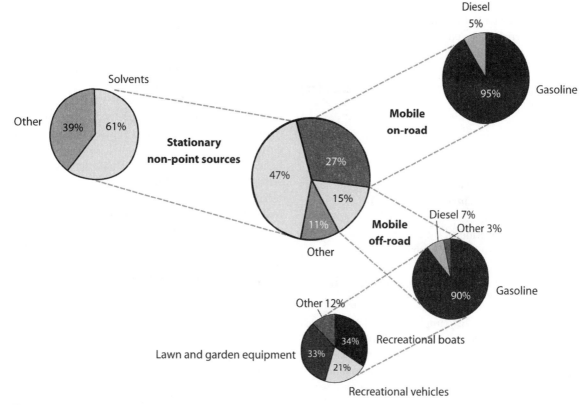

**Figure 3** Sources of VOC Emissions.

Source: U.S. Environmental Protection Agency National Emissions Inventory Database, http://www.epa.gov/air/data/neidb.html.

rules should curb air pollution in cities like Atlanta, where a drop in ozone concentrations corresponds with a drop in $NO_x$ emissions on weekends. But in VOC-limited cities like Los Angeles and Chicago, it is possible that substantial $NO_x$ reductions could worsen ozone pollution for a number of years. In effect, higher weekend ozone concentrations could become the norm during the week, as $NO_x$ emissions decline. Only after $NO_x$ reductions became very large—as high as 80 to 90 percent, according to some atmospheric modeling[41]—would ozone again diminish in a VOC-limited regime.

The weekend effect has led a number of researchers to argue that EPA and CARB should temporarily defer $NO_x$ reductions in favor of curbing VOC emissions in Los Angeles and the Bay Area. To date, however, neither agency is backing away from pursuing simultaneous $NO_x$ and VOC reductions.[42] Thus, the weekend effect remains an air pollution wild card, and identifying the appropriate approach to reduce ground-level ozone continues to pose complicated policy dilemmas.

## Overseas Ozone and Economic Growth

The fourth and final obstacle to reducing smog in the United States is ozone transported from other nations. At the time of the enactment of the 1970 Clean Air Act, scientists did not appreciate the significant contribution that transported ozone can make to U.S. air pollution. Although several studies in the mid-1970s suggested that winds were transporting acid-deposition

precursors (including $NO_x$) thousands of miles over national boundaries, it was not until the 1990s that scientists had enough information and data to track emissions reaching the United States.[43] Emissions in a number of nations can affect U.S. smog levels, but Asia appears to be the biggest foreign contributor, particularly on the West Coast during April and May. Some computer modeling suggests that emissions of ozone precursors from Asia contribute to ozone formation in the western United States[44] and that Asian air on average contributes 10 to 25 percent of the Northern Hemisphere's total springtime ground-level ozone.[45]

Meanwhile, economic growth is leading to emission increases, especially beyond America's borders. Unlike in the United States, anthropogenic emissions of ozone precursors from foreign nations are almost certain to rise in coming decades as developing countries follow a development path similar to Western nations. Economic growth in the developing world has led to large increases in the number of automobiles, trucks, and kilometers driven. These mobile sources are a ubiquitous component of emission "importation" from nations like China and Mexico. The 700 million-plus vehicles that existed worldwide in 1999 are expected to more than double by 2020. Similarly, Asian anthropogenic $NO_x$ emissions are projected to rise 50 to 100 percent from 2000 to 2020, according to some estimates.[46] Even if North America were able to eliminate all of its own anthropogenic $NO_x$ emissions, model simulations suggest that long-range transport of foreign emissions at present-day levels would maintain summer surface ozone levels of 20-40 ppb in the United States[47]—meaning that international sources could contribute a quarter to half of the emissions allowable under the new 8-hour ozone standard.

# Can Policymakers Reduce Ozone Pollution?

The decade-long stalemate in the campaign to reduce smog has three likely causes:

- the low-hanging fruit—reductions from automobiles and stationary sources—has already been picked;
- seemingly sensible policies that successfully reduce precursors can be undermined by the complex and sometimes counterintuitive chemistry of ground-level ozone; and
- emissions from outside the United States continue to rise.

For public officials, the policy implications of the ozone paradox seem clear: There is no one-size-fits-all policy solution for the problem of photochemical smog. States and localities, for example, may need additional flexibility and innovation to design State Implementation Plans. EPA officials have opted to reduce smog over the next decade chiefly by cutting $NO_x$ emissions from diesel engines, off-road equipment, and power plants. These are invaluable programs for cutting air pollution. But it is not clear that these policy prescriptions are the best strategy for every region. Both federal and state policies need to better reflect the reality of local ozone formation.

While regulatory flexibility is important, not every air basin has been as thoroughly studied as, say, that of Los Angeles or Atlanta. Thus, before EPA officials provide states with additional regulatory latitude, the federal government needs to provide technical resources and funding so that states and localities can more carefully document the changes in the $VOC/NO_x$ relationships of their airshed and model the role of transported pollution and other confounding factors. More research undoubtedly will be required in all of these areas. The need for better airshed modeling, in this case, is not simply an academic fancy but one that goes to the heart of state efforts to attain the new 8-hour ozone standard (see box).

Policy flexibility is also going to be needed to regulate off-road equipment and vehicles. Unexamined assumptions about consumer and manufacturer behavior could prove surprisingly consequential in reducing ozone pollution. Substantial tax incentives and rebates, for instance, may be necessary to encourage consumers to purchase new off-road engines. Meanwhile, current engine-scrapping programs are still far too modest to encourage many owners to remove older diesel engines from the in-use stock. Another option worth considering is the development of more after-market products to reduce emissions from existing engines.[48] EPA could thus provide or promote incentives for operators to clean up existing engines—potentially a cheaper alternative than replacing the equipment.

Federal and state officials could similarly expand the use of emission trading for off-road equipment. Emission trading has primarily been used to cut pollution at stationary point sources, but it may be possible to incorporate off-road equipment and non-point sources into trading regimes. It would therefore be valuable for EPA to explore how new information technologies, like those used in highway electronic toll devices, might help make this transformation feasible.

EPA and the U.S. Department of Energy—backed by state, local, and private sector partnerships—might also consider funding technology research and development of new motors, cleaner engines, and alternative fuels. The Department of Energy has already invested substantial research funds in these areas for on-road mobile sources and stationary sources, but it has devoted little seed money to basic research in emission reductions from off-road equipment.

Finally, policymakers need to look internationally for opportunities to reduce emissions. Whether it is emissions from Mexico or Asia, increasing economic prosperity will likely increase emissions. But there is an opportunity for countries to choose a different path, one that recognizes the growing costs associated with some of their pollution problems. To this end, the United States would be wise to work with foreign countries and invest in strategies that can reduce the growth in international emissions. Overseas emission control programs can help resolve two problems at once: reducing local and transported emissions. U.S. officials also have considerable experience reducing smog, which they can draw on in assisting developing nations. For example, the United States has the technology to collect data and understand the complex interactions among pollutants, both of which could be passed on to developing countries that do not.

The reduction of auto emissions since 1970 has justly been hailed as an environmental success story. Yet the history of motor vehicle regulation provides one last cautionary note in the ongoing battle against smog. Even successful regulation can take several decades to bear fruit—and almost inevitably there are surprises along the way. The government officials and analysts who modeled ozone scenarios in the 1970s did not foresee the popularity of SUVs and light trucks, much less the difficulties the new vehicles would create in meeting air quality standards.

Future ozone regulation is likely to engender some surprises of its own. If policymakers are fortunate, a technological breakthrough like the development of a cost-effective hydrogen fuel cell could obviate the need for large emission reductions. If they are not so lucky, an increased incidence of summer heat waves could raise ozone concentrations. While the numbers are necessarily speculative, one group of climate scientists recently projected that global warming would lead to a 4.5 percent increase in summertime ozone deaths by the 2050s in the New York metropolitan region.[49] The ozone paradox and the fight against smog may persist for many years to come.

# Notes

1. South Coast Air Quality Management District (SCAQMD), *The Southland's War on Smog* (Diamond Bar, CA, 1997), 21. The photograph and text of Nathan Louis Gordon's death certificate

## The Data Shortfall: Lessons from Houston

Modeling ozone production is notoriously difficult. The chemistry of tropospheric ozone is complex to begin with, and it is no simple task to develop a local inventory of both anthropogenic and biogenic emissions of nitrogen oxides ($NO_x$) and volatile organic compounds (VOCs). In addition to calculating these emission "inputs," modelers must also estimate the impact of meteorological conditions and transported ozone from other areas on local ozone formation. The experience of officials in Houston, Texas, provides a telling case study.

Until the late 1990s, Houston opted for VOC-only reductions, reflecting modeling that indicated $NO_x$ reductions would not help the city attain the 1-hour ozone standard. Yet Houston continued to fail to attain the 1-hour ozone standard and, in the late 1990s, shifted to a $NO_x$ reduction strategy. Houston still did not meet the standards, and in 2000, Texas and the federal government funded a comprehensive air quality study that enabled area officials to document the conditions of ozone production in the region more precisely.

Only after the 2000 air quality study was complete did officials discover that Houston's ozone problem was unique. Houston did indeed need to cut $NO_x$ as well as VOC emissions. But to be effective, the city would need to slash a particular subcategory of VOC emissions.

Not all VOCs, it turns out, are created equal. Some are more reactive than others and generate ozone far more efficiently. These highly reactive VOCs are emitted in abundance by the petrochemical industry, and previous emission inventories had underestimated the contributions of these emissions to Houston's ozone problem.[1]

---

[1]C. D. Forswall and K. E. Higgins, "Clean Air Implementation in Houston: An Historical Perspective, 1970–2005," working paper, Environmental and Energy Systems Institute, Shell Center for Sustainability, Rice University (Houston, TX, 2004), 24–25, http//www.ruf.rice.edu/~eesi/scs/SIP.pdf.

3. A. J. Haagen-Smit, "Chemistry and Physiology of Los Angeles Smog," *Industrial Engineering Chemistry* 44 (1952): 1342–46.

4. M. L. Bell et al., "Ozone and Short-term Mortality in 95 US Urban Communities, 1987–2000," *Journal of the American Medical Association* 292, no. 19 (2004): 2372–78; and B. J. Hubbell et al., "Health-Related Benefits of Attaining the 8-Hr Ozone Standard," *Environmental Health Perspectives* 113, no. 1 (2005): 73–82.

5. P. O'Driscoll, "Checking Ozone Levels Becoming Routine for Many," *USA Today,* 20 August 2001.

6. Smog is primarily made up of ground-level ozone but includes gases and particulate matter as well.

7. U.S. Environmental Protection Agency (EPA), *The Ozone Report: Measuring Progress Through 2003,* EPA 454/K-04-001 (Washington, DC, 2004), 16, http://www.epa.gov/air/airtrends/pdfs/2003ozonereport.pdf.

8. EPA, *Latest Findings on National Air Quality: 2002 Status and Trends,* EPA 454/K-03-001 (Washington, DC: 2003), 3, 8, http://www.epa.gov/airtrends/2002_airtrends_final.pdf.

9. Ibid.

10. EPA, note 6 above.

11. EPA, note 6 above.

12. D. W. Purves et al., "Human-Induced Changes in U.S. Biogenic Volatile Organic Compound Emissions: Evidence from Long-Term Forest Inventory Data," *Global Change Biology* 10 (2004): 1737–55.

13. EPA, *Final Regulatory Impact Analysis: Control of Emissions from Nonroad Diesel Engines,* EPA420-R-04-007 (Washington, DC, 2004), 26–28, http://www.epa.gov/nonroad-diesel/2004fr.htm#ria.

14. Energy Information Administration, *Country Analysis Briefs* (Washington, DC: U.S. Department of Energy), http://www.eia.doe.gov/emeu/cabs; and "Cars in China Dream Machines," *The Economist,* 4 June 2005, 25

15. Bell et al., note 4 above.

16. EPA, note 7 above.

17. EPA, note 7 above, page 17. For figures on nitrogen oxide ($NO_x$) and volatile organic compound (VOC) emissions over time, see the tables in EPA, "Air Emissions Trends—Continued Progress Through 2003," http://www.epa.gov/airtrends/econ-emissions.html.

18. Hubbell et al., note 4 above.

19. U.S. EPA National Emissions Inventory Database, http://www.epa.gov/air/data/neidb.html

20. See the table in EPA, note 7 above. In 2003, electrical utilities emitted 4.46 million tons of $NO_x$ while off-highway vehicles and equipment emitted 4.1 million tons of $NO_x$.

21. EPA, note 13 above.

22. EPA, note 13 above.

23. S. C. Davis and S. W. Diegel, *Transportation Energy Data Book, 24th Edition* (Oak Ridge, TN: Center for Transportation Analysis, Oak Ridge National Laboratory, 2004); EPA, Office of Transportation and Air Quality, Assessments and Standards Division, *Median Life, Annual Activity, and Load Factor Values for Nonroad Engine Emissions,* Modeling Report No. NR-005c (Washington, DC, 2004).

---

is not included in the online edition of the SCAQMD monograph. The Los Angeles Coroner's office took issue with Dr. Veger's listing of smog as a contributing cause of death. More than 40 years would pass before another doctor listed air pollution as a contributing cause of death on a death certificate. In September 2000, the medical coroner in Coeur d'Alene, Idaho, cited particulate pollution from agricultural field burning as a contributing cause of death in a fatal asthma attack. See D. Whitman, "Fields of Fire," *U.S. News & World Report,* 3 September 2001: 10–14.

2. The cause of deaths and subsequent illnesses in the Donora case have been the source of some controversy. See C. Bryson, "The Donora Fluoride Fog: A Secret History of America's Worst Air Pollution Disaster," *Earth Island Journal,* Fall 1998, http://www.earthisland.org/eijournal/fall98/fe_fall98donora.html; and D. Davis. *When Smoke Ran Like Water* (New York: Basic Books, 2002).

24. California Environmental Protection Agency, Air Resources Board, *California's 2001 Emission Inventory: A Review and Look to the Future* (Sacramento CA, 2001).

25. A. Kaplun, "Air Pollution: EPA Diesel Program Falling Short of Emissions Targets," *Greenwire,* 26 July 2005.

26. E. Lau and T. Bizjak, "Air Board Mandates Retrofits of Diesels," *The Sacramento Bee,* 10 December 2004.

27. Oak Ridge National Laboratory (ORNL), *Fuel Use for Off-Road Recreation: A Reassessment of the Fuel Use Model,* ORNL/TM-1999/100 (Oak Ridge, TN, 1999).

28. EPA, note 13 above.

29. Ronald Reagan, quoted in Purves et al., note 12 above.

30. R. A. Rasmussen and F. W. Went, "Volatile Organic Material of Plant Origin in the Atmosphere," *Proceedings of the National Academy of Sciences USA* 53, no. 1 (1965): 215–20.

31. A recent analysis finds that biogenic VOC emissions rose substantially in the Southeast from the mid-1980s to the mid-1990s. See Purves et al., note 12 above. Purves also summarizes recent studies on biogenic VOC emissions at page 1738.

32. From 1987 to 1997, forest cover in the United States grew nearly 10 million acres, from 736.4 million acres in 1987 to 747 million acres a decade later. For statistics on reforestation by region, see R. J. Alig et al., *Land Use Changes Involving Forestry in the United States: 1952 to 1997, with Projections to 2050,* U.S. Department of Agriculture, Forest Service, Pacific Northwest Research Station, General Technical Report PNW-GTR-587 (Washington DC, 2003), 18. For media accounts of reforestation see D. Whitman, "From Stumps, Lush Forests," *U.S. News & World Report special edition,* "The Future of Earth," Spring 2004, 77–78; and R. S. Boyd, "U.S. Forests are on a Rebound, but Can It Last?" *Seattle Times,* 13 February 2005. Although forest acreage has increased since the 1920s, much of this forest land is second growth, less diverse, and at risk due to forest health or wildfire. (Personal communication with Gerry Gray, vice president for policy, american Forests, 10 August 2005.)

33. Purves et al., note 12 above.

34. Purves et al., note 12 above.

35. For figures on VOC emissions see the tables in EPA, note 7 above. In 2003, solvent utilization emitted 4.56 million tons of VOCs; highway vehicles emitted 4.42 million tons of VOCs. The regulatory efforts of the Ozone Transport Commission (OTC) provide one illustration of the greater difficulty of reducing VOCs from solvents than from motor vehicles. OTC, a compact of Northeastern states, has promulgated a model VOC reductions rule that incorporates tougher state emissions limits than EPA rules. A number of eastern states, including New York, have adopted the tougher OTC rules in their state implementation plans (SIPs). However, the National Paint and Coating Association has argued that products formulated to meet the tightened VOC limits require more applications and increased use in warmer months. EPA's approval of New York's SIP acknowledged that "the agency is concerned that if the rule limits make it impossible for manufacturers to produce coatings that are desirable to consumers, there is a possibility that users may misuse the products by adding additional solvent, thereby circumventing the rule's intended VOC emissions reductions." In February 2005, the paint manufacturer Sherwin-Williams Co. challenged EPA's approval of New York's VOC rules in federal court. See B. German, "Paint Industry Lawsuits Target EPA Ozone Rules," *Greenwire,* 17 February 2005.

36. "Special Report on Valley Air Quality," *The Fresno Bee,* 15 December 2002, http://www.valleyairquality.com; and D. Crow, "Air Pollution Control Officer's Determination of VOC Emissions Factors for Dairies," San Joaquin Valley Unified Air Pollution Control District (Fresno, CA, 2005), http://www.valleyair.org/busind/pto/dpag/APCO%20Determination%20of%20EF_August%201_.pdf.

37. California Air Resources Board (CARB), Weekend Effect Workgroup, *ARB Report on the Ozone Weekend Effect in California: Executive Summary* (Sacramento, CA, 2003), 1, http://www.arb.ca.gov/aqd/weekendeffect/weekendeffect.htm.

38. For an important review article on the weekend effect, see J. M. Heuss et al., "Weekday/Weekend Ozone Differences: What Can We Learn from Them?" *Journal of the Air and Waste Management Association* 53, no. 7 (2003): 772–88. Heuss et al. use EPA data to chart weekday/weekend ozone differences at more than 1,200 monitoring sites across the country from 1996 to 1998. The most common pattern, found at 750 sites, was that ozone concentrations were substantially similar on weekends and weekdays. But at 131 monitoring sites, 1-hour daily maximum ozone ($O_3$) levels increased by 10 percent or more on weekends, and 200 sites saw an increase of 5–10 percent on weekends. Weekend 1-hour $O_3$ concentrations decreased at only 125 monitors. Once the new 8-hour maximums were tracked, this was true at just 62 monitors. In general, the weekend effect was evident in large cities in the coastal areas of California, the Midwest, and the Northeast Corridor. By contrast, most Southeast sites had similar or lower $O_3$ levels on weekends.

39. For data on Philadelphia, see B. K. Pun et al., "Day-of-Week Behavior of Atmospheric Ozone in Three U.S. Cities," *Journal of the Air and Waste Management Association* 53, no. 7 (2003): 789–801.

40. Both theories are discussed in Berkeley Lab, "Understanding the Weekend Effect," press release, 14 July 2004, http://www.lbl.gov/Science-Articles/Archive/sb/July-2004/4_weekend.html.

41. C. L. Blanchard and S. J. Tanenbaum, "Differences between Weekday and Weekend Air Pollution Levels in Southern California," *Journal of Air and Waste Management Association* 53, no. 7 (2003): 816–28; and E. M. Fujita et al., "Evolution of the Magnitude and Spatial Extent of the Weekend Ozone Effect in California's South Coast Air Basin, 1981–2000," *Journal of Air and Waste Management Association* 53, no. 7 (2003): 802–15.

42. For CARB's position, see note 37 above at pages 8–9. With respect to EPA's treatment of the weekend effect, see S. F. Hayward and J. Schwartz, *Emissions Down, Smog Up. Say What?* American Enterprise Institute Environmental Policy Outlook, No. 16242, posted 20 January 2004 at http://www.aei.org/publications/filter.,pubID.19746/pub_detail.asp. It is interesting to note that Blanchard and Tanenbaum and Fujita et al., note 41 above, are considerably less agnostic than CARB or EPA about the source of higher weekend $O_3$ levels. See especially the review article by Heuss et al., note 38 above. "It is prudent to conclude that $NO_x$ reduction is the primary cause of weekday/weekend $O_3$ differences," Heuss writes. These "findings suggest that $NO_x$-focused strategies to reduce $O_3$ in VOC-limited areas should be avoided or, at a minimum, evaluated carefully in photochemical models" (pages 786–87).

43. D. J. Jacob et al., Effect of Rising Asian Emissions on Surface Ozone in the United States, *Geophysical Research Letters* 26, no. 6: 711–14.

44. Ibid.

45. G. R. Carmichael, "The Impacts of Emissions from Asia on Local/Regional/Global Air Quality," Chairman's Air Pollution Seminar Series, 18 January 2001, http://www.arb.ca.gov/research/seminars/carmichael/carmichael.htm.

46. National Research Council, Global Air Quality: *An Imperative for Long-Term Observational Strategies,* (Washington, DC: National Academy Press, 2001).

47. See Jacob et al., note 43 above.

48. There is an emerging set of cost-effective retrofit technologies for small motors that include small catalytic converters and magnetic-based emission reduction devices.

49. K. Knowlton et al., "Assessing Ozone–Related Health Impacts under a Changing Climate," *Environmental Health Perspectives* 112, no. 15 (2004): 1557–63. Also see the correspondence between Knowlton et al. and critics J. Schwartz, P. Michaels, and R. Davis, "Ozone: Unrealistic Scenarios," *Environmental Health Perspectives* 113, no. 2 (2005): A86–A87. Modeling of climate change by Loretta Mickley and her colleagues in the Division of Engineering and Applied Science at Harvard University suggests that by 2050, the frequency of cold fronts bringing cool, clear weather out of Canada to the United States during summer months will decline about 20 percent by 2050. If Mickley's projections are correct, global warming could worsen ozone pollution in the Northeast and Midwest. Her modeling also suggests that ozone-alert days in the Northeast and Midwest will occur 66 percent more frequently in 2050 than in 2000, and the severity and duration of pollution episodes will increase even with constant emissions. See L. J. Mickley, "The Pollution-Climate Connection," PowerPoint presentation to the American Association for the Advancement of Science conference in Washington, DC, 19 February 2005, available at http://www.people.fas.harvard.edu/~mickley/mickley_aaas_v2.ppt. For related findings on the impact of climate change on carbon monoxide emissions and ozone concentrations, see L. J. Mickley, D. J. Jacob, B. D. Field, and D. Rind, "Effects of Future Climate Change on Regional Air Pollution Episodes in the United States," *Geophysical Research Letters* 31 (2004): L24103, doi:10.1029/2004GL021216, http://www.people.fas.harvard.edu/~mickley_2004b.pdf.

**MARK BERNSTEIN** is a senior policy analyst at the RAND Corporation in Santa Monica, California, where he specializes in energy and environmental policy. He served as a senior analyst in the White House Office of Science and Technology Policy from 1996–1998 and was also a professor at the University of Pennsylvania. He may be reached via e-mail at markb@rand.org or by phone at (310) 393-0411, ext. 6524. **DAVID WHITMAN** is a RAND consultant and a contributing editor at *U.S. News & World Report,* where he has covered environmental and energy issues. In 2004, Whitman was an Alicia Patterson Foundation Journalism Fellow. During his fellowship year, Whitman wrote a series of articles on the dangers of fine particle pollution. Whitman is the author of several books, including *The Optimism Gap: The I'm OK—They're Not Syndrome* and the *Myth of American Decline* (Walker & Company, 1998). He can be contacted by e-mail at davidwhitman.dc@verizon.net or by phone at (202) 667-0827.

From *Environment,* October 2005, pp. 29–41. Reprinted by permission of the Helen Dwight Reid Educational Foundation. Published by Heldref Publications, 1319 Eighteenth St., NW, Washington, DC 20036-1802. Copyright © 2005. www.heldref.org

# How Did Humans First Alter Global Climate?

**A bold new hypothesis suggests that our ancestors' farming practices kicked off global warming thousands of years before we started burning coal and driving cars**

WILLIAM F. RUDDIMAN

The scientific consensus that human actions first began to have a warming effect on the earth's climate within the past century has become part of the public perception as well. With the advent of coal-burning factories and power plants, industrial societies began releasing carbon dioxide ($CO_2$) and other greenhouse gases into the air. Later, motor vehicles added to such emissions. In this scenario, those of us who have lived during the industrial era are responsible not only for the gas buildup in the atmosphere but also for at least part of the accompanying global warming trend. Now, though, it seems our ancient agrarian ancestors may have begun adding these gases to the atmosphere many millennia ago, thereby altering the earth's climate long before anyone thought.

New evidence suggests that concentrations of $CO_2$ started rising about 8,000 years ago, even though natural trends indicate they should have been dropping. Some 3,000 years later the same thing happened to methane, another heat-trapping gas. The consequences of these surprising rises have been profound. Without them, current temperatures in northern parts of North America and Europe would be cooler by three to four degrees Celsius—enough to make agriculture difficult. In addition, an incipient ice age—marked by the appearance of small ice caps—would probably have begun several thousand years ago in parts of northeastern Canada. Instead the earth's climate has remained relatively warm and stable in recent millennia.

Until a few years ago, these anomalous reversals in greenhouse gas trends and their resulting effects on climate had escaped notice. But after studying the problem for some time, I realized that about 8,000 years ago the gas trends stopped following the pattern that would be predicted from their past long-term behavior, which had been marked by regular cycles. I concluded that human activities tied to farming—primarily agricultural deforestation and crop irrigation—must have added the extra $CO_2$ and methane to the atmosphere. These activities explained both the reversals in gas trends and the ongoing increases right up to the start of the industrial era. Since then,

modern technological innovations have brought about even faster rises in greenhouse gas concentrations.

> **My claim that human contributions have been ALTERING THE EARTH'S CLIMATE FOR MILLENNIA is provocative and controversial.**

My claim that human contributions have been altering the earth's climate for millennia is provocative and controversial. Other scientists have reacted to this proposal with the mixture of enthusiasm and skepticism that is typical when novel ideas are put forward, and testing of this hypothesis is now under way.

## The Current View

This new idea builds on decades of advances in understanding long-term climate change. Scientists have known since the 1970s that three predictable variations in the earth's orbit around the sun have exerted the dominant control over long-term global climate for millions of years. As a consequence of these orbital cycles (which operate over 100,000, 41,000 and 22,000 years), the amount of solar radiation reaching various parts of the globe during a given season can differ by more than 10 percent. Over the past three million years, these regular changes in the amount of sunlight reaching the planet's surface have produced a long sequence of ice ages (when great areas of Northern Hemisphere continents were covered with ice) separated by short, warm interglacial periods.

Dozens of these climatic sequences occurred over the millions of years when hominids were slowly evolving toward anatomically modern humans. At the end of the most recent glacial period, the ice sheets that had blanketed northern Europe and

## Overview/Early Global Warming

- A new hypothesis challenges the conventional assumption that greenhouse gases released by human activities have perturbed the earth's delicate climate only within the past 200 years.
- New evidence suggests instead that our human ancestors began contributing significant quantities of greenhouse gases to the atmosphere thousands of years earlier by clearing forests and irrigating fields to grow crops.
- As a result, human beings kept the planet notably warmer than it would have been otherwise—and possibly even averted the start of a new ice age.

North America for the previous 100,000 years shrank and, by 6,000 years ago, had disappeared. Soon after, our ancestors built cities, invented writing and founded religions. Many scientists credit much of the progress of civilization to this naturally warm gap between less favorable glacial intervals, but in my opinion this view is far from the full story.

In recent years, cores of ice drilled in the Antarctic and Greenland ice sheets have provided extremely valuable evidence about the earth's past climate, including changes in the concentrations of the greenhouse gases. A three-kilometer-long ice core retrieved from Vostok Station in Antarctica during the 1990s contained trapped bubbles of ancient air that revealed the composition of the atmosphere (and the gases) at the time the ice layers formed. The Vostok ice confirmed that concentrations of $CO_2$ and methane rose and fell in a regular pattern during virtually all of the past 400,000 years.

Particularly noteworthy was that these increases and decreases in greenhouse gases occurred at the same intervals as variations in the intensity of solar radiation and the size of the ice sheets. For example, methane concentrations fluctuate mainly at the 22,000-year tempo of an orbital cycle called precession. As the earth spins on its rotation axis, it wobbles like a top, slowly swinging the Northern Hemisphere closer to and then farther from the sun. When this precessional wobble brings the northern continents nearest the sun during the summertime, the atmosphere gets a notable boost of methane from its primary natural source—the decomposition of plant matter in wetlands.

After wetland vegetation flourishes in late summer, it then dies, decays and emits carbon in the form of methane, sometimes called swamp gas. Periods of maximum summertime heating enhance methane production in two primary ways: In southern Asia, the warmth draws additional moisture-laden air in from the Indian Ocean, driving strong tropical monsoons that flood regions that might otherwise stay dry. In far northern Asia and Europe, hot summers thaw boreal wetlands for longer periods of the year. Both processes enable more vegetation to grow, decompose and emit methane every 22,000 years. When the Northern Hemisphere veers farther from the sun, methane emissions start

to decline. They bottom out 11,000 years later—the point in the cycle when Northern Hemisphere summers receive the least solar radiation.

## Unexpected Reversals

Examining records from the Vostok ice core closely, I spotted something odd about the recent part of the record. Early in previous interglacial intervals, the methane concentration typically reached a peak of almost 700 parts per billion (ppb) as precession brought summer radiation to a maximum. The same thing happened 11,000 years ago, just as the current interglacial period began. Also in agreement with prior cycles, the methane concentration then declined by 100 ppb as summer sunshine subsequently waned. Had the recent trend continued to mimic older interglacial intervals, it would have fallen to a value near 450 ppb during the current minimum in summer heating. Instead the trend reversed direction 5,000 years ago and rose gradually back to almost 700 ppb just before the start of the industrial era. In short, the methane concentration rose when it should have fallen, and it ended up 250 ppb higher than the equivalent point in earlier cycles.

Like methane, $CO_2$ has behaved unexpectedly over the past several thousand years. Although a complex combination of all three orbital cycles controls $CO_2$ variations, the trends during previous interglacial intervals were all surprisingly similar to one another. Concentrations peaked at 275 to 300 parts per million (ppm) early in each warm period, even before the last remnants of the great ice sheets finished melting. The $CO_2$ levels then fell steadily over the next 15,000 years to an average of about 245 ppm. During the current interglacial interval, $CO_2$ concentrations reached the expected peak around 10,500 years ago and, just as anticipated, began a similar decline. But instead of continuing to drop steadily through modern times, the trend reversed direction 8,000 years ago. By the start of the industrial era, the concentration had risen to 285 ppm—roughly 40 ppm higher than expected from the earlier behavior.

What could explain these unexpected reversals in the natural trends of both methane and $CO_2$? Other investigators suggested that natural factors in the climate system provided the answer. The methane increase has been ascribed to expansion of wetlands in Arctic regions and the $CO_2$ rise to natural losses of carbon-rich vegetation on the continents, as well as to changes in the chemistry of the ocean. Yet it struck me that these explanations were doomed to fail for a simple reason. During the four preceding interglaciations, the major factors thought to influence greenhouse gas concentrations in the atmosphere were nearly the same as in recent millennia. The northern ice sheets had melted, northern forests had reoccupied the land uncovered by ice, meltwater from the ice had returned sea level to its high interglacial position, and solar radiation driven by the earth's orbit had increased and then began to decrease in the same way.

Why, then, would the gas concentrations have fallen during the last four interglaciations yet risen only during the current one? I concluded that something new to the natural workings of the climate system must have been operating during the past several thousand years.

# The Human Connection

The most plausible "new factor" operating in the climate system during the present interglaciation is farming. The basic timeline of agricultural innovations is well known. Agriculture originated in the Fertile Crescent region of the eastern Mediterranean around 11,000 years ago, shortly thereafter in northern China, and several thousand years later in the Americas. Through subsequent millennia it spread to other regions and increased in sophistication. By 2,000 years ago, every crop food eaten today was being cultivated somewhere in the world.

Several farming activities generate methane. Rice paddies flooded by irrigation generate methane for the same reason that natural wetlands do—vegetation decomposes in the stagnant standing water. Methane is also released as farmers burn grasslands to attract game and promote growth of berries. In addition, people and their domesticated animals emit methane with feces and belches. All these factors probably contributed to a gradual rise in methane as human populations grew slowly, but only one process seems likely to have accounted for the abruptness of the reversal from a natural methane decline to an unexpected rise around 5,000 years ago—the onset of rice irrigation in southern Asia.

Farmers began flooding lowlands near rivers to grow wet-adapted strains of rice around 5,000 years ago in the south of China. With extensive floodplains lying within easy reach of several large rivers, it makes sense that broad swaths of land could have been flooded soon after the technique was discovered, thus explaining the quick shift in the methane trend. Historical records also indicate a steady expansion in rice irrigation throughout the interval when methane values were rising. By 3,000 years ago the technique had spread south into Indochina and west to the Ganges River Valley in India, further increasing methane emissions. After 2,000 years, farmers began to construct rice paddies on the steep hillsides of Southeast Asia.

Future research may provide quantitative estimates of the amount of land irrigated and methane generated through this 5,000-year interval. Such estimates will be probably difficult to come by, however, because repeated irrigation of the same areas into modern times has probably disturbed much of the earlier evidence. For now, my case rests mainly on the basic fact that the methane trend went the "wrong way" and that farmers began to irrigate wetlands at just the right time to explain this wrong-way trend.

Another common practice tied to farming—deforestation—provides a plausible explanation for the start of the anomalous $CO_2$ trend. Growing crops in naturally forested areas requires cutting trees, and farmers began to clear forests for this purpose in Europe and China by 8,000 years ago, initially with axes made of stone and later from bronze and then iron. Whether the fallen trees were burned or left to rot, their carbon would have soon oxidized and ended up in the atmosphere as $CO_2$.

Scientists have precisely dated evidence that Europeans began growing nonindigenous crop plants such as wheat, barley and peas in naturally forested areas just as the $CO_2$ trend reversed 8,000 years ago. Remains of these plants, initially cultivated in the Near East, first appear in lake sediments in southeastern Europe and then spread to the west and north over the next several thousand years. During this interval, silt and clay began to wash into rivers and lakes from denuded hillsides at increasing rates, further attesting to ongoing forest clearance.

The most unequivocal evidence of early and extensive deforestation lies in a unique historical document—the Doomsday Book. This survey of England, ordered by William the Conqueror, reported that 90 percent of the natural forest in lowland, agricultural regions was cleared as of A.D. 1086. The survey also counted 1.5 million people living in England at the time, indicating that an average density of 10 people per square kilometer was sufficient to eliminate the forests. Because the advanced civilizations of the major river valleys of China and India had reached much higher population densities several thousand years prior, many historical ecologists have concluded that these regions were heavily deforested some two or even three thousand years ago. In summary, Europe and southern Asia had been heavily deforested long before the start of the industrial era, and the clearance process was well under way throughout the time of the unusual $CO_2$ rise.

# An Ice Age Prevented?

If farmers were responsible for greenhouse gas anomalies this large—250 ppb for methane and 40 ppm for $CO_2$ by the 1700s—the effect of their practices on the earth's climate would have been substantial. Based on the average sensitivity shown by a range of climate models, the combined effect from these anomalies would have been an average warming of almost 0.8 degree C just before the industrial era. That amount is larger than the 0.6 degree C warming measured during the past century—implying that the effect of early farming on climate rivals or even exceeds the combined changes registered during the time of rapid industrialization.

How did this dramatic warming effect escape recognition for so long? The main reason is that it was masked by natural climatic changes in the opposite direction. The earth's orbital cycles were driving a simultaneous natural cooling trend, especially at high northern latitudes. The net temperature change was a gradual summer cooling trend lasting until the 1800s.

Had greenhouse gases been allowed to follow their natural tendency to decline, the resulting cooling would have augmented the one being driven by the drop in summer radiation, and this planet would have become considerably cooler than it is now. To explore this possibility, I joined with Stephen J. Vavrus and John E. Kutzbach of the University of Wisconsin-Madison to use a climate model to predict modern-day temperature in the absence of all human-generated greenhouse gases. The model simulates the average state of the earth's climate—including temperature and precipitation—in response to different initial conditions.

# Human Disease and Global Cooling

Concentrations of $CO_2$ in the atmosphere have been climbing since about 8,000 years ago. During the past two millennia, however, that steady increase at times reversed direction, and the $CO_2$ levels fell for decades or more. Scientists usually attribute such $CO_2$ drops—and the accompanying dips in global temperature—to natural reductions in the sun's energy output or to volcanic eruptions. These factors have been regarded as major drivers of climate change over decades or centuries, but for the $CO_2$ patterns, such explanations fall short—which implies that an additional factor forced $CO_2$ levels downward. Because I had already concluded that our human ancestors had caused the slow rise in $CO_2$ for thousands of years by clearing forests for agriculture [*see main article*], this new finding made me wonder whether some kind of reversal of the ongoing clearance could explain the brief $CO_2$ drops.

The most likely root cause turns out to be disease—the massive human mortality accompanying pandemics. Two severe outbreaks of bubonic plague, the single most devastating killer in human history, correlate well with large $CO_2$ drops at approximately A.D. 540 and 1350 [*graph*]. Plague first erupted during the Roman era, with the most virulent pandemic, the Plague of Justinian, in A.D. 540 to 542. The infamous "Black Death" struck between 1347 and 1352, followed by lesser outbreaks for more than a century. Each of these pandemics killed some 25 to 40 percent of the population of Europe. An even worse catastrophe followed in the Americas after 1492 when Europeans introduced smallpox and a host of other diseases that killed around 50 million people, or about 90 percent of the pre-Columbian population. The American pandemic coincides with the largest $CO_2$ drop of all, from 1550 to 1800.

Observers at the time noted that the massive mortality rates produced by these pandemics caused widespread abandonment of rural villages and farms, leaving untended farmland to revert to the wild. Ecologists have shown that forests will reoccupy abandoned land in just 50 years. Coupled with estimates of human population and the acreage cultivated by each farmer, calculations of forest regrowth in pandemic-stricken regions indicate that renewed forests could have sequestered enough carbon to reduce concentrations of $CO_2$ in the atmosphere by the amounts observed. Global climate would have cooled as a result, until each pandemic passed and rebounding populations began cutting and burning forests anew.

*—W.F.R.*

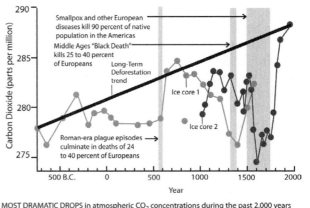

MOST DRAMATIC DROPS in atmospheric $CO_2$ concentrations during the past 2,000 years (as recorded in two Antarctic ice cores) occurred around the same periods that disease outbreaks were taking the greatest toll on human life (*grey bars*).

For our experiment, we reduced the greenhouse gas levels in the atmosphere to the values they would have reached today without early farming or industrial emissions. The resulting simulation showed that our planet would be almost two degrees C cooler than it is now—a significant difference. In comparison, the global mean temperature at the last glacial maximum 20,000 years ago was only five to six degrees C colder than it is today. In effect, current temperatures would be well on the way toward typical glacial temperatures had it not been for the greenhouse gas contributions from early farming practices and later industrialization.

I had also initially proposed that new ice sheets might have begun to form in the far north if this natural cooling had been allowed to proceed. Other researchers had shown previously that parts of far northeastern Canada might be ice covered today if the world were cooler by just 1.5 to two degrees C—the same amount of cooling that our experiment suggested has been offset by the greenhouse gas anomalies. The later modeling effort with my Wisconsin colleagues showed that snow would now persist into late summer in two areas of northeastern Canada: Baffin Island, just east of the mainland, and Labrador, farther south. Because any snow that survives throughout the summer will accumulate in thicker piles year by year and eventually become glacial ice, these results suggest that a new ice age would have begun in northeast Canada several millennia ago, at least on a small scale.

This conclusion is startlingly different from the traditional view that human civilization blossomed within a period of warmth that nature provided. As I see it, nature would have cooled the earth's climate, but our ancestors kept it warm by discovering agriculture.

## Implications for the Future

The conclusion that humans prevented a cooling and arguably stopped the initial stage of a glacial cycle bears directly on a long-running dispute over what global climate has in store for us in the near future. Part of the reason that policymakers had trouble embracing the initial predictions of global warming in the 1980s was that a number of scientists had spent the previous decade telling everyone almost exactly the opposite—that an ice age was on its way. Based on the new confirmation that orbital variations control the growth and decay of ice sheets, some scientists studying these longer-scale changes had reasonably concluded that the next ice age might be only a few hundred or at most a few thousand years away.

In subsequent years, however, investigators found that greenhouse gas concentrations were rising rapidly and that the earth's climate was warming, at least in part because of the gas increases. This evidence convinced most scientists that the relatively near-term future (the next century or two) would be dominated by global warming rather than by global cooling. This revised prediction, based on an improved understanding of the climate system, led some policymakers to discount all forecasts—whether of global warming or an impending ice age—as untrustworthy.

My findings add a new wrinkle to each scenario. If anything, such forecasts of an "impending" ice age were actually understated: new ice sheets should have begun to grow several millennia ago. The ice failed to grow because human-induced global warming actually began far earlier than previously thought—well before the industrial era.

In these kinds of hotly contested topics that touch on public policy, scientific results are often used for opposing ends. Global-warming skeptics could cite my work as evidence that human-generated greenhouse gases played a beneficial role for several thousand years by keeping the earth's climate more hospitable than it would otherwise have been. Others might counter that if so few humans with relatively primitive technologies were able to alter the course of climate so significantly, then we have reason to be concerned about the current rise of greenhouse gases to unparalleled concentrations at unprecedented rates.

The rapid warming of the past century is probably destined to persist for at least 200 years, until the economically accessible fossil fuels become scarce. Once that happens, the earth's climate should begin to cool gradually as the deep ocean slowly absorbs the pulse of excess $CO_2$ from human activities. Whether global climate will cool enough to produce the long-overdue glaciation or remain warm enough to avoid that fate is impossible to predict.

## More to Explore

**Plagues and Peoples.** William McNeill. Doubleday, 1976.

**Ice Ages: Solving the Mystery.** John Imbrie and Katherine Palmer Imbrie. Enslow, 1979.

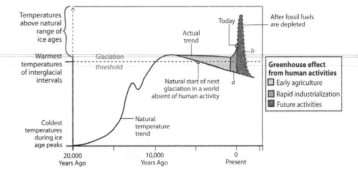

**Figure 1** Greenhouse effect from human activities has warded off a glaciation that otherwise would have begun about 5,000 years ago. Early human agricultural activities produced enough greenhouse gases to offset most of the natural cooling trend during preindustrial times, warming the planet by an average of almost 0.8 degrees Celsius. That early warming effect (*a*) rivals the 0.6 degree Celsius (*b*) warming measured in the past century of rapid industrialization. Once most fossil fuels are depleted and the temperature rise caused by greenhouse gases peaks, the earth will cool toward the next glaciation—now thousands of years overdue.

**Guns, Germs, and Steel: The Fates of Human Societies.** Jared Diamond. W. W. Norton, 1999.

**Earth's Climate: Past and Future.** William F. Ruddiman. W. H. Freeman, 2001.

**The Anthropogenic Greenhouse Era Began Thousands of Years Ago.** William F. Ruddiman in *Climatic Change,* Vol. 61, No. 3, pages 261–293; 2003.

**Deforesting the Earth: From Prehistory to Global Crisis.** Michael A. Williams. University of Chicago Press, 2003.

**Plows, Plagues, and Petroleum: How Humans Took Control of Climate.** William F. Ruddiman. Princeton University Press [in press].

**WILLIAM F. RUDDIMAN** is a marine geologist and professor emeritus of environmental sciences at the University of Virginia. He joined the faculty there in 1991 and served as department chair from 1993 to 1996. Ruddiman first began studying records of climate change in ocean sediments as a graduate student at Columbia University, where he received his doctorate in 1969. He then worked as a senior scientist and oceanographer with the U.S. Naval Oceanographic Office in Maryland and later as a senior research scientist at Columbia's Lamont-Doherty Earth Observatory.

*By Any Measure, Earth Is at...*

# The Tipping Point

**The climate is crashing, and global warming is to blame.
Why the crisis hit so soon—and what we can do about it**

JEFFREY KLUGER

No one can say exactly what it looks like when a planet takes ill, but it probably looks a lot like Earth. Never mind what you've heard about global warming as a slow-motion emergency that would take decades to play out. Suddenly and unexpectedly, the crisis is upon us.

It certainly looked that way last week as the atmospheric bomb that was Cyclone Larry—a Category 5 storm with wind bursts that reached 180 m.p.h.—exploded through northeastern Australia. It certainly looked that way last year as curtains of fire and dust turned the skies of Indonesia orange, thanks to drought-fueled blazes sweeping the island nation. It certainly looks that way as sections of ice the size of small states calve from the disintegrating Arctic and Antarctic. And it certainly looks that way as the sodden wreckage of New Orleans continues to molder, while the waters of the Atlantic gather themselves for a new hurricane season just two months away. Disasters have always been with us and surely always will be. But when they hit this hard and come this fast—when the emergency becomes commonplace—something has gone grievously wrong. That something is global warming.

The image of Earth as organism—famously dubbed Gaia by environmentalist James Lovelock—has probably been overworked, but that's not to say the planet can't behave like a living thing, and these days, it's a living thing fighting a fever. From heat waves to storms to floods to fires to massive glacial melts, the global climate seems to be crashing around us. Scientists have been calling this shot for decades. This is precisely what they have been warning would happen if we continued pumping greenhouse gases into the atmosphere, trapping the heat that flows in from the sun and raising global temperatures.

Environmentalists and lawmakers spent years shouting at one another about whether the grim forecasts were true, but in the past five years or so, the serious debate has quietly ended. Global warming, even most skeptics have concluded, is the real deal, and human activity has been causing it. If there was any consolation, it was that the glacial pace of nature would give us decades or even centuries to sort out the problem.

But glaciers, it turns out, can move with surprising speed, and so can nature. What few people reckoned on was that global climate systems are booby-trapped with tipping points and feedback loops, thresholds past which the slow creep of environmental decay gives way to sudden and self-perpetuating collapse. Pump enough $CO_2$ into the sky, and that last part per million of greenhouse gas behaves like the 212th degree Fahrenheit that turns a pot of hot water into a plume of billowing steam. Melt enough Greenland ice, and you reach the point at which you're not simply dripping meltwater into the sea but dumping whole glaciers. By one recent measure, several Greenland ice sheets have doubled their rate of slide, and just last week the journal *Science* published a study suggesting that by the end of the century, the world could be locked in to an eventual rise in sea levels of as much as 20 ft. Nature, it seems, has finally got a bellyful of us.

"Things are happening a lot faster than anyone predicted," says Bill Chameides, chief scientist for the advocacy group Environmental Defense and a former professor of atmospheric chemistry. "The last 12 months have been alarming." Adds Ruth Curry of the Woods Hole Oceanographic Institution in Massachusetts: "The ripple through the scientific community is palpable."

And it's not just scientists who are taking notice. Even as nature crosses its tipping points, the public seems to have reached its own. For years, popular skepticism about climatological science stood in the way of addressing the problem, but the naysayers—many of whom were on the payroll of energy companies—have become an increasingly marginalized breed. In a new TIME/ ABC News/ Stanford University poll, 85% of respondents agree that global warming probably is happening. Moreover, most respondents say they want some action taken. Of those polled, 87% believe the government should either encourage or require lowering of power-plant emissions, and 85% think something should be done to get cars to use less gasoline. Even Evangelical Christians, once one of the most reliable columns in the conservative base, are demanding action, most notably in February, when 86 Christian leaders formed the Evangelical Climate Initiative, demanding that Congress regulate greenhouse gases.

A collection of new global-warming books is hitting the shelves in response to that awakening interest, followed closely by TV and theatrical documentaries. The most notable of them is *An Inconvenient Truth,* due out in May, a profile of former Vice President Al Gore and his climate-change work, which is generating a lot of prerelease buzz over an unlikely topic and an equally unlikely star. For all its lack of Hollywood flash, the film compensates by conveying both the hard science of global warming and Gore's particular passion.

Such public stirrings are at last getting the attention of politicians and business leaders, who may not always respond to science but have a keen nose for where votes and profits lie. State and local lawmakers have started taking action to curb emissions, and major corporations are doing the same. Wal-Mart has begun installing wind turbines on its stores to generate electricity and is talking about putting solar reflectors over its parking lots. HSBC, the world's second largest bank, has pledged to neutralize its carbon output by investing in wind farms and other green projects. Even President Bush, hardly a favorite of greens, now acknowledges climate change and boasts of the steps he is taking to fight it. Most of those steps, however, involve research and voluntary emissions controls, not exactly the laws with teeth scientists are calling for.

Is it too late to reverse the changes global warming has wrought? That's still not clear. Reducing our emissions output year to year is hard enough. Getting it low enough so that the atmosphere can heal is a multigenerational commitment. "Ecosystems are usually able to maintain themselves," says Terry Chapin, a biologist and professor of ecology at the University of Alaska, Fairbanks. "But eventually they get pushed to the limit of tolerance."

# CO$_2$ and the Poles

As a tiny component of our atmosphere, carbon dioxide helped warm Earth to comfort levels we are all used to. But too much of it does an awful lot of damage. The gas represents just a few hundred parts per million (p.p.m.) in the overall air blanket, but they're powerful parts because they allow sunlight to stream in but prevent much of the heat from radiating back out. During the last ice age, the atmosphere's CO$_2$ concentration was just 180 p.p.m., putting Earth into a deep freeze. After the glaciers retreated but before the dawn of the modern era, the total had risen to a comfortable 280 p.p.m. In just the past century and a half, we have pushed the level to 381 p.p.m., and we're feeling the effects. Of the 20 hottest years on record, 19 occurred in the 1980s or later. According to NASA scientists, 2005 was one of the hottest years in more than a century.

It's at the North and South poles that those steambath conditions are felt particularly acutely, with glaciers and ice caps crumbling to slush. Once the thaw begins, a number of mechanisms kick in to keep it going. Greenland is a vivid example. Late last year, glaciologist Eric Rignot of the Jet Propulsion

Laboratory in Pasadena, Calif., and Pannir Kanagaratnam, a research assistant professor at the University of Kansas, analyzed data from Canadian and European satellites and found that Greenland ice is not just melting but doing so more than twice as fast, with 53 cu. mi. draining away into the sea last year alone, compared with 22 cu. mi. in 1996. A cubic mile of water is about five times the amount Los Angeles uses in a year.

Dumping that much water into the ocean is a very dangerous thing. Icebergs don't raise sea levels when they melt because they're floating, which means they have displaced all the water they're ever going to. But ice on land, like Greenland's, is a different matter. Pour that into oceans that are already rising (because warm water expands), and you deluge shorelines. By some estimates, the entire Greenland ice sheet would be enough to raise global sea levels 23 ft., swallowing up large parts of coastal Florida and most of Bangladesh. The Antarctic holds enough ice to raise sea levels more than 215 ft.

# Feedback Loops

One of the reasons the loss of the planet's ice cover is accelerating is that as the poles' bright white surface shrinks, it changes the relationship of Earth and the sun. Polar ice is so reflective that 90% of the sunlight that strikes it simply bounces back into space, taking much of its energy with it. Ocean water does just the opposite, absorbing 90% of the energy it receives. The more energy it retains, the warmer it gets, with the result that each mile of ice that melts vanishes faster than the mile that preceded it.

That is what scientists call a feedback loop, and it's a nasty one, since once you uncap the Arctic Ocean, you unleash another beast: the comparatively warm layer of water about 600 ft. deep that circulates in and out of the Atlantic. "Remove the ice," says Woods Hole's Curry, "and the water starts talking to the atmosphere, releasing its heat. This is not a good thing."

A similar feedback loop is melting permafrost, usually defined as land that has been continuously frozen for two years or more. There's a lot of earthly real estate that qualifies, and much of it has been frozen much longer than two years—since the end of the last ice age, or at least 8,000 years ago. Sealed inside that cryonic time capsule are layers of partially decayed organic matter, rich in carbon. In high-altitude regions of Alaska, Canada and Siberia, the soil is warming and decomposing, releasing gases that will turn into methane and CO$_2$. That, in turn, could lead to more warming and permafrost thaw, says research scientist David Lawrence of the National Center for Atmospheric Research (NCAR) in Boulder, Colo. And how much carbon is socked away in Arctic soils? Lawrence puts the figure at 200 gigatons to 800 gigatons. The total human carbon output is only 7 gigatons a year.

One result of all that is warmer oceans, and a result of warmer oceans can be, paradoxically, colder continents within a hotter globe. Ocean currents running between warm and cold regions serve as natural thermoregulators, distributing heat from the equator toward the poles. The Gulf Stream, carrying warmth up from the tropics, is what keeps Europe's climate relatively mild. Whenever Europe is cut off from the Gulf Stream, temperatures plummet. At the end of the last ice age, the warm current was temporarily blocked, and temperatures in Europe fell as much as 10° F, locking the continent in glaciers.

What usually keeps the Gulf Stream running is that warm water is lighter than cold water, so it floats on the surface. As it reaches Europe and releases its heat, the current grows denser and sinks, flowing back to the south and crossing under the northbound Gulf Stream until it reaches the tropics and starts to warm again. The cycle works splendidly, provided the water remains salty enough. But if it becomes diluted by freshwater, the salt concentration drops, and the water gets lighter, idling on top and stalling the current. Last December, researchers associated with Britain's National Oceanography Center reported that one component of the system that drives the Gulf Stream has slowed about 30% since 1957. It's the increased release of Arctic and Greenland meltwater that appears to be causing the problem, introducing a gush of freshwater that's overwhelming the natural cycle. In a global-warming world, it's unlikely that any amount of cooling that resulted from this would be sufficient to support glaciers, but it could make things awfully uncomfortable.

"The big worry is that the whole climate of Europe will change," says Adrian Luckman, senior lecturer in geography at the University of Wales, Swansea. "We in the U.K. are on the same latitude as Alaska. The reason we can live here is the Gulf Stream."

## Drought

As fast as global warming is transforming the oceans and the ice caps, it's having an even more immediate effect on land. People, animals and plants living in dry, mountainous regions like the western U.S. make it through summer thanks to snowpack that collects on peaks all winter and slowly melts off in warm months. Lately the early arrival of spring and the unusually blistering summers have caused the snowpack to melt too early, so that by the time it's needed, it's largely gone. Climatologist Philip Mote of the University of Washington has compared decades of snowpack levels in Washington, Oregon and California and found that they are a fraction of what they were in the 1940s, and some snowpacks have vanished entirely.

Global warming is tipping other regions of the world into drought in different ways. Higher temperatures bake moisture out of soil faster, causing dry regions that live at the margins to cross the line into full-blown crisis. Meanwhile, El Niño events—the warm pooling of Pacific waters that periodically drives worldwide climate patterns and has been occurring more frequently in global-warming years—further inhibit precipitation in dry areas of Africa and East Asia. According to a recent study by NCAR, the percentage of Earth's surface suffering drought has more than doubled since the 1970s.

## Flora and Fauna

Hot, dry land can be murder on flora and fauna, and both are taking a bad hit. Wildfires in such regions as Indonesia, the western U.S. and even inland Alaska have been increasing as timberlands and forest floors grow more parched. The blazes create a feedback loop of their own, pouring more carbon into the atmosphere and reducing the number of trees, which inhale $CO_2$ and release oxygen.

Those forests that don't succumb to fire die in other, slower ways. Connie Millar, a paleoecologist for the U.S. Forest Service, studies the history of vegetation in the Sierra Nevada. Over the past 100 years, she has found, the forests have shifted their tree lines as much as 100 ft. upslope, trying to escape the heat and drought of the lowlands. Such slow-motion evacuation may seem like a sensible strategy, but when you're on a mountain, you can go only so far before you run out of room. "Sometimes we say the trees are going to heaven because they're walking off the mountaintops," Millar says.

Across North America, warming-related changes are mowing down other flora too. Manzanita bushes in the West are dying back; some prickly pear cacti have lost their signature green and are instead a sickly pink; pine beetles in western Canada and the U.S. are chewing their way through tens of millions of acres of forest, thanks to warmer winters. The beetles may even breach the once insurmountable Rocky Mountain divide, opening up a path into the rich timbering lands of the American Southeast.

With habitats crashing, animals that live there are succumbing too. Environmental groups can tick off scores of species that have been determined to be at risk as a result of global warming. Last year, researchers in Costa Rica announced that two-thirds of 110 species of colorful harlequin frogs have vanished in the past 30 years, with the severity of each season's die-off following in lockstep with the severity of that year's warming.

In Alaska, salmon populations are at risk as melting permafrost pours mud into rivers, burying the gravel the fish need for spawning. Small animals such as bushy-tailed wood rats, alpine chipmunks and piñon mice are being chased upslope by rising temperatures, following the path of the fleeing trees. And with sea ice vanishing, polar bears—prodigious swimmers but not inexhaustible ones—are starting to turn up drowned. "There will be no polar ice by 2060," says Larry Schweiger, president of the National Wildlife Federation. "Somewhere along that path, the polar bear drops out."

## What about Us?

It is fitting, perhaps, that as the species causing all the problems, we're suffering the destruction of our habitat too, and we have experienced that loss in terrible ways. Ocean waters have warmed by a full degree Fahrenheit since 1970, and warmer water is like rocket fuel for typhoons and hurricanes. Two studies last year found that in the past 35 years the number of Category 4 and 5 hurricanes worldwide has doubled while the wind speed and duration of all hurricanes has jumped 50%. Since atmospheric heat is not choosy about the water it warms, tropical storms could start turning up in some decidedly nontropical places. "There's a school of thought that sea surface temperatures are warming up toward Canada," says Greg Holland, senior scientist for NCAR in Boulder. "If so, you're likely to get tropical cyclones there, but we honestly don't know."

# What We Can Do

So much environmental collapse happening in so many places at once has at last awakened much of the world, particularly the 141 nations that have ratified the Kyoto treaty to reduce emissions—an imperfect accord, to be sure, but an accord all the same. The U.S., however, which is home to less than 5% of Earth's population but produces 25% of $CO_2$ emissions, remains intransigent. Many environmentalists declared the Bush Administration hopeless from the start, and while that may have been premature, it's undeniable that the White House's environmental record—from the abandonment of Kyoto to the President's broken campaign pledge to control carbon output to the relaxation of emission standards—has been dismal. George W. Bush's recent rhetorical nods to America's oil addiction and his praise of such alternative fuel sources as switchgrass have yet to be followed by real initiatives.

The anger surrounding all that exploded recently when NASA researcher Jim Hansen, director of the Goddard Institute for Space Studies and a longtime leader in climate-change research, complained that he had been harassed by White House appointees as he tried to sound the global-warming alarm. "The way democracy is supposed to work, the presumption is that the public is well informed," he told TIME. "They're trying to deny the science." Up against such resistance, many environmental groups have resolved simply to wait out this Administration and hope for something better in 2009.

The Republican-dominated Congress has not been much more encouraging. Senators John McCain and Joe Lieberman have twice been unable to get through the Senate even mild measures to limit carbon. Senators Pete Domenici and Jeff Bingaman, both of New Mexico and both ranking members of the chamber's Energy Committee, have made global warming a high-profile matter. A white paper issued in February will be the subject of an investigatory Senate conference next week. A House delegation recently traveled to Antarctica, Australia and New Zealand to visit researchers studying climate change. "Of the 10 of us, only three were believers," says Representative Sherwood Boehlert of New York. "Every one of the others said this opened their eyes."

Boehlert himself has long fought the environmental fight, but if the best that can be said for most lawmakers is that they are finally recognizing the global-warming problem, there's reason to wonder whether they will have the courage to reverse it. Increasingly, state and local governments are filling the void. The mayors of more than 200 cities have signed the U.S. Mayors Climate Protection Agreement, pledging, among other things, that they will meet the Kyoto goal of reducing greenhouse-gas emissions in their cities to 1990 levels by 2012. Nine eastern states have established the Regional Greenhouse Gas Initiative for the purpose of developing a cap-and-trade program that would set ceilings on industrial emissions and allow companies that overperform to sell pollution credits to those that underperform—the same smart, incentive-based strategy that got sulfur dioxide under control and reduced acid rain. And California passed the nation's toughest automobile-emissions law last summer.

"There are a whole series of things that demonstrate that people want to act and want their government to act," says Fred Krupp, president of Environmental Defense. Krupp and others believe that we should probably accept that it's too late to prevent $CO_2$ concentrations from climbing to 450 p.p.m. (or 70 p.p.m. higher than where they are now). From there, however, we should be able to stabilize them and start to dial them back down.

That goal should be attainable. Curbing global warming may be an order of magnitude harder than, say, eradicating smallpox or putting a man on the moon. But is it moral not to try? We did not so much march toward the environmental precipice as drunkenly reel there, snapping at the scientific scolds who told us we had a problem.

The scolds, however, knew what they were talking about. In a solar system crowded with sister worlds that either emerged stillborn like Mercury and Venus or died in infancy like Mars, we're finally coming to appreciate the knife-blade margins within which life can thrive. For more than a century we've been monkeying with those margins. It's long past time we set them right.

---

With reporting by **DAVID BJERKLIE AND ANDREA DORFMAN,** New York; **DAN CRAY,** Los Angeles; **GREG FULTON,** Atlanta; **ANDREA GERLIN,** London; **RITA HEALY,** Denver and **ERIC ROSTON,** Washington

# Swift Boating, Stealth Budgeting, and Unitary Executives

JAMES HANSEN

The American Revolution launched the radical proposition that the commonest of men should have a vote equal in weight to that of the richest, most powerful citizen. Our forefathers devised a remarkable Constitution, with checks and balances, to guard against the return of despotic governance and subversion of the democratic principle for the sake of the powerful few with special interests. They were well aware of the difficulties that would be faced, however, placing their hopes in the presumption of an educated and honestly informed citizenry.

I have sometimes wondered how our forefathers would view our situation today. On the positive side, as a scientist, I like to imagine how Benjamin Franklin would view the capabilities we have built for scientific investigation. Franklin speculated that an atmospheric "dry fog" produced by a large volcano had reduced the Sun's heating of the Earth so as to cause unusually cold weather in the early 1780s; he noted that the enfeebled solar rays, when collected in the focus of a "burning glass," could "scarce kindle brown paper." As brilliant as Franklin's insights may have been, they were only speculation as he lacked the tools for quantitative investigation. No doubt Franklin would marvel at the capabilities provided by Earth-encircling satellites and super-computers that he could scarcely have imagined.

Yet Franklin, Jefferson, and the other revolutionaries would surely be distraught by recent tendencies in America, specifically the increasing power of special interests in our government, concerted efforts to deceive the public, and arbitrary actions of government executives that arise from increasing concentration of authority in a unitary executive, in defiance of the aims of our Constitution's framers. These tendencies are illustrated well by a couple of incidents that I have been involved in recently.

In the first incident, my own work was distorted for the purposes of misinforming the public and protecting special interests. In the second incident, the mission of the National Aeronautics and Space Administration (NASA) was altered surreptitiously by executive action, thus subverting constitutional division of power. These incidents help to paint a picture that reveals consequences for society far greater than simple enrichment of special interests. The effect is to keep the public in the dark about increasing risks to our society and our home planet.

The first incident prompted *New York Times* columnist Paul Krugman to argue not long ago that I must respond to "swift boaters"—those who distort the record to impugn someone's credibility. I have had reservations about doing so, stemming from the perceptive advice of Professor Henk van de Hulst, who said, when I was a post-doc at Leiden University, "Your success will depend upon choosing what not to work on." Unfortunately, given the shrinking fuse on the global warming time bomb, Krugman is probably right: we cannot afford the luxury of ignoring swift boaters and focusing only on science.

Pat Michaels, a swift boater to whom Krugman refers, is sometimes described as a "contrarian." Contrarians address global warming as if they were lawyers, not scientists. A lawyer's job often is to defend a client, not seek the truth. Instead of following Richard Feynman's dictum on scientific objectivity ("The only way to have real success in science . . . is to describe the evidence very carefully without regard to the way you feel it should be"), contrarians present only evidence that supports their desired conclusion.

Skepticism, an inherent aspect of scientific inquiry, should be carefully distinguished from contrarianism. Skepticism, and the objective weighing of evidence, are essential for scientific success. Skepticism about the existence of global warming and the principal role of human-made greenhouse gases has diminished as empirical evidence and our understanding have advanced. However, many aspects of global warming need to be understood better, including the best ways to minimize climate change and its consequences. Legitimate skepticism will always have an important role to play.

However, hard-core global warming contrarians have an agenda other than scientific truth. Their target is the public. Their goal is to create an impression that global warming or its causes are uncertain. Debating a contrarian leaves an impression with today's public of an argument among theorists. Sophistical contrarians do not need to win the scientific debate to advance their cause.

## Science Fiction

Consider, for example, Pat Michaels' deceit (in a 2000 article in *Social Epistemology*) in portraying climate "predictions" that I made in 1988 as being in error by "450 percent." This distortion is old news, but by sheer repetition has become received wisdom among climate-change deniers. In fact, science fiction writer

Michael Crichton was duped by Michaels, although Crichton reduced my "error" to "wrong by 300 percent" in his 2004 novel *State of Fear.*

People acquainted with this topic are aware that Michaels, in comparing global warming predictions made with the GISS (Goddard Institute for Space Studies) climate model with observations, played a dirty trick by showing model calculations for only one of the three scenarios (not predictions!) that I presented in 1988. Here's why this trick has a big impact.

The three scenarios (see figure, next page) were intended to bracket the range of likely future climate forcings (changes imposed on the Earth's energy balance that tend to alter global temperature either way). Scenario C had the smallest greenhouse gas forcing: it extended recent greenhouse gas growth rates to the year 2000 and thereafter kept greenhouse gas amounts constant, i.e., it assumed that after 2000 human sources of these gases would be just large enough to balance removal of these gases by the "sinks." Scenario B continued approximately linear growth of greenhouse gases beyond 2000. Scenario A showed exponential growth of greenhouse gases and included a substantial allowance for trace gases that were suspected of increasing but were unmeasured.

Scenarios A, B, and C also differed in their assumptions about future volcanic eruptions. Scenarios B and C included occasional eruptions of large volcanoes, at a frequency similar to that of the real world in the previous few decades. Scenario A, intended to yield the largest plausible warming, included no volcanic eruptions, as it is not uncommon to have no large eruptions for extended periods, such as the half century between the Katmai eruption in 1912 and the Agung eruption in 1963.

Multiple scenarios are used to provide a range of plausible climate outcomes, but also so that we can learn something by comparing real-world outcomes with model predictions. How well the model succeeds in simulating the real world depends upon the realism of both the assumed forcing and the climate sensitivity (the global temperature response to a standard climate forcing) of the model.

As it turned out, in the real world the largest climate forcing in the decade after 1988, by far, was caused by the Mount Pinatubo volcanic eruption, the greatest volcanic eruption of the past century. Forcings are measured in watt-years per square meter (W-yr/m$^2$) averaged over the surface of the Earth (1 W-yr/m$^2$ is a heating of 1 W/m$^2$ over the entire planet maintained for one year). The small particles injected into the Earth's stratosphere by Pinatubo reflected sunlight back to space, causing a negative (cooling) climate forcing of about $-5$ W-yr/m$^2$. In contrast, the added greenhouse gas climate forcings ranged from about $+1.6$ W-yr/m$^2$ in scenario C to about $+2.3$ W-yr/m$^2$ in scenario A.

So of the four scenarios (A, B, C, and the real world) only scenario A had no large volcanic eruption. The volcanic activity modeled in scenarios B and C was somewhat weaker than in the real world and was misplaced by a few years, but by good fortune it was such as to have a cooling effect pretty similar to that of Pinatubo. Despite the fact that scenario A omitted the largest climate forcing, Michaels chose to compare scenario A—and *only* scenario A—with the real world. Is this a case of scientific idiocy or is there something else at work? Perhaps Michaels is just not very interested in learning about the real world.

Although less important for the temperature change between 1997 and 1988 that Michaels examined, measured real-world greenhouse gas changes in carbon dioxide ($CO_2$), methane ($CH_4$), nitrous oxide ($N_2O$), and chlorofluorocarbons (CFCs) yielded a forcing similar to those in scenarios B and C. The reason for the slow real-world growth rate was that both $CH_4$ and $CO_2$ growth rates decreased in the early 1990s (the slowdowns may have been associated with Pinatubo; in any case the $CO_2$ growth rate has subsequently accelerated rapidly).

An astute reader may wonder why the world showed any warming during the period 1988–97, given that the negative (cooling) forcing by Pinatubo exceeded the positive (warming) forcing by greenhouse gases added in that period. The reason is that the climate system was also being pushed by the planetary "energy imbalance" that existed in 1988. The climate system had not yet fully responded to greenhouse gases added to the atmosphere before then. The observed continued decadal warming, despite the very large negative volcanic forcing, provides some confirmation of that planetary energy imbalance.

## Noise and Distortion

Michaels' trick of comparing the real world only with the inappropriate scenario A accounts for his specious, incorrect conclusions. However, a second unscientific aspect of his method is also worth pointing out.

Scientists seek to learn something by comparing the real world with climate model calculations. Climate sensitivity is of special interest, as future climate change depends strongly upon it. In principal, we can extract climate sensitivity if we have accurate knowledge of the net forcing that drove climate change, and the global temperature change that occurred in response to that change. However, even if these demanding conditions are met, it is necessary to compare the magnitude of the calculated changes with the magnitude of "noise," including errors in the measurements and chaotic (unforced) variability in the model and real-world climate changes.

If Michaels had examined the noise question he would have realized that a nine-year change is insufficient to determine the real-world temperature trend or distinguish among the model runs. Even the period 1988–2005 is too brief for most purposes. Within several years the differences among scenarios A, B, and C, and comparisons with the real world, will become more meaningful.

Michaels' latest tomfoolery, repeated on several occasions, is the charge that I approve of exaggeration of potential consequences of future global warming. This is more unadulterated hogwash. Michaels quotes me as saying, "Emphasis on extreme scenarios may have been appropriate at one time, when the public and decision-makers were relatively unaware of the global warming issue."

What trick did Michaels use to create the impression that I advocate exaggeration? He took the above sentence out of context from a paragraph in which I was being gently critical of a tendency of Intergovernmental Panel on Climate Change climate simulations to emphasize only cases with very large increases of climate forcings. My entire paragraph (from a June

2003 presentation to the Council on Environmental Quality) read as follows:

> *Summary opinion re scenarios. Emphasis on extreme scenarios may have been appropriate at one time, when the public and decision-makers were relatively unaware of the global warming issue, and energy sources such as "synfuels," shale oil, and tar sands were receiving strong consideration. Now, however, the need is for demonstrably objective climate forcing scenarios consistent with what is realistic under current conditions. Scenarios that accurately fit recent and near future observations have the best chance of bringing all of the important players into the discussion, and they also are what is needed for the purpose of providing policy-makers the most effective and efficient options to stop global warming.*

Would an intelligent reader who read the entire paragraph (or even the entire sentence; by chopping off half of the sentence Michaels brings quoting-out-of-context to a new low) infer that I was advocating exaggeration? On the contrary. Perhaps I should take it as a compliment that anyone would search my writing so hard to find something that can be quoted out of context.

Having taken this trouble to refute Michaels' claims, I still wonder about the wisdom of arguing with contrarians as a strategy. Many of them, including Michaels, receive support from special interests such as fossil fuel and automotive companies. It is understandable that special interests gravitated, early on, to scientists who had a message they preferred to hear. But now that global warming and its impacts are clearer, it is time for business people to reconsider their position—and scientists, rather than debating contrarians, may do better to communicate with business leaders. The latter did not attain their positions without being astute and capable of changing. We need to make clear to them the legal and moral liabilities that accrue with continued denial of global warming. It is time for business leaders to chuck contrarians and focus on the business challenges and opportunities.

## Stealth Budgets & Unitary Executives

The second incident involved NASA's budget. Many people are aware that something bad happened to the NASA Earth Science budget this year, yet the severity of the cuts and their long-term implications are not universally recognized. In part this is because of a stealth budgeting maneuver.

When annual budgets for the coming fiscal year are announced, the differences in growth from the previous year, for agencies and their divisions, are typically a few percent. An agency with +3 percent growth may crow happily, in comparison to agencies receiving +1 percent. Small differences are important because every agency has fixed costs (civil service salaries, buildings, other infrastructure), so new programs or initiatives are strongly dependent upon any budget growth and how that growth compares with inflation.

When the administration announced its planned fiscal 2007 budget, NASA science was listed as having typical changes of 1 percent or so. However, Earth Science research actually had a staggering reduction of about 20 percent from the 2006 budget. How could that be accomplished? Simple enough: reduce the 2006 research budget retroactively by 20 percent! One-third of the way into fiscal year 2006, NASA Earth Science was told to go figure out how to live with a 20-percent loss of the current year's funds.

The Earth Science budget is almost a going-out-of-business budget. From the taxpayers' point of view it makes no sense. An 80-percent budget must be used mainly to support infrastructure (practically speaking, you cannot fire civil servants; buildings at large facilities such as Goddard Space Flight Center will not be bulldozed to the ground; and the grass at the centers must continue to be cut). But the budget cuts wipe off the books most planned new satellite missions (some may be kept on the books, but only with a date so far in the future that no money

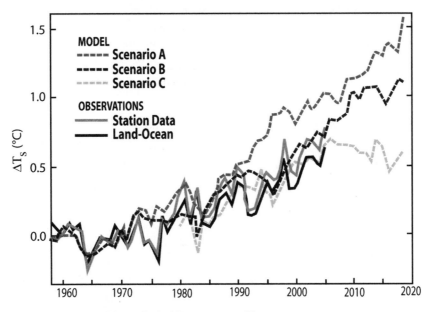

**Figure 1**  Annual Mean Global Temperature Change.

needs to be spent now), and support for contractors, young scientists, and students disappears, with dire implications for future capabilities.

Bizarrely, this is happening just when NASA data are yielding spectacular and startling results. Two small satellites that measure the Earth's gravitational field with remarkable precision found that the mass of Greenland decreased by the equivalent of 200 cubic kilometers of ice in 2005. The area on Greenland with summer melting has increased 50 percent, the major ice streams on Greenland (portions of the ice sheet moving most rapidly toward the ocean and discharging icebergs) have doubled in flow speed, and the area in the Arctic Ocean with summer sea ice has decreased 20 percent in the last 25 years.

One way to avoid bad news: stop the measurements! Only hitch: the first line of the NASA mission is "to understand and protect our home planet." Maybe that can be changed to "...protect special interests' backside."

I should say that the mission statement *used* to read "to understand and protect our home planet." That part has been deleted—a shocking loss to me, as I had been using the phrase since December 2005 to justify speaking out about the dangers of global warming. The quoted mission statement had been constructed in 2001 and 2002 via an inclusive procedure involving representatives from the NASA Centers and e-mail interactions with NASA employees. In contrast, elimination of the "home planet" phrase occurred in a spending report delivered to Congress in February 2006, the same report that retroactively slashed the Earth Science research budget. In July 2006 I asked dozens of NASA employees and management people (including my boss) if they were aware of the change. Not one of them was. Several expressed concern that such management changes by fiat would have a bad effect on organization morale.

The budgetary goings-on in Washington have been noted, e.g., in editorials of *The Boston Globe:* "Earth to NASA: Help!" (June 15, 2006) and "Don't ask; don't ask" (June 22), both decrying the near-termination of Earth measurements. Of course, the *Globe* might be considered "liberal media," so their editorials may not raise many eyebrows.

But it is conservatives and moderates who should be most upset, and I consider myself a moderate conservative. When I was in school we learned that Congress controlled the purse strings; it is in the Constitution. But it does not really seem to work that way, not if the Bush administration can jerk the science budget the way they have, in the middle of a fiscal year no less. It seems more like David Baltimore's "Theory of the Unitary Executive" (the legal theory that the president can do pretty much whatever he wants) is being practiced successfully. My impression is that conservatives and moderates would prefer that the government work as described in the Constitution, and that they prefer to obtain their information on how the Earth is doing from real observations, not from convenient science fiction.

Congress is putting up some resistance to the budget manipulation. The House restored a fraction of the fiscal year 2007 cuts to science and is attempting to restore planning for some planetary missions. But the corrective changes are moderate. You may want to check your children's textbooks for the way the U.S. government works. If their books still say that Congress controls the purse strings, some updating is needed.

## The NASA Mission
**To understand and protect our home planet, To explore the universe and search for life, To inspire the next generation of explorers . . . as only NASA can.**

But may it be that this is all a bad dream? I will stand accused of being as wistful as the boy who cried out, "Joe, say it ain't so!" to the fallen Shoeless Joe Jackson of the 1919 Chicago Black Sox, yet I maintain the hope that NASA's dismissal of "home planet" is not a case of either shooting the messenger or a too-small growth of the total NASA budget, but simply an error of transcription. Those who have labored in the humid, murky environs of Washington are aware of the unappetizing forms of life that abound there. Perhaps the NASA playbook was left open late one day, and by chance the line "to understand and protect our home planet" was erased by the slimy belly of a slug crawling in the night. For the sake of our children and grandchildren, let us pray that this is the true explanation for the devious loss, and that our home planet's rightful place in NASA's mission will be restored.

**JAMES HANSEN** is an adjunct professor at the Columbia University Earth Institute and director of NASA's Goddard Institute for Space Studies in New York. He expresses his opinions here as a private citizen under the protection of the First Amendment.

# Glossary

*This glossary of environmental terms is included to provide you with a convenient and ready reference as you encounter general computer terms that are unfamiliar or require a review. It is not intended to be comprehensive, but, taken together with the many definitions included in the articles, it should prove to be quite useful.*

## A

**Abiotic.** Without life; any system characterized by a lack of living organisms.

**Absorption.** Incorporation of a substance into a solid or liquid body.

**Acid.** Any compound capable of reacting with a base to form a salt; a substance containing a high hydrogen ion concentration (low pH).

**Acid Rain.** Precipitation containing a high concentration of acid.

**Adaptation.** Adjustment of an organism to the conditions of its environment, enabling reproduction and survival.

**Additive.** A substance added to another in order to impart or improve desirable properties or suppress undesirable ones.

**Adsorption.** Surface retention of solid, liquid, or gas molecules, atoms, or ions by a solid or liquid.

**Aerobic.** Environmental conditions where oxygen is present; aerobic organisms require oxygen in order to survive.

**Aerosols.** Tiny mineral particles in the atmosphere onto which water droplets, crystals, and other chemical compounds may adhere.

**Air Quality Standard.** A prescribed level of a pollutant in the air that should not be exceeded.

**Alcohol Fuels.** The processing of sugary or starchy products (such as sugar cane, corn, or potatoes) into fuel.

**Allergens.** Substances that activate the immune system and cause an allergic response.

**Alpha Particle.** A positively charged particle given off from the nucleus of some radioactive substances; it is identical to a helium atom that has lost its electrons.

**Ammonia.** A colorless gas comprised of one atom of nitrogen and three atoms of hydrogen; liquefied ammonia is used as a fertilizer.

**Anthropocentric.** Considering humans to be the central or most important part of the universe.

**Aquaculture.** Propagation and/or rearing of any aquatic organism in artificial "wetlands" and/or ponds.

**Aquifers.** Porous, water-saturated layers of sand, gravel, or bedrock that can yield significant amounts of water economically.

**Atom.** The smallest particle of an element, composed of electrons moving around an inner core (nucleus) of protons and neutrons. Atoms of elements combine to form molecules and chemical compounds.

**Atomic Reactor.** A structure fueled by radioactive materials that generates energy, usually in the form of electricity; reactors are also utilized for medical and biological research.

**Autotrophs.** Organisms capable of using chemical elements in the synthesis of larger compounds; green plants are autotrophs.

## B

**Background Radiation.** The normal radioactivity present; coming principally from outer space and naturally occurring radioactive substances on Earth.

**Bacteria.** One-celled microscopic organisms found in the air, water, and soil. Bacteria cause many diseases of plants and animals; they also are beneficial in agriculture, decay of dead matter, and food and chemical industries.

**Benthos.** Organisms living on the bottom of bodies of water.

**Biocentrism.** Belief that all creatures have rights and values and that humans are not superior to other species.

**Biochemical Oxygen Demand (BOD).** The oxygen utilized in meeting the metabolic needs of aquatic organisms.

**Biodegradable.** Capable of being reduced to simple compounds through the action of biological processes.

**Biodiversity.** Biological diversity in an environment as indicated by numbers of different species of plants and animals.

**Biogeochemical Cycles.** The cyclical series of transformations of an element through the organisms in a community and their physical environment.

**Biological Control.** The suppression of reproduction of a pest organism utilizing other organisms rather than chemical means.

**Biomass.** The weight of all living tissue in a sample.

**Biome.** A major climax community type covering a specific area on Earth.

**Biosphere.** The overall ecosystem of Earth. It consists of parts of the atmosphere (troposphere), hydrosphere (surface and ground water), and lithosphere (soil, surface rocks, ocean sediments, and other bodies of water).

**Biota.** The flora and fauna in a given region.

**Biotic.** Biological; relating to living elements of an ecosystem.

**Biotic Potential.** Maximum possible growth rate of living systems under ideal conditions.

**Birthrate.** Number of live births in one year per 1,000 midyear population.

**Breeder Reactor.** A nuclear reactor in which the production of fissionable material occurs.

# C

**Cancer.** Invasive, out-of-control cell growth that results in malignant tumors.

**Carbon Cycle.** Process by which carbon is incorporated into living systems, released to the atmosphere, and returned to living organisms.

**Carbon Monoxide (CO).** A gas, poisonous to most living systems, formed when incomplete combustion of fuel occurs.

**Carcinogens.** Substances capable of producing cancer.

**Carrying Capacity.** The population that an area will support without deteriorating.

**Chlorinated Hydrocarbon Insecticide.** Synthetic organic poisons containing hydrogen, carbon, and chlorine. Because they are fat-soluble, they tend to be recycled through food chains, eventually affecting nontarget systems. Damage is normally done to the organism's nervous system. Examples include DDT, Aldrin, Deildrin, and Chlordane.

**Chlorofluorocarbons (CFCs).** Any of several simple gaseous compounds that contain carbon, chlorine, fluorine, and sometimes hydrogen; they are suspected of being a major cause of stratospheric ozone depletion.

**Circle of Poisons.** Importation of food contaminated with pesticides banned for use in this country but made here and sold abroad.

**Clear-Cutting.** The practice of removing all trees in a specific area.

**Climate.** Description of the long-term pattern of weather in any particular area.

**Climax Community.** Terminal state of ecological succession in an area; the redwoods are a climax community.

**Coal Gasification.** Process of converting coal to gas; the resultant gas, if used for fuel, sharply reduces sulfur oxide emissions and particulates that result from coal burning.

**Commensalism.** Symbiotic relationship between two different species in which one benefits while the other is neither harmed nor benefited.

**Community Ecology.** Study of interactions of all organisms existing in a specific region.

**Competitive Exclusion.** Resulting from competition; one species forced out of part of an available habitat by a more efficient species.

**Conservation.** The planned management of a natural resource to prevent overexploitation, destruction, or neglect.

**Conventional Pollutants.** Seven substances (sulfur dioxide, carbon monoxide, particulates, hydrocarbons, nitrogen oxides, photochemical oxidants, and lead) that make up the largest volume of air quality degradation, as identified by the Clean Air Act.

**Core.** Dense, intensely hot molten metal mass, thousands of kilometers in diameter, at Earth's center.

**Cornucopian Theory.** The belief that nature is limitless in its abundance and that perpetual growth is both possible and essential.

**Corridor.** Connecting strip of natural habitat that allows migration of organisms from one place to another.

**Crankcase Smog Devices (PCV System).** A system, used principally in automobiles, designed to prevent discharge of combustion emissions into the external environment.

**Critical Factor.** The environmental factor closest to a tolerance limit for a species at a specific time.

**Cultural Eutrophication.** Increase in biological productivity and ecosystem succession resulting from human activities.

# D

**Death Rate.** Number of deaths in one year per 1,000 midyear population.

**Decarbonization.** To remove carbon dioxide or carbonic acid from a substance.

**Decomposer.** Any organism that causes the decay of organic matter; bacteria and fungi are two examples.

**Deforestation.** The action or process of clearing forests without adequate replanting.

**Degradation (of water resource).** Deterioration in water quality caused by contamination or pollution that makes water unsuitable for many purposes.

**Demography.** The statistical study of principally human populations.

**Desert.** An arid biome characterized by little rainfall, high daily temperatures, and low diversity of animal and plant life.

**Desertification.** Converting arid or semiarid lands into deserts by inappropriate farming practices or overgrazing.

**Detergent.** A synthetic soaplike material that emulsifies fats and oils and holds dirt in suspension; some detergents have caused pollution problems because of certain chemicals used in their formulation.

**Detrivores.** Organisms that consume organic litter, debris, and dung.

**Dioxin.** Any of a family of compounds known chemically as dibenzo-p-dioxins. Concern about them arises from their potential toxicity as contaminants in commercial products. Tests on laboratory animals indicate that it is one of the more toxic anthropogenic (man-made) compounds.

**Diversity.** Number of species present in a community (species richness), as well as the relative abundance of each species.

**DNA (Deoxyribonucleic Acid).** One of two principal nucleic acids, the other being RNA (Ribonucleic Acid). DNA contains information used for the control of a living cell. Specific segments of DNA are now recognized as genes, those agents controlling evolutionary and hereditary processes.

**Dominant Species.** Any species of plant or animal that is particularly abundant or controls a major portion of the energy flow in a community.

# Glossary

**Drip Irrigation.** Pipe or perforated tubing used to deliver water a drop at a time directly to soil around each plant. Conserves water and reduces soil waterlogging and salinization.

## E

**Ecological Density.** The number of a singular species in a geographical area, including the highest concentration points within the defined boundaries.

**Ecological Succession.** Process in which organisms occupy a site and gradually change environmental conditions so that other species can replace the original inhabitants.

**Ecology.** Study of the interrelationships between organisms and their environments.

**Ecosystem.** The organisms of a specific area, together with their functionally related environments; considered as a definitive unit.

**Ecotourism.** Wildlife tourism that could damage ecosystems and disrupt species if strict guidelines governing tours to sensitive areas are not enforced.

**Edge Effects.** Change in ecological factors at the boundary between two ecosystems. Some organisms flourish here; others are harmed.

**Effluent.** A liquid discharged as waste.

**El Niño.** Climatic change marked by shifting of a large warm water pool from the western Pacific Ocean toward the East.

**Electron.** Small, negatively charged particle; normally found in orbit around the nucleus of an atom.

**Eminent Domain.** Superior dominion exerted by a governmental state over all property within its boundaries that authorizes it to appropriate all or any part thereof to a necessary public use, with reasonable compensation being made.

**Endangered Species.** Species considered to be in imminent danger of extinction.

**Endemic Species.** Plants or animals that belong or are native to a particular ecosystem.

**Environment.** Physical and biological aspects of a specific area.

**Environmental Impact Statement (EIS).** A study of the probable environmental impact of a development project before federal funding is provided (required by the National Environmental Policy Act of 1968).

**Environmental Protection Agency (EPA).** Federal agency responsible for control of air and water pollution, radiation and pesticide problems, ecological research, and solid waste disposal.

**Erosion.** Progressive destruction or impairment of a geographical area; wind and water are the principal agents involved.

**Estuary.** Water passage where an ocean tide meets a river current.

**Eutrophic.** Well nourished; refers to aquatic areas rich in dissolved nutrients.

**Evolution.** A change in the gene frequency within a population, sometimes involving a visible change in the population's characteristics.

**Exhaustible Resources.** Earth's geologic endowment of minerals, nonmineral resources, fossil fuels, and other materials present in fixed amounts.

**Extinction.** Irrevocable elimination of species due to either normal processes of the natural world or through changing environmental conditions.

## F

**Fallow.** Cropland that is plowed but not replanted and is left idle in order to restore productivity mainly through water accumulation, weed control, and buildup of soil nutrients.

**Fauna.** The animal life of a specified area.

**Feral.** Refers to animals or plants that have reverted to a noncultivated or wild state.

**Fission.** The splitting of an atom into smaller parts.

**Floodplain.** Level land that may be submerged by floodwaters; a plain built up by stream deposition.

**Flora.** The plant life of an area.

**Flyway.** Geographic migration route for birds that includes the breeding and wintering areas that it connects.

**Food Additive.** Substance added to food usually to improve color, flavor, or shelf life.

**Food Chain.** The sequence of organisms in a community, each of which uses the lower source as its energy supply. Green plants are the ultimate basis for the entire sequence.

**Fossil Fuels.** Coal, oil, natural gas, and/or lignite; those fuels derived from former living systems; usually called nonrenewable fuels.

**Fuel Cell.** Manufactured chemical systems capable of producing electrical energy; they usually derive their capabilities via complex reactions involving the sun as the driving energy source.

**Fusion.** The formation of a heavier atomic complex brought about by the addition of atomic nuclei; during the process there is an attendant release of energy.

## G

**Gaia Hypothesis.** Theory that Earth's biosphere is a living system whose complex interactions between its living organisms and nonliving processes regulate environmental conditions over millions of years so that life continues.

**Gamma Ray.** A ray given off by the nucleus of some radioactive elements. A form of energy similar to X-rays.

**Gene.** Unit of heredity; segment of DNA nucleus of the cell containing information for the synthesis of a specific protein.

**Gene Banks.** Storage of seed varieties for future breeding experiments.

**Genetic Diversity.** Infinite variation of possible genetic combinations among individuals; what enables a species to adapt to ecological change.

**Geothermal Energy.** Heat derived from the Earth's interior. It is the thermal energy contained in the rock and fluid (that fills the fractures and pores within the rock) in the Earth's crust.

**Germ Plasm.** Genetic material that may be preserved for future use (plant seeds, animal eggs, sperm, and embryos).

**Global Warming.** An increase in the near surface temperature of the Earth. Global warming has occurred in the distant past as the result of natural influences, but the term is most often used to refer to the warming predicted to occur as a result of increased emissions of greenhouse gases. Scientists generally agree that the Earth's surface has warmed by about 1 degree Fahrenheit in the past 140 years.

**Green Revolution.** The great increase in production of food grains (as in rice and wheat) due to the introduction of high-yielding varieties, to the use of pesticides, and to better management techniques.

**Greenhouse Effect.** The effect noticed in greenhouses when shortwave solar radiation penetrates glass, is converted to longer wavelengths, and is blocked from escaping by the windows. It results in a temperature increase. Earth's atmosphere acts in a similar manner.

**Gross National Product (GNP).** The total value of the goods and services produced by the residents of a nation during a specified period (such as a year).

**Groundwater.** Water found in porous rock and soil below the soil moisture zone and, generally, below the root zone of plants. Groundwater that saturates rock is separated from an unsaturated zone by the water table.

## H

**Habitat.** The natural environment of a plant or animal.

**Habitat Fragmentation.** Process by which a natural habitat/landscape is broken up into small sections of natural ecosystems, isolated from each other by sections of land dominated by human activities.

**Hazardous Waste.** Waste that poses a risk to human or ecological health and thus requires special disposal techniques.

**Herbicide.** Any substance used to kill plants.

**Heterotroph.** Organism that cannot synthesize its own food and must feed on organic compounds produced by other organisms.

**Hydrocarbons.** Organic compounds containing hydrogen, oxygen, and carbon. Commonly found in petroleum, natural gas, and coal.

**Hydrogen.** Lightest-known gas; major element found in all living systems.

**Hydrogen Sulfide.** Compound of hydrogen and sulfur; a toxic air contaminant that smells like rotten eggs.

**Hydropower.** Electrical energy produced by flowing or falling water.

## I

**Infiltration.** Process of water percolation into soil and pores and hollows of permeable rocks.

**Intangible Resources.** Open space, beauty, serenity, genius, information, diversity, and satisfaction are a few of these abstract commodities.

**Integrated Pest Management (IPM).** Designed to avoid economic loss from pests, this program's methods of pest control strive to minimize the use of environmentally hazardous, synthetic chemicals.

**Invasive.** Refers to those species that have moved into an area and reproduced so aggressively that they have replaced some of the native species.

**Ion.** An atom or group of atoms, possessing a charge; brought about by the loss or gain of electrons.

**Ionizing Radiation.** Energy in the form of rays or particles that have the capacity to dislodge electrons and/or other atomic particles from matter that is irradiated.

**Irradiation.** Exposure to any form of radiation.

**Isotopes.** Two or more forms of an element having the same number of protons in the nucleus of each atom but different numbers of neutrons.

## K

**Keystone Species.** Species that are essential to the functioning of many other organisms in an ecosystem.

**Kilowatt.** Unit of power equal to 1,000 watts.

## L

**Leaching.** Dissolving out of soluble materials by water percolating through soil.

**Limnologist.** Individual who studies the physical, chemical, and biological conditions of aquatic systems.

## M

**Malnutrition.** Faulty or inadequate nutrition.

**Malthusian Theory.** The theory that populations tend to increase by geometric progression (1, 2, 4, 8, 16, etc.) while food supplies increase by arithmetic means (1, 2, 3, 4, 5, etc.).

**Metabolism.** The chemical processes in living tissue through which energy is provided for continuation of the system.

**Methane.** Often called marsh gas ($CH_4$); an odorless, flammable gas that is the major constituent of natural gas. In nature it develops from decomposing organic matter.

**Migration.** Periodic departure and return of organisms to and from a population area.

**Monoculture.** Cultivation of a single crop, such as wheat or corn, to the exclusion of other land uses.

**Mutation.** Change in genetic material (gene) that determines species characteristics; can be caused by a number of agents, including radiation and chemicals, called mutagens.

## N

**Natural Selection.** The agent of evolutionary change by which organisms possessing advantageous adaptations leave more offspring than those lacking such adaptations.

**Niche.** The unique occupation or way of life of a plant or animal species; where it lives and what it does in the community.

# Glossary

**Nitrate.** A salt of nitric acid. Nitrates are the major source of nitrogen for higher plants. Sodium nitrate and potassium nitrate are used as fertilizers.

**Nitrite.** Highly toxic compound; salt of nitrous acid.

**Nitrogen Oxides.** Common air pollutants. Formed by the combination of nitrogen and oxygen; often the products of petroleum combustion in automobiles.

**Nonrenewable Resource.** Any natural resource that cannot be replaced, regenerated, or brought back to its original state once it has been extracted, for example, coal or crude oil.

**Nutrient.** Any nutritive substance that an organism must take in from its environment because it cannot produce it as fast as it needs it or, more likely, at all.

## O

**Oil Shale.** Rock impregnated with oil. Regarded as a potential source of future petroleum products.

**Oligotrophic.** Most often refers to those lakes with a low concentration of organic matter. Usually contain considerable oxygen; Lakes Tahoe and Baikal are examples.

**Organic.** Living or once fliving material; compounds containing carbon formed by living organisms.

**Organophosphates.** A large group of nonpersistent synthetic poisons used in the pesticide industry; include parathion and malathion.

**Ozone.** Molecule of oxygen containing three oxygen atoms; shields much of Earth from ultraviolet radiation.

## P

**Particulate.** Existing in the form of small, separate particles; various atmospheric pollutants are industrially produced particulates.

**Peroxyacyl Nitrate (PAN).** Compound making up part of photochemical smog and the major plant toxicant of smog-type injury; levels as low as 0.01 ppm can injure sensitive plants. Also causes eye irritation in people.

**Pesticide.** Any material used to kill rats, mice, bacteria, fungi, or other pests of humans.

**Pesticide Treadmill.** A situation in which the cost of using pesticides increases while the effectiveness decreases (because pest species develop genetic resistance to the pesticides).

**Petrochemicals.** Chemicals derived from petroleum bases.

**pH.** Scale used to designate the degree of acidity or alkalinity; ranges from 1 to 14; a neutral solution has a pH of 7; low pHs are acid in nature, while pHs above 7 are alkaline.

**Phosphate.** A phosphorous compound; used in medicine and as fertilizers.

**Photochemical Smog.** Type of air pollution; results from sunlight acting with hydrocarbons and oxides of nitrogen in the atmosphere.

**Photosynthesis.** Formation of carbohydrates from carbon dioxide and hydrogen in plants exposed to sunlight; involves a release of oxygen through the decomposition of water.

**Photovoltaic Cells.** An energy-conversion device that captures solar energy and directly converts it to electrical current.

**Physical Half-Life.** Time required for half of the atoms of a radioactive substance present at some beginning to become disintegrated and transformed.

**Phytoplankton.** That portion of the plankton community comprised of tiny plants (e.g., algae, diatoms).

**Pioneer Species.** Hardy species that are the first to colonize a site in the beginning stage of ecological succession.

**Plankton.** Microscopic organisms that occupy the upper water layers in both freshwater and marine ecosystems.

**Plutonium.** Highly toxic, heavy, radioactive, manmade, metallic element. Possesses a very long physical half-life.

**Pollution.** The process of contaminating air, water, or soil with materials that reduce the quality of the medium.

**Polychlorinated Biphenyls (PCBs).** Poisonous compounds similar in chemical structure to DDT. PCBs are found in a wide variety of products ranging from lubricants, waxes, asphalt, and transformers to inks and insecticides. Known to cause liver, spleen, kidney, and heart damage.

**Population.** All members of a particular species occupying a specific area.

**Predator.** Any organism that consumes all or part of another system; usually responsible for death of the prey.

**Primary Production.** The energy accumulated and stored by plants through photosynthesis.

## R

**Rad (Radiation Absorbed Dose).** Measurement unit relative to the amount of radiation absorbed by a particular target, biotic or abiotic.

**Radioactive Waste.** Any radioactive by-product of nuclear reactors or nuclear processes.

**Radioactivity.** The emission of electrons, protons (atomic nuclei), and/or rays from elements capable of emitting radiation.

**Rain Forest.** Forest with high humidity, small temperature range, and abundant precipitation; can be tropical or temperate.

**Recycle.** To reuse; usually involves manufactured items, such as aluminum cans, being restructured after use and utilized again.

**Red Tide.** Population explosion or bloom of minute single-celled marine organisms (dinoflagellates), which can accumulate in protected bays and poison other marine life.

**Renewable Resources.** Resources normally replaced or replenished by natural processes; not depleted by moderate use.

**Riparian Water Right.** Legal right of an owner of land bordering a natural lake or stream to remove water from that aquatic system.

## S

**Salinization.** An accumulation of salts in the soil that could eventually make the soil too salty for the growth of plants.

**Sanitary Landfill.** Land waste disposal site in which solid waste is spread, compacted, and covered.

**Scrubber.** Antipollution system that uses liquid sprays in removing particulate pollutants from an airstream.

**Sediment.** Soil particles moved from land into aquatic systems as a result of human activities or natural events, such as material deposited by water or wind.

**Seepage.** Movement of water through soil.

**Selection.** The process, either natural or artificial, of selecting or removing the best or less-desirable members of a population.

**Selective Breeding.** Process of selecting and breeding organisms containing traits considered most desirable.

**Selective Harvesting.** Process of taking specific individuals from a population; the removal of trees in a specific age class would be an example.

**Sewage.** Any waste material coming from domestic and industrial origins.

**Smog.** A mixture of smoke and air; now applies to any type of air pollution.

**Soil Erosion.** Detachment and movement of soil by the action of wind and moving water.

**Solid Waste.** Unwanted solid materials usually resulting from industrial processes.

**Species.** A population of morphologically similar organisms, capable of interbreeding and producing viable offspring.

**Species Diversity.** The number and relative abundance of species present in a community. An ecosystem is said to be more diverse if species present have equal population sizes and less diverse if many species are rare and some are very common.

**Strip Mining.** Mining in which Earth's surface is removed in order to obtain subsurface materials.

**Strontium-90.** Radioactive isotope of strontium; it results from nuclear explosions and is dangerous, especially for vertebrates, because it is taken up in the construction of bone.

**Succession.** Change in the structure and function of an ecosystem; replacement of one system with another through time.

**Sulfur Dioxide (SO₂).** Gas produced by burning coal and as a by-product of smelting and other industrial processes. Very toxic to plants.

**Sulfur Oxides (SOx).** Oxides of sulfur produced by the burning of oils and coal that contain small amounts of sulfur. Common air pollutants.

**Sulfuric Acid (H₂ SO₄).** Very corrosive acid produced from sulfur dioxide and found as a component of acid rain.

**Sustainability.** Ability of an ecosystem to maintain ecological processes, functions, biodiversity, and productivity over time.

**Sustainable Agriculture.** Agriculture that maintains the integrity of soil and water resources so that it can continue indefinitely.

# T

**Technology.** Applied science; the application of knowledge for practical use.

**Tetraethyl Lead.** Major source of lead found in living tissue; it is produced to reduce engine knock in automobiles.

**Thermal Inversion.** A layer of dense, cool air that is trapped under a layer of less-dense warm air (prevents upward flowing air currents from developing).

**Thermal Pollution.** Unwanted heat, the result of ejection of heat from various sources into the environment.

**Thermocline.** The layer of water in a body of water that separates an upper warm layer from a deeper, colder zone.

**Threshold Effect.** The situation in which no effect is noticed, physiologically or psychologically, until a certain level or concentration is reached.

**Tolerance Limit.** The point at which resistance to a poison or drug breaks down.

**Total Fertility Rate (TFR).** An estimate of the average number of children that would be born alive to a woman during her reproductive years.

**Toxic.** Poisonous; capable of producing harm to a living system.

**Tragedy of the Commons.** Degradation or depletion of a resource to which people have free and unmanaged access.

**Trophic.** Relating to nutrition; often expressed in trophic pyramids in which organisms feeding on other systems are said to be at a higher trophic level; an example would be carnivores feeding on herbivores, which, in turn, feed on vegetation.

**Turbidity.** Usually refers to the amount of sediment suspended in an aquatic system.

# U

**Uranium 235.** An isotope of uranium that when bombarded with neutrons undergoes fission, resulting in radiation and energy. Used in atomic reactors for electrical generation.

# Z

**Zero Population Growth.** The condition of a population in which birthrates equal death rates; it results in no growth of the population.

# Index

## A

ABI Research, 120
Actus Lend Lease, 124
affluence, 6.40, 36, 40, 47, 52
Africa: agroforestry in, 81; drylands in,
    151; environmental data collection in, 7;
    organic farming in, 67; privatization in, 28
agriculture: climate change and, 79–82; early,
    and climate change, 216–220; organic
    farming, 66–70
agroforestry, 81
Allen, Hartwell, 79–80
Allen, Tony, 75
Alley, William M., 162–170
Alliance for Environmental Innovation, 86
Alliance to Save Energy (ASE), 90
Amazon rainforest, 175–180; family forestry
    in, 179, 180; future of, 179–180; history of
    timber industry in, 175–176; smallholder
    forestry in, 178–179; Trans-Amazon
    highway in, 177–178, 179
American Association for the Advancement of
    Science, 59
American Electric Power Company, 117, 146
American Revolution, 225
Anderson, Ewan, 73
Anderson, Terry, 25, 26
Annan, Kofi, 59, 60
Antweiler, Werner, 25
appropriate situational tailoring, 64
aquifers, 162–170
Arrow, Kenneth, 5
Asia: monsoon in, 79; ozone pollution in,
    211, 212
Aspen Institute, 8
assured security, 62
Austin Energy, 88
Ausubel, Jesse, 12, 13

## B

Baden, John, 26
Badgley, Catherine, 67, 68
ballast water, and invasive alien species,
    130, 131, 134, 136
banana, 188, 189, 190, 195
Barnett, Harold, 12
Battelle, 116, 117–118
Bell, Michelle, 207–208
Bender, Marty, 81–82
Bernstam, Mikhail, 25
Bernstein, Mark, 205–213
Bingaman, Jeff, 224
biodiversity: buying, 11; in drylands, 153;
    invasive alien species and, 129–137;
    markets for services, 139–146; organic
    farming and, 68
biodiversity credits, 142
biofuels, 69, 117
biological oxygen demand (BOD), 24
biotechnology, 39–40

bird mortality, wind turbines and, 111
birthrate, decline in, 18
Blair, Tony, 59, 60
Blix, Hans, 59–60
Bloom, Barry R., 46
Boehlert, Sherwood, 224
Bongaarts, John, 53
Borer, Douglas A., 71–77
Borlaug, Norman, 66
Borroni-Bird, Chris, 121, 122
Bosch, Carl, 67
Boserup, Ester, 155
Bowden College, 173
BP, 10, 14, 96, 123, 124
Brazil, 142, 146. *See also* Amazon rainforest
Bruce, Judith, 53
Brundtland, Gro Harlem, 4, 5
Buchmann, Stephen, 173
Bunch, Roland, 70
Bush, George H. W., 9, 14
Bush, George W., 5, 7, 9, 59, 63, 64, 77, 121,
    123, 224

## C

Cairncross, Frances, 20
California: conservation banking in, 144;
    energy pricing in, 124; green marketing in,
    84, 87; groundwater in, 163–164; hydrogen
    highway in, 121–122; invasive alien
    species in, 129, 130, 131, 132, 135; smog
    in, 13, 205, 208, 209, 210; wind power in,
    103, 104, 107, 109, 110, 111, 112
California Air Resources Board (CARB),
    206, 209
Cambodia, 80
Cambridge University, 10, 15, 66
Canada, 129, 132, 195, 216, 219, 222, 223
Cape Wind Project, 109, 112
carbon dioxide: capturing and storing, 9; early
    farming and, 216–219; emissions of, 45,
    47, 224; hydrogen production and, 120;
    during last ice age, 222; and plant growth,
    80; and polar ice, 221, 222–223; reducing
    emissions of, 7–8, 96–101, 119, 146
carbon tax, 9, 124
Carbon Trust, 8–9
Cato Institute, 15
Catskill Mountains, 172, 173
cell phones, 20
Centre for Risk Analysis, 5
CFL. *See* compact fluorescent light
Chameides, Bill, 221
Chapin, Terry, 222
Chengjian, Wu, 12
Chevron-Texaco, 146
Chicago Council on Foreign Relations, 62–63
China: CFC-free refrigerators in, 88;
    economic growth in, 6; energy efficiency
    in, 101; logging in, 15; ozone pollution in,
    211; pollution in, 11; vehicles in, 13
chlorofluorocarbons (CFCs), 9–10, 84, 88

Churchill, Winston, 7, 9, 10
Clean Air Act, 14, 205, 208, 211
Clean Development Mechanism (CDM), 145
Clean Power Now, 111
cleantech, 85
Clemson University, 27
Climate, Community and Biodiversity
    Alliance, 145
climate change, 8, 221–224; abrupt, 61;
    accelerated, 60, 221; and agriculture,
    79–82; biodiversity conservation for
    mitigating, 145; contrarians on, 225; and
    drought, 223; early, 216–220; at end
    of last ice age, 79; errors in predictions
    about, 225–227; and flora and fauna, 223;
    and hurricanes, 223; Pentagon report on,
    59, 60–61; and polar ice, 222–223, 228;
    as threat, 62, 63, 64; uncertainty over,
    7–8, 10
climate policy, 8–9
climate sensitivity, 226
Clinton, Bill, 64
coal gasification, 117
coffee, 81, 188–191, 193, 194
Cohen, Joel E., 46
Columbia University, 16–17
commercialism, attitudes toward, 36–39
Commoner, Barry, 52
communications, 19–20
community-based forestry initiatives, 11
community-driven conservation, 145
compact fluorescent light (CFL), 83, 84, 86,
    97
competitive society, 40
composite materials, 87
comprehensive operational integration, 64
connectivity, communications and, 19–20
conservation: community-driven, 145; in
    drylands, 152; financial incentives for,
    140–141; new market solutions for,
    141–143
conservation banking, 142, 144
conservation finance, 139, 140
Conservation International, 11, 139, 141
consumerism, attitudes toward, 36–39
consumption, 53–55; definition of, 54; energy,
    115–118; and population, 54–55, 56–57;
    reducing, 55–57; trends in, 54–55
contrarianism, 225
Convention on Biological Diversity (1992), 134
Convention to Combat Desertification, 151
Copeland, Brian R., 25
Cornell University, 173
corruption, 20, 27
Corsell, Peter, 124
Costa Rica: extinction in, 223; pesticide use
    in, 188–197
Cox, Craig, 110
crisis prevention, 64
CropLife International, 193
cultural modernization, 18–19, 21
Curry, Ruth, 221, 222

# Index

# Index

# Test Your Knowledge Form

We encourage you to photocopy and use this page as a tool to assess how the articles in *Annual Editions* expand on the information in your textbook. By reflecting on the articles you will gain enhanced text information. You can also access this useful form on a product's book support Web site at *http://www.mhcls.com/online/*.

NAME:                                                                    DATE:

_____

TITLE AND NUMBER OF ARTICLE:

_____

BRIEFLY STATE THE MAIN IDEA OF THIS ARTICLE:


_____

LIST THREE IMPORTANT FACTS THAT THE AUTHOR USES TO SUPPORT THE MAIN IDEA:




_____

WHAT INFORMATION OR IDEAS DISCUSSED IN THIS ARTICLE ARE ALSO DISCUSSED IN YOUR TEXTBOOK OR OTHER READINGS THAT YOU HAVE DONE? LIST THE TEXTBOOK CHAPTERS AND PAGE NUMBERS:




_____

LIST ANY EXAMPLES OF BIAS OR FAULTY REASONING THAT YOU FOUND IN THE ARTICLE:



_____

LIST ANY NEW TERMS/CONCEPTS THAT WERE DISCUSSED IN THE ARTICLE, AND WRITE A SHORT DEFINITION:

# We Want Your Advice

ANNUAL EDITIONS revisions depend on two major opinion sources: one is our Advisory Board, listed in the front of this volume, which works with us in scanning the thousands of articles published in the public press each year; the other is you—the person actually using the book. Please help us and the users of the next edition by completing the prepaid article rating form on this page and returning it to us. Thank you for your help!

## ANNUAL EDITIONS: Environment 07/08

### ARTICLE RATING FORM

Here is an opportunity for you to have direct input into the next revision of this volume.
We would like you to rate each of the articles listed below, using the following scale:

1. **Excellent: should definitely be retained**
2. **Above average: should probably be retained**
3. **Below average: should probably be deleted**
4. **Poor: should definitely be deleted**

Your ratings will play a vital part in the next revision.
Please mail this prepaid form to us as soon as possible.
Thanks for your help!

| RATING | ARTICLE | RATING | ARTICLE |
|---|---|---|---|
| | 1. How Many Planets? A Survey of the Global Environment | | 15. Hydrogen: Waiting for the Revolution |
| | 2. Five Meta-Trends Changing the World | | 16. Sunrise for Renewable Energy? |
| | 3. Globalization's Effects on the Environment | | 17. Strangers in Our Midst: The Problem of Invasive Alien Species |
| | 4. Do Global Attitudes and Behaviors Support Sustainable Development? | | 18. Markets for Biodiversity Services: Potential Roles and Challenges |
| | 5. The Climax of Humanity | | 19. Dryland Development: Success Stories from West Africa |
| | 6. Population and Consumption: What We Know, What We Need to Know | | 20. Tracking U.S. Groundwater: Reserves for the Future? |
| | 7. A New Security Paradigm | | 21. How Much Is Clean Water Worth? |
| | 8. Can Organic Farming Feed Us All? | | 22. Searching for Sustainability: Forest Policies, Small-holders, and the Trans-Amazon Highway |
| | 9. Where Oil and Water Do Mix: Environmental Scarcity and Future Conflict in the Middle East and North Africa | | 23. Agricultural Pesticides in Developing Countries |
| | 10. The Irony of Climate | | 24. Smog Alert: The Challenges of Battling Ozone Pollution |
| | 11. Avoiding Green Marketing Myopia | | 25. How Did Humans First Alter Global Climate? |
| | 12. More Profit with Less Carbon | | 26. The Tipping Point |
| | 13. Wind Power: Obstacles and Opportunities | | 27. Swift Boating, Stealth Budgeting, and Unitary Executives |
| | 14. Personalized Energy: The Next Paradigm | | |

NO POSTAGE
NECESSARY
IF MAILED
IN THE
UNITED STATES

## BUSINESS REPLY MAIL
FIRST CLASS MAIL PERMIT NO. 551 DUBUQUE IA

POSTAGE WILL BE PAID BY ADDRESSEE

**McGraw-Hill Contemporary Learning Series**
2460 KERPER BLVD
DUBUQUE, IA 52001-9902

---

# ABOUT YOU

Name                                                    Date

Are you a teacher? ☐ A student? ☐
Your school's name

Department

Address                          City            State            Zip

School telephone #

# YOUR COMMENTS ARE IMPORTANT TO US!

Please fill in the following information:
For which course did you use this book?

Did you use a text with this ANNUAL EDITION? ☐ yes ☐ no
What was the title of the text?

What are your general reactions to the Annual Editions concept?

Have you read any pertinent articles recently that you think should be included in the next edition? Explain.

Are there any articles that you feel should be replaced in the next edition? Why?

Are there any World Wide Web sites that you feel should be included in the next edition? Please annotate.

May we contact you for editorial input? ☐ yes ☐ no
May we quote your comments? ☐ yes ☐ no